The Beginning of A New Dawn For Humanity

(Introduction Into The World of Micro- And Macro- Molecular Chemistry)

Ω

Sunny N. E. Omorodion

Professor of Chemical Engineering

authorHOUSE®

AuthorHouse™ UK
1663 Liberty Drive
Bloomington, IN 47403 USA
www.authorhouse.co.uk
Phone: 0800.197.4150

Published by AuthorHouse 02/01/2017

ISBN: 978-1-5246-6750-4 (sc)
ISBN: 978-1-5246-6749-8 (e)

Print information available on the last page.

THE BEGINNING OF A NEW DAWN FOR HUMANITY

[Introduction to the world of Micro- and Macro- Molecular Chemistry]

Page Number

PREFACE

This is the beginning of a new science in which drastic changes are bound to take place virtually in all disciplines including Religion. The materials contained herein are new challenges we are bound to face as we advance into the third millennium. Since antiquity, the developments of Science and associated disciplines have been very obscure, despite the fact that the world "seems" to have been advancing technologically and in all respects. The world of advancement has been that filled with *illusions and ignorance* and for one to continue to live in this kind of world is unacceptable. It has been a world where all what we have been doing technologically, scientifically, sociologically, and in all others, have been *ninety five percent ART and five percent SCIENCE,* just analogous to our inabilities of not using *ninety five percent of the brains* given to all humans.

We must have the abilities to ask questions and provide solutions to them regardless the operating conditions. We have no knowledge about mechanism of reactions or for all systems. Unfortunately, we think we know, without realizing that the more we know, we then begin to see that we know nothing. If we knew, the need for the use of ***trials and errors, guess work, semblance, copying, postulations, hypothesis, necessary and unnecessary assumptions, accidental discoveries (wherein the discoverer knows nothing of what discovered), and so on would not have arisen.*** So far, we have lived with these methods for so long, something to be expected. But now, the time for change has arrived, because we have begun to see better to the point, where it is clear that so many things have been missing, for which for example we cannot even explain as simple as it may seem, how heat is generated during Combustion, a phenomenon used every nano second universally everywhere! How can one continue to live in such a world?

Unknown is that *over ninety percent of all research works carried out in all disciplines universally since antiquity have no correct explanations.* Without Mathematics, Physics, and Chemistry – the great trios called Natural Sciences, nothing exists in humanity. They were there before humans and all other things were created. Mathematics being the natural language of communication in the real and imaginary domains, should apply to all disciplines whether Arabic, Edo, English, French, German, Greek, Latin, History, Law or not. Physics being the study of the Forces of Nature, that is, the senses (Sight, Sound, Smell, Taste and Feelings) in the real and imaginary domains, should apply to all disciplines whether Arabic, Edo, English, French, German, Greek, Latin, History, Law or not. Chemistry being the study of the Laws of Nature in the real and imaginary domains, is the greatest of the trios. Based on the laws, it applies to all disciplines. Though these laws are not stated herein, they are enormous to comprehend. It is only inside CHEMISTRY, you see the fundamentals of Mathematics, Physics, Engineering, Medicine, Social Sciences, the Arts, and indeed all disciplines including in particular RELIGION. Hence, these materials including those in THE NEW FRONTIERS of twelve volumes where the Laws have begun to be stated, must be read by all regardless your discipline. However, a solid secondary background in Chemistry is important, because CHEMISTRY is the MOTHER of all disciplines, since all other disciplines are inside CHEMISTRY based on the manners by which Chemical, Polymeric and Nuclear reactions take place.

Indeed, this introduction to Chemistry contained herein, very short book is herein used to launch the publications of the first six Volumes of the NEW FRONTIERS along with applications to Control in Engineering and Mathematical Numerical methods for Engineers in two books. The last section of the sixth volume also contains applications to Solar energy in Electrical engineering and Dominance theory in Nature with respect to Medical Sciences. The present text like others were developed using universal data. They mark the beginning of a new era wherein the present text is just an introduction to Chemistry and related disciplines such as Natural Sciences, Chemical Engineering, Polymer Engineering and Sciences, Zoological Engineering, Botanical Engineering, Medical Sciences, Social Sciences, and more. We have only just begun, since all disciplines are inter-related.

Acknowledgements: The first acknowledgement goes to my students of more than thirty years who have bothered me with questions, the solutions of which cannot be found anywhere universally. Secondly my acknowledgment goes to greatest journals universally (Nature, American Chemical Society, European Journal of Polymer Science and more) that advised that such works be published as text-books for several reasons. My greatest acknowledgment goes to ALMIGHTY INFINITE GOD for giving me the ability to use the "eye of the needle" to "SEE" better.

University of Benin Sunny N. E. Omorodion
ETG

ABOUT THE AUTHOR

Sunny N. E. Omorodion has been a teacher mostly all his life. He started his teaching career at the age of nineteen in a High school teaching students many of whom were older than him. After graduating from the University of Ibadan with a **BSc (Hons)** in Chemistry at the age of twenty three, he left for Canada after teaching in two High schools again, to acquire another bachelor **(B Eng.)** in Chemical Engineering at the University of Alberta, since his dream career was Chemical Engineering since he was a child. In the same university, he acquired two Bachelors in one year in Mathematics and Physics, since when a student at Ibadan he only needed one year and two years to complete degrees in Mathematics and Physics respectively. It was the civil war in the country that made him to study Chemistry at a time when Chemical Engineering did not exist as a discipline in any of the Nigerian Universities, something which can be said to be a blessing in a different way, since in the process he was fully exposed to Mathematics and Physics at the tertiary level, while graduating with only Chemistry.

During the acquisition of four Bachelor degrees, he left for McMaster University in Ontario, Canada to acquire M Eng. and Ph. D degrees in an area which is a hybrid of Chemistry, Mathematics, Physics and Chemical Engineering-**Polymer Engineering**. He then worked in an industry (Polysar-Sania, Ontario, Canada) for about two years, before coming back to his teaching career at the University of Benin. After about twenty years of service as teacher/consultant, he left on sabbatical leave and leave of absence to teach at three universities- University of Regina, Saskatchewan, Canada, University of Windsor, Ontario, Canada, and University of Toledo, Ohio, USA. Introduction of three new courses at post-graduate level along with the teaching of other courses in Canada and USA made him one of the best professors in all the universities. Presently, he is now back to University of Benin to complete the cycle of one of the stages of life.

Sunny's research interests include Chemical and Polymer reaction Engineering very different from what exists in Present-day Science and Engineering, Environmental Science and Engineering with respect to Pollution Prevention, Waste Management, Enzymatic Chemistry and Engineering, Energy Sources and Conservation, Unit Separations and Process Control of Industrial Systems. Based on "***The New Science***", some research works which were thought not to be possible, have been made possible, such as oxidation of propane to propanol, polymerization of some monomers, which could not previously be polymerized to give useful products and so on. These are works which cannot be published without introducing ***"The New Science"*** in ***"The New Frontiers"*** universally.

Sunny is a member of American Institute of Chemical Engineers since 1972, Canadian Society of Chemical Engineers since 1974, Chemical Society of Canada since 1972, American Chemical Society since 2002, American Association for the Advancement of Science since 1996, African Academy of Science since 1990, Nigerian Society of Chemical Engineers since 1988, Polymer Institute of Nigeria since 1990, The Association of Professional Engineers and Geoscientists of Saskatchewan, since 2002. Because he has been so involved in writing, spending at least fourteen hours every day since Jan 1st 1992, he has not been an active member of these bodies and even rejected serving the State Gov. as Commissioner and other appointments. Sunny is a Fellow of some Professional bodies such as Institute of Industrial Administration since 2008, American and Cambridge Biographical centers and Professional bodies since 1997, and Strategic Institute for Natural Resources and Human Development since 2012. Sunny has won more than twenty International awards with respect to Who is Who from ABI, IBC, The Marquis and more.

We are TENANTS not only in this WORLD, but also in our PHYSICAL BODY, for there is no DEATH of the BEING, BUT THE DEATH of the PHYSICAL BODY.

The Author.

Chapter 1
New Classifications for Radicals and Their Impacts

Herein, one looks at the outer sphere of an atom where radicals as opposed to electrons exist. With the new definitions provided for mathematics, physics, and chemistry, the three most important subjects in humanity, new concepts in chemistry have begun to be introduced. These include operating conditions, states of existences, new classifications for mechanisms of systems and stage-wise operations. These resulted in establishing new classification for radicals.

These new concepts are the *missing links* that search for solutions to questions of *fear of the unknown.* Applications are provided to distinguish between the mechanisms of combustion and oxidation, starting with the use of hydrogen, aliphatic and aromatic hydrocarbons, providing origin of one of the sources of lung cancer. With hydrocarbons, methane is the only environmentally friendly one. With hydrogen, no battery is required when used as one of the sources for energy. For the first time, the structures of oxidizing oxygen and ozone are provided.

For the first time, the states of existence of all the components of air are also provided. In providing additional supports for the new foundations, the examples provided are too numerous to list. However, the worthiest of note is provision of mechanisms of chemical reactions that are completely different from what is known today universally. With the new understanding of mechanisms of systems, new doors are bound to be opened.

1.0 Introduction

There is a need to bridge the gap between the ***art and science*** and ***arts and sciences,*** since they all depend on the atom. Unlike what is known today since the development of the **Periodic Table,** which indeed is the organizing principle of chemistry, the *electrons* only reside *inside the nucleus* of an atom. There are many schools of thought with respect to the existence of an electron and where it is located. Some scientists believe electrons don't exist in an atom; others think they exist *outside the nucleus* (Thomson's corpuscle hypothesis), and others think they exist inside the nucleus as reflected by some classifications of the subatomic elements in the nucleus. George Francis FitzGerald (1851–1901), an Irish *physicist,* suggested that Thomson's "corpuscles" making up the cathode ray were actually *free electrons.* Dying thirty-nine years before Thomson, he did not live to continue the fight. Ironically, an English physicist bearing both names - Sir Thomson George Paget (1892–1975), shared with C.J. Davisson in 1937 the

Nobel Prize in physics for the simultaneous independent discovery of diffraction phenomena in the electron. Paget's previous works included *The Atom* (1930), *The Wave Mechanics of Free Electrons* (1930), *Theory and Practice of Electron Diffraction* (with William Cochrane, 1939), and *The Inspiration of Science* (1961).[1–19] One can imagine the heavy "state of confusion" above, despite the enormous abundance of data.

What today are called electrons when used for the outer sphere of an atom are indeed *radicals,* as will be explained and supported via applications. Hence, what today are called *electronic configurations* that display how radicals are arranged on the outer sphere of an atom are *radical configurations*. Electronic configurations, which will be shown downstream, are based on the arrangement of the subatomic particles in the nucleus of every atom. This understanding here, though very simple and heavily supported with current developments based on research and observations, will greatly affect many things, as will shortly be shown, based on understanding how nature operates. All the new concepts described herein are largely based on the use of current universal literature data in the sciences, engineering, and the arts—too numerous to comprehend and list herein.

In atoms, chemical reactions based largely on chemistry (the laws) take place in the outer sphere, while nuclear reactions based largely on physics (the forces) take place in the nucleus. It is in the nucleus where something close to electrons (which indeed are Gamma rays), called *negatrons – the beta rays* (negative), exist inside matter. On the other side of the nucleus called antimatter, exists the opposite of negatrons and these known to the world are called *positrons* (positive). In atomic units, the masses of these sub-particles are 0.0. (See appendix 1.). However, today, based on current developments in conjunction with Einstein's theory of relativity, positron (with positive charge) is said to have a mass that is the same as the negatron (with negative charge) of 511 KeV/c^2.[20–24] This is an abnormal correlation. The remaining three subatomic particles on the matter side have positive masses "in our gravitational world" while the other three in the antimatter side have negative masses "in our gravitational world," such as *the protons and antiprotons*[25, 26] have atomic masses of +1 and -1 respectively. Under vacuum however, the situation may be different. All eight subatomic particles are primary and secondary, some of which gave birth to other short-lived subatomic particles, such as muons, peons, kaons, lambda, and more. Knowing the mechanisms by which chemical, polymeric, and nuclear reactions take place, have been the missing links in understanding how nature operates.

Today, very little is known about the mechanisms of systems in particular chemical reactions (biochemical or enzymatic, non-biochemical, and polymeric), despite the abundance of data universally, most of which are available and in order. In recent years, there have been foci in research developments for energy resources, genetics for food, cloning, diseases and protein/biomedical technologies, environmental concerns with respect to pollution control, and much more. Important data, mostly experimental in character, are gathered for them without understanding their full mechanisms of operations. For lack of this, so many things cannot be clearly explained and indeed are not in place. For example, as simple as it may seem, what today is universally called ***genetic engineering*** is a small subset of ***enzymatic engineering.*** Its name should be ***genetic enzymatic engineering*** since what carries out chemical reactions, such as oxidation, hydrogenation, dehydrogenation, hydrolysis, combustion, stages of reproduction using genes and skeletal foundations in living systems in a very narrow standard range of operating conditions are ***enzymes.*** *These are unique types of optically active very specific catalysts that carry blueprints, such as oxidation centres, hydrogenation centres and more on gene backbones, just like other enzymes with these centres placed on other backbones.* With the new concepts and use of universal data, *the molecular structures* of these enzymes have begun to be known in the New Frontiers.

For the lack of knowledge of the mechanisms of systems and structures of many compounds and elements, so many definitions as above are not in place, and when not in place, any forward movement is *an illusionary process*. This has generally been the case for almost all disciplines. All disciplines indeed depend on the *natural sciences*. Knowing the actual definitions of the *natural sciences* composed essentially of *chemistry, physics, and mathematics* will help to bridge many gaps. However, in the process, much of the past works must be revisited, because the wheel will eventually be turned backwards by no one but NATURE dismantling and rebuilding for the purpose of providing orderliness.

Based on what is going on universally, Chemistry deals with the study of the *laws of nature* based on the manners by which chemical reactions take place under *different operating conditions*, microscopically and macroscopically in the *real and imaginary domains*. These laws, unlike material laws, have no exceptions. Physics deals with the study of the *forces of nature*. The natural forces in question are obviously the five senses: sight, sound, smell, taste, and feeling, all also in the *real and imaginary domains*. Of all the five parts (indeed ten) of Physics, only those of smell and taste are lagging behind. Mathematics deals with the study of the *natural language of communication* for all systems in the *real and imaginary domains*. Unlike universal languages and dialects, it has no beginning and no end and cannot be changed or killed. It is in chemistry that one sees what mathematics and physics are, because of the nature of atoms. These new definitions gave birth to these new concepts and vice versa contained herein.

In support of these new concepts, one will begin by showing the mechanisms of combustion and oxidation, an area very important to chemical engineers. There are many ways by which energy can be provided to run systems: chemical (use of fuels, fossils, and their equivalents), electrical (greatest life source, the best of which are nuclear and solar), and mechanical (wind, waterfalls, and other forces). However, the present focus based on the outer sphere of an atom will be fuels (one of the main current sources in our world with respect to transformation to other forms of energy), beginning with the first members from the petroleum industry: methane and ethane (the aliphatic, both of which are rarely used in gasoline), benzene, and toluene (the aromatics used in many types of gasoline).[27] This will be compared with the case of H_2, which is about to be used to replace or complement hydrocarbon and other fuels. In recent years, H_2 has largely been used for space programs with very little attentions to surface uses.[28–34] Unknown is that H_2 offers great advantages with respect to little or no impact to environmental pollutions.[35–37] So also is methane. As opposed to the use of *abnormal operating conditions,* such as flames under very high temperatures and pressures as used in space programs, herein one is going to use H_2 catalysts and other chemical methods in the domains of *normal to mild operating conditions*.

Worthy of note is that the overall equations for, for example, the combustion of aliphatic and aromatic hydrocarbons have never been clearly and properly represented in many texts.[38, 39] Most people don't know what combustion and oxidation are, or if they know, they are ignorant of the distinctions between them. Indeed, present-day science doesn't know the difference between molecular oxygen and oxidizing oxygen in terms of their molecular structures and more, except in terms of characters/behaviours—all from experimental observations. With the new concepts resulting from the use of universal data, past and present, this will now be unquestionably explained. However, new forms of researches will now emerge based on our abilities to copy nature, that which is engineering applying four basic fundamental principles:

- the natural sciences (mathematics, physics, and chemistry),
- the social sciences (sociology, anthropology, economics, accounting, and business administration),

- the arts (English, history, law, art), and
- imaginative capabilities.

The last is most unique to all engineers. This is now becoming obvious with the emergence of Nano-Sciences and Technologies, Remote control, Genetic Enzymatic engineering and more.

1.1 THE NEW CONCEPTS

The three new concepts that follow herein, form in part the basic foundations for all disciplines. The fourth concept which indeed is number one in the list- ***Operating Conditions*** latently presented herein is the most important, for without it, the three concepts herein and more other concepts to follow, do not exist and in its absence, nothing takes place in Nature.

1.1.1 **States of Existences**

Everything existing has fixed states of existences (like fingerprints). In general, Figure 1.1 shows the three major kinds of states of existences. The real finger-print is *EQUILIBRIUM STATE OF EXISTENCE.*

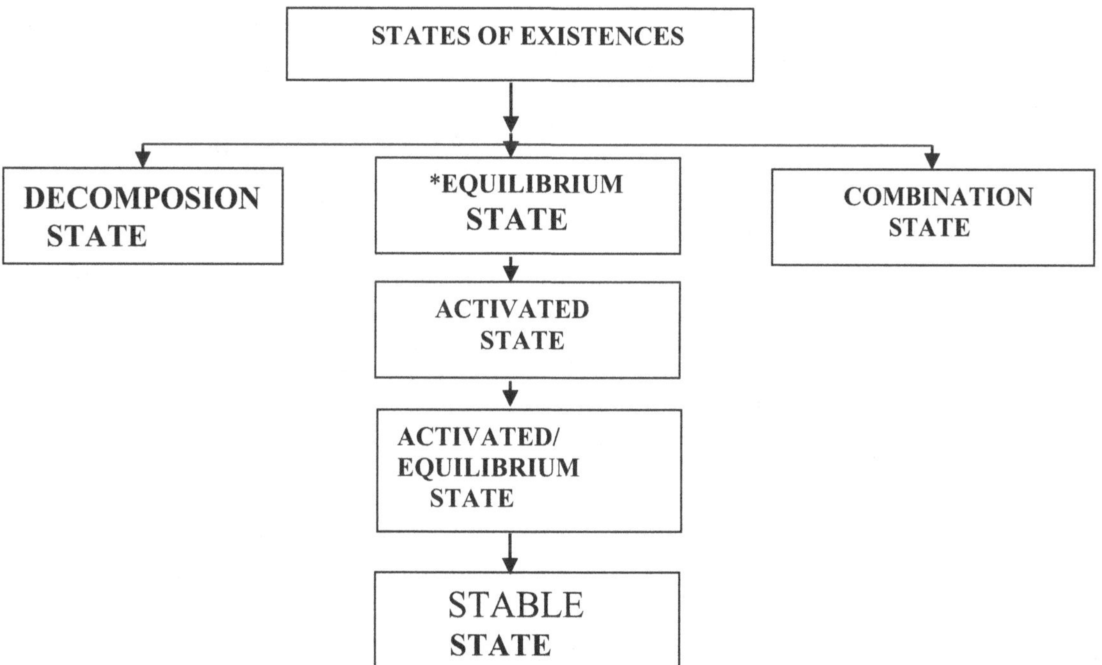

Figure 1.1 Types of States of Existences

The three states of existences are Equilibrium, Decomposition and Combination. If anything is not in Equilibrium or cannot be in Equilibrium state, it can be in Activated or Activated/Equilibrium states if **it has activation center(s)**. Presence of Activation centers is the basic origin of Life, since that is where one can see what is called a male or female. If not Activated or in Equilibrium, it can be Stable. For anything like chemical compounds or molecules, these states are fixed and different for each existence and for each compound. Consider a system containing only two compounds, A and B. Before they can react with each other to give **a productive or non-productive** stage or stages, one of them or both of them must first exist in Equilibrium or Decomposition state, otherwise there will be no reaction. If all remain in Stable States such as the components of air, they can either **dissolve or miscibilize** or remain inactive with themselves as will be shown.

Some compounds can readily exist in Equilibrium state, while others cannot. To put some in Equilibrium state may require for example special **operating conditions** such as high temperatures (for example with propylene) or a special neighbour such as **a Passive catalyst** (For ėxample hydrogen catalyst for hydrogen), noting that there are Active and Passive catalysts and ***Enzymes are Active in character***. In polymerization reactions, parts of the so-called catalysts, which indeed are Initiators, are part from the beginning of the polymeric product formed. There are others that remain in Equilibrium state all the time (in the absence of any passive catalysts) to make us smell and perceive them as will become clear. Shown below are three compounds, which fixed states of existences will shortly be shown.

H - H	H - O -H	*O* ‖ :*C* –□←——*Vacant Orbital*	
(A) Hydrogen Molecule: H_2	(B) Water H_2O	(C) Carbon monoxide CO	1.1

It is important to note the real structure of (C) above wherein a vacant orbital in the last shell is clearly shown with two paired unbonded radicals in another orbital. In this state, it is non-poisonous. It has two different types of activation centers of different capacities- *the π-bond type* of activation center in C = O and *the vacant orbital/paired unbonded radicals* on the carbon center.

To keep (A) in equilibrium state of existence, requires the use of some special catalysts that cannot form hydrides (such as hydrogen hydride (H_2), sodium hydride (NaH), aluminium hydride (AlH_3)) with hydrogen. These catalysts are today called H_2-catalysts[40] and can only be found amongst some ***Transition transitional non-ionic metallic compounds*** (Group VIIIA elements). [Note the use of double *Transition*. The first use shows that these metals are Transition metals. The second use indicates that they are the boundary between ***Transition metals*** and ***Non-Ionic-non-Transition metals*** such as Zinc, Copper, Carbon, Iodine and more.] Why they are called H_2 catalysts is unknown today. However, what these passive catalysts do is that their pores break the covalent bond in the H_2 molecule into the male and female free-radical parts (not charged parts) under Equilibrium conditions. Chargedly, it is impossible.

(B) is very unique and presently very little is known about it. It can readily exist in Equilibrium state or be suppressed based on the Operating Conditions. (C) cannot exist in Equilibrium state. It can only be activated. Unique to (C) is that the π-bond activation center in the double bond (unlike in CO_2 or alkenes) can only be activated at temperatures above 1000^0C, while the second activation center can readily be activated due to the presence of a vacant orbital in the boundary of the carbon center. When activated by heat via that center, it becomes a poison. Depending on the operating conditions, these compounds can be attacked or they do the attack by existing in Equilibrium state to lead to a productive or non-productive stage(s).

1.1.2 **Stage-wise Operations**

All operations in nature whether real or imaginary or for positive or negative developments, take place in a stage or stages. Each stage contains one or more STEP(s). This is based on the manners by which chemical reactions take place. Same applies to nuclear reactions noting that our present focus is on the outer sphere of an atom. In the Appendix, one will see the case for nuclear reactions. This concept (Stage-wise operations) is not new to for example chemical engineers in Unit Separation Processes.[41] A system can be seen as a Single-stage or Multi-stage process depending on the type, and focus or foci. The stages are fixed, because you cannot by-pass a stage and go to another stage as we do in our world.

In chemical reactions, there are countless numbers of single-stage processes; others with only two stages; others with many stages depending on the types of participants and other operating conditions. While a single stage can be productive or non-productive, a multi-stage process is productive. In a multi-stage process where products are desired, one cannot neglect intermediate stages to get to the last stage such as making a four-stage chemical reaction a two-stage process by introducing the concept of for example <u>rate determining steps</u> as done in kinetic studies.[42] By neglecting the two intermediate stages, one cannot get to what is desired. All the stages are interconnected. For example, considering the combustion of paraffin in Aliphatic <u>linear normal</u> hydrocarbon families, methane combusts in four stages, ethane combusts in eight or more stages, propane combusts in twelve or more stages, butane combusts in sixteen or more stages and so on, all under different operating conditions based largely on their physical states. It is easier for gases than for liquids and in turn than for solids. How and where energy in form of heat is released in the stages is important to know. One cannot make the four stages for methane a one stage or two-stage process in modelling for any reason when Mathematics is there. ***Nature is never in a hurry.*** It is only we humans that are in a hurry.

For a stage to exist as a stage, it must be that which can be interpreted into simple mathematical equation(s) without the need for necessary or unnecessary assumptions, trials and errors, guess-work and so on. What this means will shortly become obvious after introducing all these new concepts. Multi-stage systems can be in series or in parallel or both. There are many chemical reactions that take place in series; many that take place in parallel and many that take place in series and parallel, noting that TIME the fourth dimension, is an imaginary independent variable. ***We live in a four dimensional world in TIME and SPACE.*** Space of three dimensions cartesianly or cylindrically or spherically are *real Independent variables*. Time of one dimension is an *imaginary independent variable.* Space unlike Time can also be *Dependent variables*, such as in velocity.

1.1.3 New Classifications for Mechanisms of System

Though the present focus is chemical in character that which is on the boundary of the outer sphere of an atom, the same applies to all other systems, since all systems, are atomic build-ups of same or different atoms.

Just like states of existences, the mechanisms by which all systems (governments, chemical plants, humans, universities, chemical reactions, and so on) operate, are of ***three main kinds*** as shown in Figure 1.2.

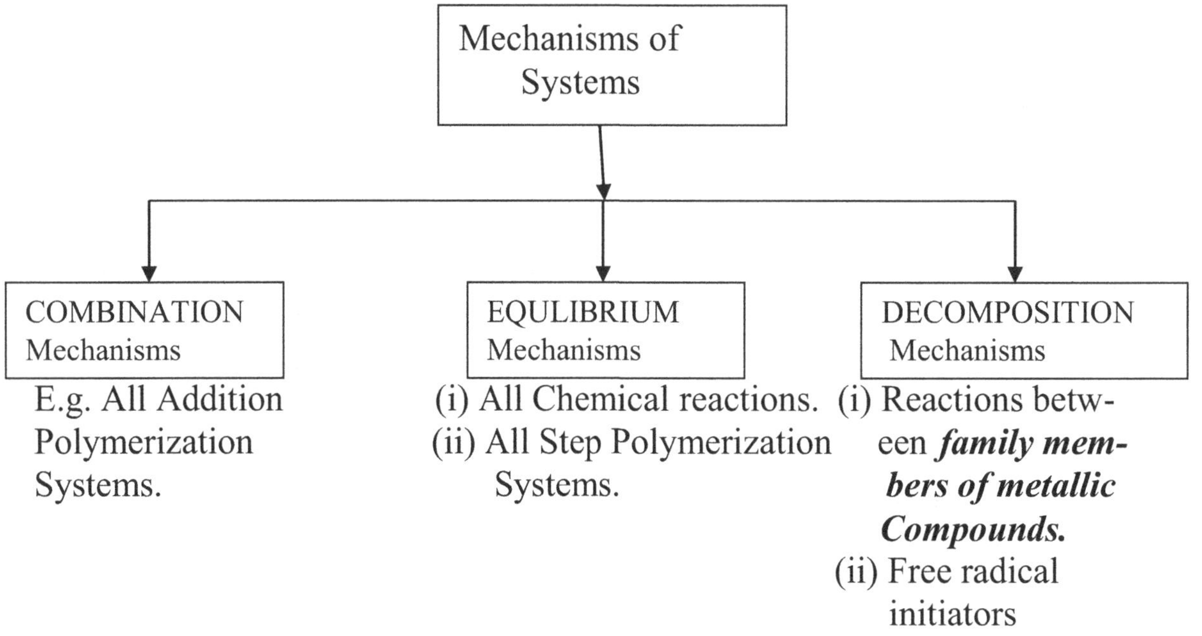

Figure 1.2 Classifications of Mechanisms of Systems

Note that the examples shown below the figure above, use only chemical and polymeric reactions. All the three mechanisms lead to positive or negative results, depending on the ***operating conditions*** and the type of products desired and much more. Decomposition should not be seen as something negative. Useful products can be obtained under such mechanism. Combination mechanism like all the other mechanisms, take place in living and non-living systems.

Based on universal data, unlike what is presently known, metals can also and indeed be ***reclassified*** as-

(a) Ionic Non-Transition metals (Groups 1A and IIA),
(b) Ionic Transition metals (Group IIIA),
(c) Non-ionic Transition metals (Groups IVA, VA, VIA and VIIA),
(d) Non-ionic Transition-transition metals (Groups VIIIA), and
(e) *Non-ionic Non-Transition metals (Groups IB, IIB, IIIB and some of Groups IVB, VB, VIB, and VIIB).*

These make up the group of metallic compounds alluded to in Figure 1.2 under Decomposition mechanisms.

All systems whether mechanical, electrical, chemical or otherwise, operate using one or more of the mechanisms above in a stage or many stages. In a single-stage system, only one mechanism operates. In multi-stage systems, one or more mechanisms can be involved. In polymerization reactions, one, two or three mechanisms are involved. Combination mechanism takes place in countless number of stages, while the other two take place in a single stage or in many stages. There are also *different types of kinds* of Equilibrium, Decomposition and Combination mechanisms.

1.2 New Classifications for Radicals

For the first time since antiquity, a Radical is being differently defined. It is well known that ions are of two characters- the cations (positive ionic charges) and the anions (negative ionic charges). Based on the *second law of Nature that which is the law of duality*, the same for ions applies to radicals. Shown in Figure 1.3, is the new classification for radicals compared with present-day Science (Old Frontiers), which classify them as radicals or free radicals. It is true that they are highly reactive and unstable. It is also true that they are produced by cleavage of a covalent bond. It is also finally true that if the term ***free*** is dropped in referring to free- radicals, this could lead to confusion.[43, 44] Based on the ways Nature operates, there are **Free and Non-Free-Radicals** each with their male and female counterparts. Radicals are indeed weightless sub-atomic particles outside the Nucleus of an Atom carrying latent energy in them, just almost like those in the Nucleus but different, all of them with their own identities. These radicals are not electrons or Thomson's corpuscles, which are present as one of the main ***eight*** sub-atomic particles in the nucleus (see Appendix (I)). This indeed is new to present-day Science. It is not based on hypothesis or assumptions or postulations or otherwise, but on current universal observations, developments and data.

The first law of Nature has different interpretations in many or all disciplines. In Chemical Engineering with respect to material conservation as also applies to energy conservation, everything that goes in comes out under ***steady state conditions.*** Under ***unsteady state conditions*** there is growth materially or acquired cooling/heating energy-wise, that is, accumulation (a function of Time). In Chemistry, that first law for all conservations independent of all different types of operating conditions is **to-be-free and not-to-be-free**. Based on this law, there are two classes of Radicals, the Free radicals and the Non-free-radicals. It is the operating conditions and type of species however, that makes them to be manifested.

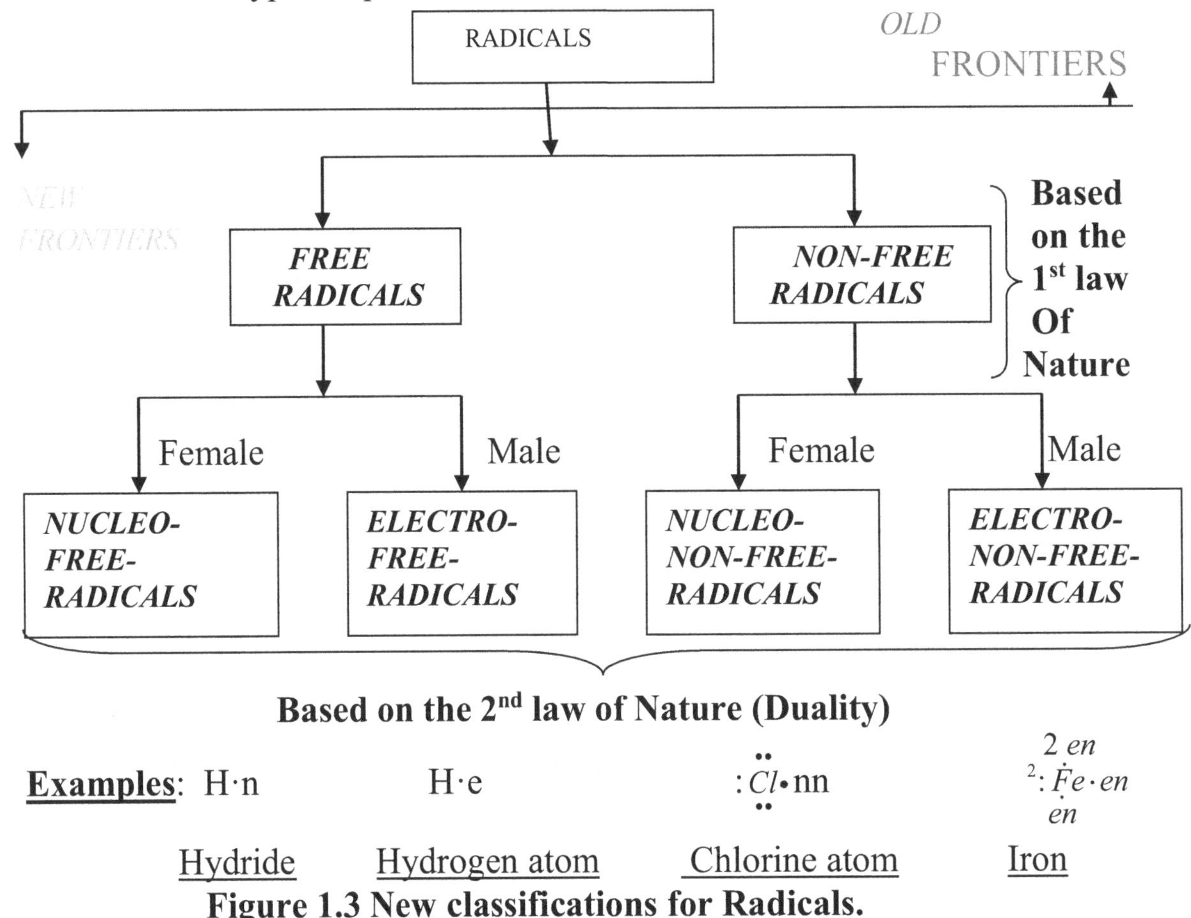

Figure 1.3 New classifications for Radicals.

Every atom has a **DOMAIN** and a **BOUNDARY**, just like every other thing. These are their LIMITATIONS for which reason everything in life has limitations. The domain is marked by the distance from the Nucleus to the Boundary. Shown in Figure 1.4, are the domains and boundaries of H and Cl atoms.

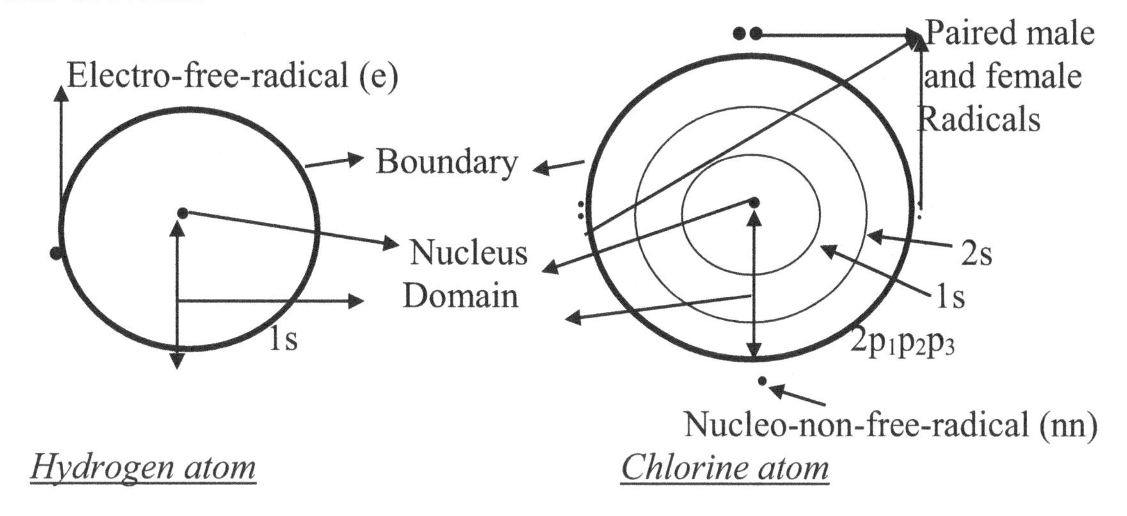

Hydrogen atom *Chlorine atom*

Figure 1.4 Domains and Boundaries of H and Cl atoms.

The boundary of H atom, its last shell (1s), cannot carry more than two radicals no matter the operating conditions. In the last shell of H, there is only one radical. Hence, it is called an **electro-free-radical** (the male character), since H is electro-positive. The boundary of Cl cannot carry more than eight radicals no matter the operating conditions. In the last shell of chlorine, there are seven radicals, three paired as male and female radicals with opposite spin and one alone. The single one radical is not free, because it can interchange with *any one of its like in the other three in view of the dynamic character of Nature a statement of fact that is not new.*[45] Because chlorine is electronegative, that single radical is female and called a **nucleo-non-free-radical.** The equilibrium states of existence of hydrogen and chlorine molecules are as follows-

$$H_2 \underset{\text{Other chemical Means}}{\overset{H_2\ Cat.\ or}{\rightleftharpoons}} H\cdot e \quad + \quad H\cdot n$$

Stable — (Male) **Electro-free-radical** The Atom — (Female) **Nucleo-free-radical** The Hydride **1.2**

$$Cl_2 \rightleftharpoons :\ddot{\underset{..}{Cl}}\cdot en \quad + \quad :\ddot{\underset{..}{Cl}}\cdot nn$$

Stable — (Male) **Electro-non-free-radical (en)** A chloride — (Female) **Nucleo-non-free-radical (nn)** The Atom **1.3**

While hydrogen catalyst is required to keep hydrogen molecule in equilibrium state of existence, nothing other than temperature and a type of neighbour (i.e., another compound) around is needed for chlorine. A neighbour such as oxidizing oxygen will stabilize it, since oxidizing oxygen molecule is always in Equilibrium state of existence. It will suppress that of Cl_2.

It is such type of state of Existence (Equilibrium) of different capacities that partly determines so many characters of a compound- such as ***smell (toxicity, [not aromaticity]), flammability, volatility, taste (sweetness, bitterness, etc.), and much more***. Notice the type of **half-sided headed reversed arrows** used to denote Equilibrium state of existence. It is *neither here nor there* and not as shown below as currently used with cations and anions, noting that ***ions*** are one of the four kinds of charges. This is new to humanity, because all the charges are uniquely different as will shortly be shown. Universally, all charges are thought to be *ionic in character!*

Wrong representation of Equilibrium State of Existence

$$H_2 \rightleftarrows \underset{Male}{H \cdot e} + \underset{Female}{H \cdot n}$$

↓

Full-sided
Headed reversed
Arrows 1.4

The last equation as written above is also not a complete representation of reversible reactions. Indeed, based on present-day Science, it is difficult to distinguish between Reversible and Equilibrium equations radically or ionically in many respects including their manners of representations symbolically and mathematically. These can lead to confusions. ***Secondly, in Equilibrium states of existence, male and female radicals are the components held together and these are fixed***. For example, considering glycerol, a non-symmetric compound, none of what is shown below is the real Equilibrium state of existence.

$$H - \underset{CH_2OH}{\overset{OH}{\underset{|}{\overset{|}{C}}}} - CH_2OH \rightleftharpoons H \cdot e + n \cdot \underset{CH_2OH}{\overset{OH}{\underset{|}{\overset{|}{C}}}} - CH_2OH \qquad 1.5a$$

$$\xrightleftharpoons{OR} H\cdot e \quad + \quad HO-\overset{\overset{\displaystyle nn\cdot CHOH}{|}}{\underset{\underset{\displaystyle CH_2OH}{|}}{C}}-H \qquad 1.5b$$

$$\xrightleftharpoons{OR} H\cdot e + HOCH_2-\overset{\overset{\displaystyle H}{|}}{\underset{\underset{\displaystyle O.nn}{|}}{C}}-CH_2OH \qquad 1.5c$$

$\xrightleftharpoons{OR}$ *More* (Cases where H.e is held.) 1.5d

$$\xrightleftharpoons{OR} HO\cdot nn + e\cdot \overset{\overset{\displaystyle CH_2OH}{|}}{\underset{\underset{\displaystyle CH_2OH}{|}}{C}}-H \qquad 1.6a$$

$$\xrightleftharpoons{OR} HO\cdot nn + e\cdot CH_2-\overset{\overset{\displaystyle H}{|}}{\underset{\underset{\displaystyle OH}{|}}{C}}-CH_2OH \qquad 1.6b$$

ETC.

The search for the real one like with other compounds entailed a vigorous process, for which most of its reactions (e.g. to give fats and oils and much more) were revisited and the only one was found. The only one state for glycerol is that represented below.

$$HO-\overset{\overset{\displaystyle H}{|}}{\underset{\underset{\displaystyle H}{|}}{C^*}}-\overset{\overset{\displaystyle H}{|}}{\underset{\underset{\displaystyle OH}{|}}{C}}-\overset{\overset{\displaystyle H}{|}}{\underset{\underset{\displaystyle H}{|}}{C}}-OH \xrightleftharpoons[\text{Existence of Glycerol}]{\text{Equilibrium State of}} \underline{H}\cdot e + nn.O-\overset{\overset{\displaystyle H}{|}}{\underset{\underset{\displaystyle H}{|}}{C^*}}-\overset{\overset{\displaystyle H}{|}}{\underset{\underset{\displaystyle OH}{|}}{C}}-\overset{\overset{\displaystyle H}{|}}{\underset{\underset{\displaystyle H}{|}}{C}}-OH \qquad 1.7$$

* Central C atom in charge. $\underline{H} \equiv$ The H held as a male in its Equilibrium state of existence.

All others apart from Equation 1.7 are Combination states *when a single double-headed arrow in the opposite direction is used.* For Decomposition state, which indeed is cleavage of a covalent bond under some forces, it is not reversed but with a single double-headed arrow with OH group being the component held like an alcohol (See Equation 1.6b with change in arrow). For symmetric molecules, the cleavage gives two identical radicals under Decomposition state of existence, while for non-symmetric molecules, it gives two opposite radicals. The state shown in Equation 1.7 makes this polyalcohol look like an acid when in Equilibrium state of existence.

The Decomposition and Combination states of existence of H_2 and Cl_2 both of which are symmetric are as follows.

Decomposition State of Existence

$$H_2 \xrightarrow[\text{Into Two Like Poles}]{\text{A Molecule breaking}} H\cdot e \quad + \quad H\cdot e$$

(Two Atoms)

$$Cl_2 \xrightarrow[\text{Breaking into two like Poles}]{\text{A Molecule}} Cl\bullet nn + Cl\bullet nn$$

(Two Atoms) 1.8a

(One full-sided headed right directional arrow)

Wrong Decomposition State representation

$$H_2 \xrightarrow[\text{Possible}]{\text{Not}} H\cdot e \ + \ H\cdot n \qquad \text{(Equilibrium state character)}$$

$$Cl_2 \xrightarrow[\text{Possible}]{\text{Not}} Cl\bullet en + Cl\bullet nn \qquad \text{(Equilibrium State character)} \qquad 1.8b$$

Combination State of Existences

(Two males) $H\bullet e + H\bullet e \longrightarrow H_2$

(Two females) $H\bullet n + H\bullet n \longrightarrow H_2$ 1.9a

(Two males) $Cl\bullet en + Cl\bullet en \longrightarrow Cl_2$

(Two females) $Cl\bullet nn + Cl\bullet nn \longrightarrow Cl_2$

Note here, that ***like poles or radicals*** are combining to form a molecule *in support of one of the laws of Metaphysics, but against one the laws of Physics.* Both are true depending on what is being done. While this law of Metaphysics applies in Equilibrium mechanism for example as will shortly be explained, its Physics law (i.e., like poles repel while unlike poles attract) does apply under all real operating conditions and visibly too. In Equilibrium mechanism, it partly makes a stage to become productive in the last step only as will shortly be seen.

Possible or Impossible representation of Combination State of Existence

$H\cdot e \ + \ H\cdot n \xrightarrow[Possible]{Not} H_2$ (Equilibrium State character)

$Cl{\bullet}en + Cl{\bullet}nn \xrightarrow[Possible]{Not} Cl_2$ (Equilibrium State character) 1.9b

[Unlike Poles]

The first reaction above for example is only possible in the presence of ionic and non-ionic metallic hydrides, since their Equilibrium states of Existence are greater than that of H_2. Their presence will suppress that of hydrogen to make it form a stable molecule of hydrogen as shown above (In Equation 1.9b); unless a passive catalyst for hydrogen is present in the system. This is made possible, because these metals are more electropositive than H which is also a metal.

$$LiH > NaH > KH > AlH_3 >>>>> H_2$$

$$E.g.\ \ NaH \underset{of\ Existence}{\overset{Equilibrium\ State}{\rightleftharpoons}} Na\cdot e \ + \ H{\bullet}n$$

Order of Equilibrium States of Existence of Hydrides 1.9c

For the second reaction above (in Equation 1.9b), only presence of specific interhalogen compounds gaseous in character can suppress the Equilibrium state of existence of Cl_2 during Combination.

$$Br_2 > BrCl > Cl_2 >>>>>> F_2$$ 1.9d

(Always Stable)

Order of Equilibrium States of Existence of some Halogens

Against one of the laws of Physics, that is like poles repel and unlike poles attract, notice that in Combination state of existence, two like poles are attracting to form a molecule as already shown in Equation 1.9a. This takes place at the end of a stage or the last step in a productive stage only in Equilibrium Mechanisms. ***This basically has to do with existence of a situation where radicals can only attract between like and unlike, but cannot repel.*** This as will become clear is not the case with Charges. *Two unlike poles in chemical reactions do not attract to form a molecule, unless when suppressed and no other state exists for the compound as already shown above.* This is not an exception in the laws of Nature. For another example, consider non-symmetric molecules such as sodium chloride (NaCl) and aluminium chloride ($AlCl_3$). Na and Al cannot carry female character in the presence of chlorine, being electropositive.

Hence

$$NaCl \rightleftharpoons \underset{(Male)}{Na{\bullet}e} + \underset{(Female-chloride)}{Cl{\bullet}nn}$$

EQUILIBRIUM STATE OF EXISTENCE OF SODIUM CHLORIDE

$$\underset{State}{\overset{Impossible}{\rightleftharpoons}} \{\underset{(Does\ not\ exist)}{Na{\bullet}n} + \underset{(Exists)}{Cl{\bullet}en}\} - Cannot\ exist \qquad 1.10a$$

Na•n can never exist as a female. Ionic solid metals like Na (as applies to all the solids in the Groups IA, IIA and IIIA elements) cannot carry female characters such as in alloys, whereas Al can in ***alloys and coordinated compounds.***

NaCl ⟶ Na•e + Cl•nn (Not Na•n + Cl•en)
Decomposition State of existence of Sodium chloride 1.10b

Na•e + Cl•nn ⟶ NaCl (Suppressed by the presence of e.g. H_2O)
Combination State of existence of Sodium chloride 1.10c

$AlCl_3 \rightleftharpoons Cl_2Al{\bullet}e + Cl{\bullet}nn$
Equilibrium State of existence of Aluminium chloride 1.11a

$AlCl_3 \longrightarrow Cl_2Al{\bullet}e + Cl{\bullet}nn$ or none at all.
Decomposition State of existence of Aluminium chloride 1.11b

$Cl{\bullet}nn + e{\bullet}\ AlCl_2 \longrightarrow AlCl_3$ (Suppressed by the presence of e.g. HCl)

And $Cl{\bullet}\ en + n{\bullet}AlCl_2 \longrightarrow AlCl_3$ (No Suppression needed)
Combination States of existence of Aluminium chloride 1.11c

$AlR_3 \rightleftharpoons R{\bullet}e + n{\bullet}\ AlR_2$
Equilibrium State of existence of Aluminium tri-alkyl 1.11d

$AlR_3 \longrightarrow R_2Al{\bullet}e + R{\bullet}n$
Decomposition State of existence of Aluminium trialkyl 1.11e

In Equation 1.11d above, notice that it is Aluminium that is carrying the female radical in the finger print (Equilibrium state of existence) of aluminium tri-alkyl. The alkyl group such as C_4H_9 is carrying the male radical. It is true that the friends you keep know you; hence the great influence of specific neighbours or components in suppressing or enhancing the equilibrium state of existence of a compound in particular those for which only one type exists for all the states of existence. Table 1.1 shows the states of existence of the components of clean air.

Table 1.1 **States of Existence of the Components of Air**

Components	Stable State	Activated State	Equilibrium State	Decomposition State	Combination State
Oxygen O_2 (21%)	O=O (Double Bonds, one π and one σ)	O_2 $\overset{Heat}{\rightleftharpoons}$ en•O-O•nn	Oxidizing* Molecular Oxygen O ⬭ O ⇌ 2nn•O•en Oxidizing Oxygen element.	O_2 ⟶ 2nn•O•nn Atomic Oxygen	2nn•O•nn ⟶ O_2 OR en•O-O•nn $\xrightarrow{Deactivation}$ O_2 + Heat
Hydrogen H_2 (<1%)	H—H (One σ bond)	No Activation Center.	H_2 $\underset{Catalyst}{\overset{H_2}{\rightleftharpoons}}$ H•e + H•n	H_2 ⟶ 2H•e Hydrogen atoms	2H•e OR 2H•n ⟶ H_2
Nitrogen N_2 (78%)	N≡N (Triple Bonds, two π s and one σ)	N_2 $\overset{Heat}{\rightleftharpoons}$ $en \cdot \underset{..}{N} = \ddot{N} \cdot nn$	Nitrogenizing* Molecular Nitrogen N ⬭ N $\underset{Conditions}{\overset{Drastic}{\rightleftharpoons}}$ $\underset{nn}{\overset{nn}{\dot{N}}} \cdot en + nn \cdot \underset{en}{\overset{en}{\dot{N}}}$ *Their existence is deadly* Nitrogen elements.	N_2 ⟶ $2\underset{nn}{\overset{nn}{\dot{N}}} \cdot nn$ Nitrogen atoms	$2\underset{nn}{\overset{nn}{\dot{N}}} \cdot nn$ ⟶ N_2 OR $nn \cdot \underset{..}{N} = \ddot{N} \cdot en$ $\xrightarrow{Deactivation}$ N_2 + Heat
Carbon Dioxide CO_2 (0%- ?" One of Green-House Gases")	O=C=O (Two π and two σ Bonds)	CO_2 $\overset{Heat}{\rightleftharpoons}$ e•C(=O) - O•nn	None	CO_2 → C(=O) + en•O•nn	e•C(=O) - O•nn $\xrightarrow{Deactivation}$ CO_2 + Heat

Inert Gases (<1%)	All Stable	None	None	None	None

*These do not exist in Air. Only oxidizing oxygen "seems to be well known" best obtained from oxidizing agents such as fuming nitric acid, hydrogen peroxide, vanadium pentoxide and more. ***All the components of air in the environment are in Stable State of Existence; otherwise there will be no life.*** Internally, they become actively used for regeneration of some of the Blueprints on the enzymes and much more based on the operating conditions. Note that water vapor has not been included in the Table above for specific reasons.

From what has been shown so far yet so little, one can begin to observe that the Periodic Table (the most complete of which is that which contains the electro-negativity, the atomic number and the radical configuration of the last shell of the atom or element, that which is the boundary) now has completely new interpretations for its atoms. It is now for example the real structures of all elements and compounds such as ***Carbon black, Activated Carbon, Coal, Coke, Charcoal, Zinc dust, Zinc metal, Oxygen atom, Oxygen molecule, Oxidizing oxygen element, Oxidizing oxygen molecule, Ozone, Different types of Sulfur elements, Phosphorus elements, Ammonium hydroxide, Aluminium hydroxide, and much more can clearly be known and explicitly shown. Without the use of universal data based on researches and emperisms, the developments of these new concepts would have been impossible. Therefore, the universal data are indeed as important as the current developments of these new concepts.*** However, there is need to now start applying the new concepts based on current level of advancements in our world in order to give supports to the new concepts and foundations being laid, since it will affect the structural foundations of many compounds and many things as we know them today. Many old structures with exact molecular compositions will have new structures as opposed to the currently known ones. For these reasons, the real chemical bonds will be identified all based on the new concepts. Note that so far, no mentions of CHARGES, which abundantly exist, have been made. These are contained in the next chapter.

1.3 Some Important Examples of Applications of the New Concepts

Shown below are some very important applications of the new concepts, wherein the one and only one mechanism for the examples are provided.

1.3.1 Combustion of Methane

Combustion involves the reaction between the component to be combusted and Molecular oxygen of Air and nothing else.

Stage1:

$$CH_4 \xrightleftharpoons[\text{Existence Of Methane At any Temperature}]{\text{Equilibrium State Of}} H\cdot e + n\cdot CH_3$$

$$H\cdot e + O=O \xrightleftharpoons[\text{Oxygen (Heat required)}]{\text{Activation of}} H\cdot e + nn\cdot\ddot{\underset{\cdot\cdot}{O}}-\ddot{\underset{\cdot\cdot}{O}}\cdot en \rightleftharpoons H-O-O\cdot en$$

$$(A)$$

$$(A) + n\cdot CH_3 \xrightarrow{\text{Combination Step}} H-O-O-CH_3 \qquad 1.12a$$

$$\underline{Overall\ Equation}: \underset{(Air)}{CH_4 + O_2} \longrightarrow \underset{(B)}{H-O-O-CH_3} \qquad 1.12b$$

In this first stage for this case, the oxygen molecule is activated via its Activation center (the π-bond) and its activation depends on the operating conditions in which temperature is very important. Note the Combination State of Existence of the peroxide (B). Its real Equilibrium state of Existence is shown in the next stage. For some compounds such as water in place of methane above, the stage is non-productive. For some other compounds such as some ether, the stage is productive and ends there to give a peroxide. For others as the case above, the peroxide formed is unstable under the operating conditions, ***wherein the energy carried by it must be released.*** Hence, the second stage follows, that in which the peroxide is decomposed via Equilibrium mechanism and not Decomposition mechanism. What a stage via Decomposition mechanism looks like will be shown very shortly downstream. Unlike in Equilibrium mechanisms, the first step in the stage begins with the decomposition symbol [$\longrightarrow$] and the last step could end with Combination or Equilibrium symbol. The first step above began with Equilibrium state of existence of one the components and this makes it to be so classified as Equilibrium mechanism. It continues in that state in all the steps until the last step where the product can either be formed or not formed. The last step is the most important step in a stage via Equilibrium mechanism.

Stage 2:

$$H-O-O-CH_3 \underset{\text{Existence of the peroxide}}{\overset{\text{Equilibrium State of}}{\rightleftharpoons}} H-O\cdot nn + en\cdot OCH_3$$

$$en\cdot O-\underset{H}{\overset{H}{C}}-H \underset{\text{Release}}{\overset{\text{Energy}}{\rightleftharpoons}} H\cdot e + \underset{H}{\overset{H}{O=C}} + Energy\ Release$$

$$(Formaldehyde)$$

$$H\cdot e + nn\cdot OH \xrightarrow[\text{by presence of the energy Or product formed}]{\text{Combination state suppressed}} H_2O \qquad 1.13a$$

$$\underline{Overall\ Equation}: CH_4 + O_2 \longrightarrow H_2O + H_2C=O + Energy\ Release \qquad 1.13b$$

As shown in the second step above, it is forbidden to find oxygen carrying an electro-non-free-radical in the presence of hydrogen and carbon, which are more electropositive or male in character than O on the molecular species (CH_3O). Under usual conditions as will be seen, the O center should be carrying a nucleo-non-free-radical. Since the oxygen center has to carry an electro-non-free-radical based on the fixed Equilibrium state of Existence of the peroxide here (as fixed by NATURE), that unusual group in the second step of the second stage releases energy to form a stable molecule (formaldehyde) and hydrogen atom, that is, H carrying an electro-free radical. The very important concept to realize herein is the structures of groups of compounds. Under this condition, the energy (Heat) released is enormous, but extremely small compared to that in the Nucleus (Electromagnetic). There are different kinds of ENERGIES of different types. The Energy above is HEAT generated from chemical reactions. The Equilibrium state of Existence of water is suppressed by the presence of energy and not the formaldehyde, since water forms Para-formaldehydes with formaldehydes at specific operating conditions. If there is no oxygen in the system, the combustion ends here. It is important to note what an Equilibrium stage looks like. In a stage, there could be two, three, four or more steps up to twelve for specific reasons as will be shown in the New Frontiers. The first step begins with equilibrium symbol ($\rightleftharpoons$) and the last step ends with combination symbol ($\longrightarrow$) to indicate that the stage is productive. If the last step had ended with equilibrium symbol, then the stage is ***reactive but non-productive***. As will shortly be shown, this indicates SOLUBILITY or INSOLUBILITY, mathematics being the Natural language of communication. With more oxygen in the system, combustion continues as follows with the formaldehyde.

Stage 3:

$$H_2C=O \xrightleftharpoons[\text{Existence of formaldehyde}]{\text{Activation State of}} H\cdot e + n\cdot\overset{\displaystyle H}{\overset{|}{C}}=O$$

$$H\cdot e + O=O \xrightleftharpoons[\text{Oxygen}]{\text{Activation of}} \underset{(A)}{H-O-O\cdot en}$$

$$(A) + n\cdot\overset{\displaystyle H}{\overset{|}{C}}=O \xrightarrow{\text{Combination}} \underset{(Another\ peroxide)}{H-O-O-\overset{\displaystyle H}{\overset{|}{C}}=O} \qquad 1.14$$

Another peroxide is formed here with no release of energy. The peroxide formed, being unstable decomposes next as follows.

Stage 4:

$$H-O-O-\overset{\displaystyle H}{\overset{|}{C}}=O \xrightleftharpoons[\text{Existence}]{\text{Equilibrium State of}} H-\ddot{\underset{\cdot\cdot}{O}}\cdot nn + \underset{(C)}{en\cdot O-\overset{\displaystyle H}{\overset{|}{C}}=O}$$

$$(C) \xrightleftharpoons[\text{Release}]{\text{Energy}} CO_2 + H\cdot e + Energy$$

$$H\cdot e + nn\cdot OH \xrightarrow{\text{Combination}} H_2O \qquad 1.15a$$

$$\underline{Overall\ Equation}: CH_4 + 2O_2 \longrightarrow CO_2 + 2H_2O + Energy \qquad 1.15b$$

At the end, water, carbon dioxide and energy are produced. Two moles of oxygen were required to fully combust one mole of the methane. Four moles of oxygen will be required to combust two moles of methane and so on. If only one mole of oxygen had been used, according to Stage 2, only water, formaldehyde and less energy will be produced. If half a mole of oxygen was used, combustion will still take place, but with less energy released, leaving behind large fraction of methane with only two stages. If two moles of both are used, only two stages each in parallel will similarly be obtained, provided all the methane is in Equilibrium state of existence. On molar basis, full combustion takes place (i.e. hundred percent conversions) when exact molar ratios and the right operating conditions have been used. When zero mole of oxygen is used, no combustion takes place. One can observe the positive and negative significance of the number zero here, the most important of all numbers. For without zero, no other number exists.

Note that CO_2 is produced instantaneously here without the need for deactivation, because when activated, the C center carries the electro-free-radical (i.e., male) while the O center carries the nucleo-non-free-radical (i.e., female). It is non-free-, because of the presence of paired unbonded radicals on the oxygen center. When an activated CO_2 is deactivated, energy is released, just as when the stable CO_2 is activated, energy must be added. Without the methane existing in Equilibrium state of Existence, its combustion is impossible since the oxygen molecule is in stable state of existence under normal and mild operating conditions. Energy is produced in Stages 2 and 4. Via researches, that amount of energy is relatively known.[38] For the first time, one can begin to see some of the ways by which energy in form of Heat is released in Chemical systems.

Notice so far, that the reactions during combustion are radical in character. No charges have been shown. Indeed. while water (H_2O) can exist in Equilibrium state of existence ionically (one of four types of charges as will shortly be shown) and radically, methane, ethane or any of the Hydrocarbon family of compounds can exist in Equilibrium State of existence only radically. As will shortly be shown, while the H center can carry an ionic charge, the C center cannot carry an ionic charge, but only covalent, polar and electrostatic charges, all of which cannot be isolated, unlike the ionic charges or radicals. Indeed, in general, all compounds exist in Equilibrium state of existence only radically (free or non-free) where possible. Only the ***Polar/Ionic compounds*** as will shortly be shown can in addition exist in Equilibrium state of existence ionically, with no productive stage ever made possible.

1.3.2 **Combustion of Ethane**

Stage 1:

$$C_2H_6 \underset{\text{Existence at above Melting Point}}{\overset{\text{Equilibrium State of}}{\rightleftharpoons}} H\cdot e + n\cdot C_2H_5$$

$$H\cdot e + O=O \underset{\text{Oxygen}}{\overset{\text{Activation of}}{\rightleftharpoons}} H-O-O\cdot en$$

$$H-O-O\cdot en + n\cdot C_2H_5 \xrightarrow{\text{Combination State}} H-O-O-C_2H_5 \quad (A\ peroxide) \qquad 1.16$$

This like the case of methane, decomposes in the next stage to give acetaldehyde and water.

Stage 2:

$$H-O-O-\underset{H}{\overset{H}{C}}-\underset{H}{\overset{H}{C}}-H \underset{\text{Existence of the Peroxide}}{\overset{\text{Equilibrium State of}}{\rightleftharpoons}} H-O\cdot nn + en\cdot O-\underset{H}{\overset{H}{C}}-\underset{H}{\overset{H}{C}}-H \quad (A)$$

$$(A) \underset{\text{Release}}{\overset{\text{Energy}}{\rightleftharpoons}} H\cdot e + O=\underset{H}{\overset{CH_3}{C}} + Energy \quad (Acetaldehyde)$$

$$H\cdot e + nn\cdot O-H \xrightarrow{\text{Combination}} H_2O \qquad 1.17a$$

$$\underline{Overall\ Equation}: C_2H_6 + O_2 \longrightarrow H_2O + H_3CCHO + Energy \qquad 1.17b$$

Because the oxygen center is carrying an electro-non-free- radical in the presence of hydrogen and carbon, hence energy is released in that step with the instantaneous formation of acetaldehyde. It is H as opposed to the CH_3 group that is released, since CH_3 is more *radical-pushing* than H. If there is enough oxygen in the system, combustion of the acetaldehyde continues as follows.

Stage 3:

$$CH_3CHO \underset{\text{Existence of the Unactivated Aldehyde}}{\overset{\text{Equilibrium state of}}{\rightleftharpoons}} H\cdot e + O=\overset{CH_3}{C}\cdot n$$

$$H\cdot e + O=O \underset{\text{Oxygen}}{\overset{\text{Activation of}}{\rightleftharpoons}} H-O-O\cdot en$$

$$H-O-O\cdot en + n\cdot\overset{CH_3}{C}=O \xrightarrow{\text{Combination}} H-O-O-\overset{CH_3}{C}=O \quad (B)-A\ peroxide \qquad 1.18$$

In this stage, no energy is released. Instead a very little or no fraction from the second stage is used to activate the molecular oxygen from air. Acetaldehyde has two types of Equilibrium states of Existence, their existence of which is based on the operating conditions (Temperature, Pressure, and catalysts) – *Activated/Equilibrium state of Existence*, *Activated state of*

Existence and *Equilibrium state of Existence* as shown below.

$$H_3CCHO \underset{\text{State of Existence}}{\overset{\text{Unactivated state of Equilibrium}}{\rightleftharpoons}} H \cdot e + H_3C - \overset{\overset{O}{\|}}{C} \cdot n \qquad 1.19a$$

$$H_3CCHO \underset{\text{State of Existence}}{\overset{\text{Activated State of Equilibrium}}{\rightleftharpoons}} H \cdot e + e \cdot \underset{\underset{n \cdot CH_2}{|}}{\overset{\overset{H}{|}}{C}} - O \cdot nn \qquad 1.19b$$

$$H_3CCHO \underset{\text{Existence}}{\overset{\text{Activated State of}}{\rightleftharpoons}} e \cdot \underset{\underset{H}{|}}{\overset{\overset{CH_3}{|}}{C}} - \ddot{\underset{\cdot\cdot}{O}} \cdot nn \qquad 1.19c$$

The Hydrogen in the second equation is transfer species of the first kind of the the first type as will shortly be explained.

It was the first case that was used in Stage 3 above. The peroxide, (B), formed in Stage 3 is very unstable and decomposes as follows.

<u>Stage 4</u>:

$$H_3C - \overset{\overset{O}{\|}}{C} - O - O - H \underset{\text{of Existence}}{\overset{\text{It's Equilibrium State}}{\rightleftharpoons}} 2H - O \cdot en + 2nn \cdot O - \overset{\overset{O}{\|}}{C} - CH_3$$

$$2nn \cdot O - \overset{\overset{O}{\|}}{C} - CH_3 \rightleftharpoons 2nn \cdot O - \overset{\overset{O}{\|}}{C} \cdot e + 2n \cdot CH_3$$

$$2H - O \cdot en \underset{\text{Release}}{\overset{\text{Energy}}{\rightleftharpoons}} 2H \cdot e + \underset{(\text{Oxidising Oxygen molecule})}{\underline{O_2}} + Energy$$

$$2H \cdot e + 2n \cdot CH_3 \underset{\text{Existence of } CH_4}{\overset{\text{Equilibrium State of}}{\rightleftharpoons}} 2CH_4$$

$$2e \cdot \underset{\underset{O}{\|}}{C} - O \cdot nn \xrightarrow[\text{Energy Release}]{\text{Deactivation}} 2CO_2 + Energy \qquad 1.20a$$

$$\underline{\text{Overall Equation}}: 2C_2H_6 + \underset{(Air)}{4O_2} \longrightarrow \underset{(Methane)}{\underline{2CH_4}} + 2CO_2 + 2H_2O + \underset{(\text{Oxidising Oxygen Molecule})}{\underline{O_2}} + Energy \qquad 1.20b$$

At this point, the followings are worthy of note-

i. For the partial combustion of ethane to be completed unlike the case of methane, two moles of ethane are required with four moles of molecular oxygen from air.

ii. The "oxygen molecule" produced is not the same as the molecular oxygen from the air we breathe. This is oxidizing oxygen molecule, a ringed very unstable molecule that all the time is always in Equilibrium state of Existence unlike molecular oxygen in the Air we breathe (See Table I). Unlike other rings, there are different types depending on the size of the ring since the amount of strain energy (SE) in the ring is dependent of the size for this imaginary family of rings. Oxygen-oxygen containing rings none of which exists are different from sulfur-sulfur containing rings in which rhombic sulfur an eight-membered ring is the only one known to exist. All these (why some rings exist and some don't and much more) will be shown in the New Frontiers. The molecular oxidizing oxygen two -membered ring cannot exist permanently, because of the amount of strain energy in the ring. It is there, but imaginarily. Hence, the oxidizing oxygen molecule is always in Equilibrium state of existence all the time. It only exists as an element of Oxygen atom. It is the only environmentally unfriendly product above.

iii. In the fourth stage, energy is released in two steps – Steps 3 and 5 of the stage. Energy is required for Activation and released for Deactivation. Less energy is released when HO•en decomposes than when CH_3CO•en decomposes instantaneously under equilibrium conditions, because of the presence of both C and H in the latter with only H in the former. This is clearly distinguished by the difference in Equilibrium states of Existences of the peroxides from formaldehyde and from acetaldehyde.

iv. With presence of more molecular oxygen in the system, the methane produced cannot continue combusting in a similar manner as was the case above, because of the presence of oxidizing oxygen. The Equilibrium state of existence of the oxidizing oxygen molecule is so strong to suppress that of methane. With its presence, the methane can no longer exist in Equilibrium state of existence. If it was not suppressed then this would have brought the number of stages to eight stages.

The next stage which involves the oxidation of the methane therefore follows as shown below. For years since the development of Science, oxidation of alkanes to alcohols has never been reported. But here for the first time, it can be observed that it can be oxidized, once a strong oxidizing agent is used.

Stage 5:

$$\underset{--}{O_2} \xrightleftharpoons[\text{Oxygen molecule}]{\text{Equilibruim State Existence of Oxidising}} 2en \cdot \ddot{\underset{..}{O}} \cdot nn$$

$$2en \cdot O \cdot nn + \underset{(Stabilized)}{2CH_4} \xrightleftharpoons{\text{Oxidation}} 2H - O \cdot nn + 2e \cdot CH_3$$

$$\longrightarrow \quad 2CH_3OH \qquad 1.21a$$

$$\underline{Overall\ Equation}: \underset{--}{O_2} + 2CH_4 \longrightarrow 2CH_3OH \qquad 1.21b$$

With the formation of methanol, combustion proceeds as shown below, since there is no more oxidizing oxygen in the system anymore.

Stage 6:

$$H_3COH \xrightleftharpoons[\text{of Existence}]{\text{Equilibrium State}} H \bullet e \quad + \quad nn \bullet OCH_3$$

$$H \bullet e \quad + \quad O = O \xrightleftharpoons[(Heat)]{\text{Activation of } O_2} \underset{(A)}{H - O - O \bullet en}$$

$$(A) \quad + \quad nn \bullet OCH_3 \longrightarrow \underset{(B)}{H - O - O - O - CH_3} \qquad 1.22a$$

$$\underline{Overall\ equation}: \quad H_3COH \quad + \quad O_2 \longrightarrow H - O - O - O - CH_3 \qquad 1.22b$$

Worthy of note is the important fact that methanol like ethanol, propanol and higher alcohols, are ACIDS, based on their Equilibrium states of existence shown above. It took years based on the wrong foundations to find and confirm that it is the H atom that is held as opposed to OH group.

(B), a peroxide of an alcohol is quite unstable and this decomposes via Equilibrium mechanism as follows in the next stage.

Stage 7:

$$(B) \xrightleftharpoons[\text{of existence}]{\text{Equilibrium State}} H-O-O\bullet nn \quad + \quad en\bullet O-\overset{H}{\underset{H}{\overset{|}{\underset{|}{C}}}}-H$$

(C)

$$(C) \rightleftharpoons H\bullet e \quad + \quad H_2C=O$$

(Formaldehyde)

$$H-O-O\bullet nn \; + \; H\bullet e \longrightarrow H-O-O-H \qquad 1.23a$$

$$\underline{Overall\ equation}: \quad (B) \longrightarrow H-O-O-H \; + \; H_2C=O \qquad 1.23b$$

PARTIAL COMBUSTION

$$\underline{Overall\ overall\ equation}: \; 2C_2H_6 \; + \; 6O_2 \longrightarrow 2CO_2 \; + \; 2H_2O \; + \; Energy \; + \; 2H-O-O-H \; + \; 2H_2C=O$$

(Hydrogen peroxide) (Formaldehyde)

1.24

This is then followed by the immediate decomposition of the hydrogen peroxide in the next stage to give Energy, water and oxidizing oxygen molecule.

Stage 8:

$$2H-O-O-H \xrightleftharpoons[\text{Existence of Hyrogen peroxide}]{\text{Equilibrium State of}} 2H-O\cdot en + 2nn\cdot O-H$$

$$2H-O\cdot en \xrightleftharpoons[\text{Release}]{\text{Energy}} 2H\cdot e + \underline{O_2} \; + \; Energy$$

$$2H\cdot e + 2nn\cdot O-H \xrightarrow[\text{presence of } \underline{O_2}]{\text{Suppressed by the}} 2H_2O \qquad 1.25a$$

$$\underline{Overall\ Equation}: 2HOOH \longrightarrow 2H_2O + \underline{O_2} \; + \; Energy \qquad 1.25b$$

This may then be followed with the combustion of the formaldehyde in the presence of more oxygen in the system via same Equilibrium mechanism in two stages as shown below and indeed already shown during the combustion of methane. Notice the number of moles involved in all the stages.

Stage 9a:

$$H_2C=O \xrightleftharpoons[\text{Existence of formaldehyde}]{\text{Activation State of}} H\cdot e + n\cdot \overset{H}{\overset{|}{C}}=O$$

$$H\cdot e + O=O \xrightleftharpoons[\text{Oxygen}]{\text{Activation of}} \underset{(A)}{H-O-O\cdot en}$$

$$A) + n\cdot \overset{H}{\overset{|}{C}}=O \xrightarrow{\text{Combination}} \underset{(Another\ peroxide)}{H-O-O-\overset{H}{\overset{|}{C}}=O} \qquad 1.26a$$

$$\underline{Overall\ equation}: 2H_2C=O + 2O_2 \longrightarrow 2HOOCHO \qquad 1.26b$$

Stage 10a:

$$H-O-O-\overset{H}{\overset{|}{C}}=O \xrightleftharpoons[\text{Existence}]{\text{Equilibrium State of}} H-\ddot{\underset{\cdot\cdot}{O}}\cdot nn + \underset{(C)}{en\cdot O-\overset{H}{\overset{|}{C}}=O}$$

$$(C) \xrightleftharpoons[\text{Release}]{\text{Energy}} CO_2 + H\cdot e + Energy$$

$$H\cdot e + nn\cdot OH \xrightarrow{\text{Combination}} H_2O \qquad 1.27a$$

$$\underline{Overall\ Equation}: 2H_2C=O + 2O_2 \longrightarrow 2CO_2 + 2H_2O + Energy \qquad 1.27b$$

OR

Stage 9b:

$$\underset{--}{O_2} \xrightleftharpoons[\text{Oxygen molecule}]{\text{Equilibruim State Existence of Oxidising}} 2en\cdot\ddot{\underset{\cdot\cdot}{O}}\cdot nn$$

$$2en\cdot O\cdot nn + \underset{(Stabilized)}{2H_2C=O} \xrightleftharpoons{\text{Oxidation}} 2H-O\cdot nn + 2e\cdot \overset{H}{\overset{|}{C}}=O$$

$$\longrightarrow 2O=CHOH \qquad 1.28a$$

$$\underline{Overall\ Equation}: \underset{--}{O_2} + 2H_2C=O \longrightarrow 2O=CHOH \qquad 1.28b$$

Stage10b:

$$O=CHOH \xrightleftharpoons[\text{of Existence}]{\text{Equilibrium State}} H\bullet e + n\bullet \overset{O}{\overset{\|}{C}}-OH$$

$$H\bullet e + O=O \xrightleftharpoons[(\textbf{Heat})]{\text{Activation of } O_2} \underset{(A)}{H-O-O\bullet en}$$

$$(A) + n\bullet \overset{O}{\overset{\|}{C}}-OH \longrightarrow \underset{(B)}{H-O-O-\overset{O}{\overset{\|}{C}}-OH} \qquad 1.29a$$

$$\underline{Overall\ equation}:\ 2O=CHOH + 2O_2 \longrightarrow 2H-O-O-\overset{O}{\overset{\|}{C}}-OH \qquad 1.29b$$

Stage 11:

$$2H-O-O-\overset{O}{\overset{\|}{C}}-OH \xrightleftharpoons[\text{of existence}]{\text{Equilibrium State}} 2H-O\bullet en + \underset{(A)}{2nn\bullet O-\overset{O}{\overset{\|}{C}}-OH}$$

$$2H-O\bullet en \rightleftharpoons 2H\bullet e + \underline{O_2}$$

$$(A) \rightleftharpoons 2HO\bullet nn + \underset{(B)}{2e\bullet \overset{O}{\overset{\|}{C}}-O\bullet nn}$$

$$2H\bullet e + 2nn\bullet OH \rightleftharpoons 2H_2O$$

$$(B) \xrightarrow[\text{Release of Heat}]{\text{Deactivation}} 2CO_2 + Energy \qquad 1.30a$$

$$\underline{Overall\ equation}:\ 2O=CHOH + 2O_2 \longrightarrow 2H_2O + 2CO_2 + \underline{O_2} + Energy$$

1.30b

FULL COMBUSTION (But incomplete in Stage 9b-see downstream)

$$\underline{Final\ Overall\ equation}: 2C_2H_6 + 8O_2 \longrightarrow 4CO_2 + 6H_2O + \underline{\underline{O_2}}\ (Oxidizing\ oxygen\ molecular) + Energy(From\ five\ Stages) \qquad 1.30c$$

Stages 9a and 10a would have been favoured if hydrogen peroxide had not decomposed. However, being more unstable than formaldehyde, it was the first to decompose to release its energy content, water and oxidizing oxygen molecule which is always in Equilibrium state of existence all the time. Hence, Stages 9b, 10b and 11 are the favoured stages. ***Thus, while methane is the only member of the family of Alkanes that undergoes full combustion with molecular oxygen of the AIR, all the other members in the family undergo both combustion and oxidation with molecular oxygen.*** The reason is because methane is the first member of the family, something which is always unique with all first members of families of compounds- Alkenes, alkynes, Cyclic members, Aldehydes, Ketones, Carboxylic acids and so on, as will be shown in the New Frontiers.

It is generally believed or reported that hydrocarbons in the presence of strong oxidizing agents such as chromium trioxide in concentrated sulfuric acid at 150^0C or oxygen at 600^0C burn to carbon dioxide and water[46].

$$CH_4 + 2O_2 \longrightarrow CO_2 + 2H_2O$$

$$C_2H_6 + 3\frac{1}{2}O_2 \longrightarrow 2CO_2 + 3H_2O$$

$$C_3H_8 + 5O_2 \longrightarrow 3CO_2 + 4H_2O$$

$$C_nH_{2n+2} + \frac{3n+1}{2}O_2 \longrightarrow nCO_2 + (n+1)H_2O \qquad 1.31$$

From the statement above, one can observe that combustion of hydrocarbons is quite different from oxidation of hydrocarbons particularly with respect to the operating conditions in particular temperature. But the equations above are not correct. On the other hand, CO is well known to be one of the products of combustion of hydrocarbons. The presence of CO is not reflected in the equations above. Then how is the CO produced? There have been so many different schools of thought with respect to combustion and oxidation of hydrocarbons. So many schools have observed the presence of "***oxygen***" as one of the products during the combustion of hydrocarbons. Hence in the Equations above, one can observe the presence of fractional moles of oxygen, such as three and half, without realizing that the oxygen produced is completely different from Molecular oxygen. ***Otherwise, why do we have what are called OXIDIZING AGENTS, when molecular oxygen in the AIR we breathe is in abundance?*** We must ask questions though it may look stupid.

On the whole, eleven stages were ***tentatively*** observed during the combustion of ethane. If methanol had not been formed, then it would have been eight stages. If hydrogen peroxide had not decomposed, then it would have been ten stages. Indeed, oxidation took place, because what was produced in Stage 4 is not molecular oxygen, but oxidizing oxygen. Interestingly enough, the oxidizing oxygen was finally produced in the last stage- the eleventh stage, with no oxidable compound left to oxidize. Water cannot be oxidized. So also, is carbon dioxide.

While the combustion of hydrocarbons other than methane with molecular oxygen cannot fully take place without oxidation, oxidation of hydrocarbons without molecular oxygen can fully take place with release of full energy. In Stage 5 of Equation 1.21a, no energy was released. Now let us begin with the full oxidation of the methane in the presence of oxidizing agents such as chromium trioxide in concentrated sulfuric acid as reflected in Equation 1.31.

1.3.3 Oxidation of Methane

Stage 1:

$$\underline{\underline{O_2}} \xrightleftharpoons[\text{Oxygen molecule}]{\text{Equilibruim State Existence of Oxidising}} 2en\cdot\ddot{\underset{\cdot\cdot}{O}}\cdot nn$$

$$2en\cdot O\cdot nn + \underset{(Stabilized)}{2CH_4} \xrightleftharpoons{\text{Oxidation}} 2H-O\cdot nn + 2e\cdot CH_3$$

$$\longrightarrow 2CH_3OH \qquad 1.32a$$

$$\underline{Overall\ Equation}: \underline{\underline{O_2}} + 2CH_4 \longrightarrow 2CH_3OH \qquad 1.32b$$

Stage 2a:

$$\underset{--}{O_2} \xrightleftharpoons[\text{Oxygen molecule}]{\text{Equilibruim State Existence of Oxidising}} 2en\cdot\ddot{\underset{\cdot\cdot}{O}}\cdot nn$$

$$2en\cdot O\cdot nn + \underset{(Stabilized)}{2C\underline{H}_3OH} \xrightleftharpoons{\text{Oxidation}} 2H - O\cdot nn + 2e\cdot CH_2OH$$

$$\longrightarrow 2HOCH_2OH \qquad 1.33a$$

$$\underline{Overall\ Equation}: \underset{--}{O_2} + 2HOCH_2OH \longrightarrow 2HOCH_2OH \qquad 1.33b$$

Stage2b:

$$\underset{--}{O_2} \xrightleftharpoons[\text{Oxygen molecule}]{\text{Equilibruim State Existence of Oxidising}} 2en\cdot\ddot{\underset{\cdot\cdot}{O}}\cdot nn$$

$$en\cdot O\cdot nn + \underset{(Stabilized)}{2CH_3O\underline{H}} \xrightleftharpoons{\text{Oxidation}} 2H - O\cdot nn + 2CH_3O\bullet en$$

$$2CH_3O\bullet en \rightleftharpoons 2H\bullet e + \underset{(A)}{2\,\overset{H}{\underset{H}{|}}C = O} + Energy$$

$$2H\bullet e + 2nn\bullet OH \longrightarrow 2H_2O \qquad 1.34a$$

$$\underline{Overall\ Equation}: \underset{--}{O_2} + 2CH_3OH \longrightarrow 2H_2O + 2H_2C = O + Energy \qquad 1.34b$$

While Stage 2a looks favoured, based on the H atom required to be abstracted, it is Stage 2b that is favoured, since OH group is more radical-pushing (i.e. richer or stronger in capacity) than CH_3 group. With continued presence of oxidizing agent, whether it is Stage 2a or Stage 2b, the same final product will be obtained. Continuing from Stage 2b however, the followings are obtained.

Stage 3:

$$\underset{--}{O_2} \xrightleftharpoons[\text{Oxygen molecule}]{\text{Equilibruim State Existence of Oxidising}} 2en\cdot\ddot{\underset{..}{O}}\cdot nn$$

$$\underset{(Stabilized)}{en\cdot O\cdot nn + \quad 2O=CH_2} \xrightleftharpoons{\text{Oxidation}} 2H-O\cdot nn \;+\; \underset{(B)}{2e\cdot \overset{O}{\overset{\|}{C}}-H}$$

$$\longrightarrow 2HO-\overset{H}{\overset{|}{C}}=O \qquad 1.35a$$

$$\underline{Overall\ Equation}: \underset{--}{O_2} + 2O=CH_2 \longrightarrow 2HO-CHO \qquad 1.35b$$

Stage 4:

$$\underset{--}{O_2} \xrightleftharpoons[\text{Oxygen molecule}]{\text{Equilibruim State Existence of Oxidising}} 2en\cdot\ddot{\underset{..}{O}}\cdot nn$$

$$\underset{(Stabilized)}{en\cdot O\cdot nn + \quad 2HO-CHO} \xrightleftharpoons{\text{Oxidation}} 2H-O\cdot nn \;+\; \underset{(B)}{2e\bullet \overset{OH}{\overset{|}{C}}=O}$$

$$(B) \rightleftharpoons \underset{(C)}{2\,e\bullet \overset{O\bullet nn}{\overset{|}{C}}=O} \;+\; H\bullet e$$

$$2H\bullet e \;+\; 2nn\bullet OH \rightleftharpoons 2H_2O$$

$$(C) \xrightarrow[\text{Release of Energy}]{\text{Deactivation}} 2CO_2 \;+\; Energy \qquad 1.36a$$

$$\underline{Overall\ Equation}: \underset{--}{O_2} + 2HO-CHO \longrightarrow 2H_2O \;+\; 2CO_2 \;+\; Energy \qquad 1.36b$$

FROM OXIDATION (FULL)

$$\underline{Overall\ overall\ equation}: 2CH_4 \;+\; 4\underset{--}{O_2} \xrightarrow{\text{Oxidation}} 4H_2O \;+\; 2CO_2 \;+\; Energy(2)$$

$$(From\ four\ Stages)1.36c$$

FROM COMBUSTION (FULL)

$$\underline{Overall\ overall\ equation}: 2CH_4 \;+\; 4O_2 \xrightarrow{\text{Combustion}} 4H_2O \;+\; 2CO_2 \;+\; Energy(2)$$

$$(From\ four\ Stages)1.36d$$

During full oxidation of methane, same number of stages obtained during combustion of methane (See Equations 1.12a to 1.16b) can also be observed here, with the same products and different amounts of energies. Energy was released here via deactivations different from the ways it was released during combustion. Energies were released in the same two types of stages- the second and the fourth stage. Notice that without using two moles of methane, oxidation may not readily take place if the oxidizing agent is of the type where oxidizing oxygen has to be used in situ in a stage. For this type, as will be shown in the New Frontiers, one mole of methane can be used.

How the chromium trioxide breaks down to give the oxidizing oxygen molecules in the presence of concentrated sulfuric acids, will be shown downstream in the New Frontiers. All the reactions take place radically. If the operating conditions are higher than expected, in Stage 3 of Equation 1.35a, carbon monoxide could be released and remain in the system if no oxidizing oxygen are available in the system as shown below.

Stage 3 of Equation 1.35a <u>at higher operating conditions</u>

<u>Stage 3:</u>

$$\underset{--}{O_2} \underset{\text{Oxygen molecule}}{\overset{\text{Equilibruim State Existence of Oxidising}}{\rightleftharpoons}} 2en\cdot\ddot{\underset{\cdot\cdot}{O}}\cdot nn$$

$$\underset{(Stabilized)}{en\cdot O\cdot nn + 2O=CH_2} \overset{\text{Oxidation}}{\rightleftharpoons} 2H-O\cdot nn + \underset{(B)}{2e\cdot \overset{O}{\overset{\|}{C}}-H}$$

$$(B) \rightleftharpoons 2H\bullet e + \underset{(A)}{2n\bullet\overset{O}{\overset{\|}{C}}\bullet e}$$

$$2H\bullet e + 2nn\bullet OH \rightleftharpoons 2H_2O$$

$$(A) \xrightarrow[\text{release of Energy}]{\text{Deactivation}} 2:C=O + Energy \qquad 1.37a$$

$$\underline{Overall\ Equation}: \underset{--}{O_2} + 2O=CH_2 \longrightarrow 2H_2O + 2CO + Energy \qquad 1.37b$$

<u>Stage 4:</u>

$$\underset{--}{O_2} \underset{\text{Oxygen molecule}}{\overset{\text{Equilibruim State Existence of Oxidising}}{\rightleftharpoons}} 2en\cdot\ddot{\underset{\cdot\cdot}{O}}\cdot nn$$

$$\underset{(Stabilized)}{en\cdot O\cdot nn + 2C=O} \overset{\text{Oxidation}}{\rightleftharpoons} \underset{(B)}{2e\bullet\overset{O}{\overset{\|}{C}}-O\bullet nn}$$

$$(B) \xrightarrow[\text{Release of Energy}]{\text{Deactivation}} 2CO_2 + Energy \qquad 1.38a$$

$$\underline{Overall\ Equation}: \underset{--}{O_2} + 2CO \longrightarrow 2CO_2 + Energy \qquad 1.38b$$

From the two stages above, one can observe that if there is no oxidizing oxygen in the system, CO_2 will not be produced in Stage 4. ***For the first time, one has begun to show the origin of carbon monoxide both during combustion and oxidation, but less in oxidation than in combustion.*** When it takes place during combustion, such as in Stage 9b of Equation 1.28a, CO and water are produced in that stage and the stage becomes the last stage (Nine stages) with the following final overall equation.

<u>FROM COMBUSTION (FULL)</u>

$$\underline{Final\ Overall\ equation}: 2C_2H_6 + 6O_2 \longrightarrow 2CO_2 + 6H_2O + 2CO + Energy(From\ five\ Stages) \qquad 1.39$$

This equation complements the final equation of Equation 1.30c for the combustion of ethane at high operating condition with nine stages as opposed to eleven stages when CO is not formed. With nine stages, only 6 moles of O_2 as opposed to 8 moles when there is no CO as product. Note that CO like oxidizing oxygen, but unlike CO_2 is also environmentally unfriendly

When the mechanisms of how chemical reactions take place (which indeed is HOW NATURE OPERATES) are not known, then one can begin to see why the oxidation of alkanes to give alcohols has always been an impossible task. Indeed, NOTHING is known universally about the wonders of NATURE. But we think we know. E.g., why do we use oxidizing agents?

1.3.4 Oxidation of Ethane

Methane is the first member of the Alkane family followed by ethane. Because of the weight and number of elements carried by them, at normal operating conditions, they exist as gases. As such, the Equilibrium state of existence of ethane is very strong, but far less than that of oxidizing oxygen molecule. Hence, its ability to exist in Equilibrium state of existence is suppressed, far more so than methane whose Equilibrium state of existence is stronger than that of ethane.

Stage1:

$$\underset{--}{O_2} \xrightleftharpoons[\text{Oxygen molecule}]{\text{Equilibruim State Existence of Oxidising}} 2en\cdot\ddot{\underset{..}{O}}\cdot nn$$

$$\underset{(Stabilized)}{2en\cdot O\cdot nn + \quad 2C_2H_6} \xrightleftharpoons{\text{Oxidation}} 2H-O\cdot nn \; + \; 2e\cdot C_2H_5$$

$$2e\bullet C_2H_5 \rightleftharpoons 2H\bullet e \; + \; \underset{(A)}{2e\bullet \overset{H}{\underset{H}{C}} - \overset{H}{\underset{H}{C}}\bullet n}$$

$$2H\bullet e \; + \; 2nn\bullet OH \rightleftharpoons 2H_2O$$

$$(A) \xrightarrow[\text{Release of Energy}]{\text{Deactivation}} 2H_2C=CH_2 \; + \; Energy \qquad 1.40a$$

$$\underline{Overall\ Equation}:\ \underset{--}{O_2} + 2C_2H_6 \longrightarrow 2H_2O \; + \; 2H_2C=CH_2 \; + \; Energy \quad 1.40b$$

Worthy of note here, is that based on the operating conditions for the oxidation, ethanol could not be produced here, because in the third step of the stage, the ethyl (Real name as will be shown herein or in the New

Frontiers is *ethylane*) group carrying an electro-free-radical is forced to release H•e (Hydrogen atom) immediately to give (A) which latter on deactivates to release energy to complete the stage with a single double headed symbol to indicate formation of product(s)- water and ethylene (Real name is ethene, because there is a compound called ethylene in ***the Carbene family of Hydrocarbon family tree*** as to be seen in the New Frontiers).

With methane, nothing was there to be released, except if methylene, the first member of the Carbene family was formed. If formed, the same product will still remain to be formed as shown below.

Stages 1 and 2 for methane in place of Stages 1 and 2 of Equations 1.32a and 1.34a respectively

Stage1:

$$\underline{\underline{O_2}} \xrightleftharpoons[\text{Oxygen molecule}]{\text{Equilibruim State Existence of Oxidising}} 2en\cdot\ddot{\underset{\cdot\cdot}{O}}\cdot nn$$

$$2en\cdot O\cdot nn + \underset{(Stabilized)}{2CH_4} \xrightleftharpoons{\text{Oxidation}} 2H-O\cdot nn + 2e\cdot CH_3$$

$$2e\bullet CH_3 \rightleftharpoons 2H\bullet e + \underset{(A)-Methylene}{2e\bullet \overset{H}{\underset{H}{C}}\bullet n}$$

$$2H\bullet e + 2nn\bullet OH \rightleftharpoons 2H_2O$$

$$(A) \xrightarrow[\text{Release of Energy}]{\text{Deactivation}} 2:CH_2\,(A\,Carbene) + Energy \quad (1.32a)^{**}$$

$$\underline{Overall\ Equation}: \underline{\underline{O_2}} + 2CH_4 \longrightarrow 2H_2O + 2:CH_2 + Energy \quad (1.32b)^{**}$$

Stage 2:

$$\underline{\underline{O_2}} \xrightleftharpoons[\text{Oxygen molecule}]{\text{Equilibruim State Existence of Oxidising}} 2en\cdot\ddot{\underset{\cdot\cdot}{O}}\cdot nn$$

$$2en\cdot O\cdot nn + \underset{(Stabilized)}{2:CH_2} \xrightleftharpoons{\text{Oxidation}} \underset{(B)}{2e\bullet\overset{H}{\underset{H}{C}}-O\bullet nn}$$

$$\xrightarrow[\text{Release of Energy}]{\text{Deactivation}} 2H_2C=O + Energy \quad (1.34a)^{**}$$

$$\underline{Overall\ Equation}: \underline{\underline{O_2}} + 2H_2C: \longrightarrow 2H_2C=O + Energy \quad (1.34b)^{**}$$

From here, stages 3 and 4 of Equations 1.37a and 1.38a follow in place of Stages 3 and 4 of Equations 1.35a and 1.36a respectively. Thus, to produce these alcohols, very low operating conditions and the exact molar ratios of oxidizing oxygen and the alkane must be used. Nevertheless, based on what has been shown during the combustion of ethane in the second step of Stage 5 of Equation 1.21a, if methanol was not formed, the combustion of the ethane beyond that stage would still have been possible. CO will still remain produced with number of stages reduced from nine to eight stages as shown below. Stage 5 has been replaced by Equation 1.32a** above. In Stage 6, the methylene is combusted to form a three-membered ring with two oxygen atoms too strained to exist. This decomposes in the seventh stage to give formaldehyde and oxidizing oxygen. In the eighth stage, CO is produced.

Continuation of the Final stages in combustion of Ethane

Stage 5: Stage 1 of Equation (1.32a)**

Stage 6:

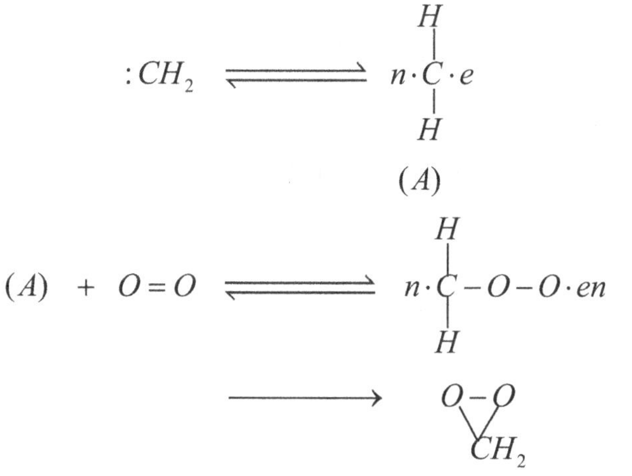

$$:CH_2 \rightleftharpoons n\cdot\overset{H}{\underset{H}{C}}\cdot e \quad (A)$$

$$(A) + O=O \rightleftharpoons n\cdot\overset{H}{\underset{H}{C}}-O-O\cdot en \longrightarrow \underset{CH_2}{O-O} \quad (B) \qquad (1.22)a^{**}$$

$$\underline{Overall\ equation}:\ 2H_2C + 2O_2 \longrightarrow 2(B) \qquad (1.22)b^{**}$$

Stage 7:

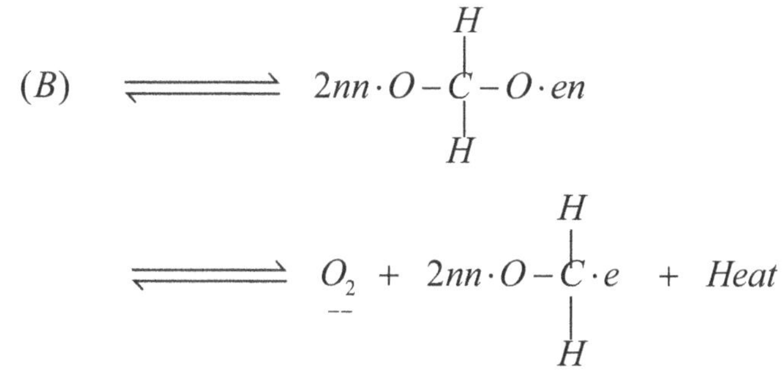

$$(B) \rightleftharpoons 2nn\cdot O-\overset{H}{\underset{H}{C}}-O\cdot en$$

$$\rightleftharpoons \underline{\underline{O_2}} + 2nn\cdot O-\overset{H}{\underset{H}{C}}\cdot e + Heat \quad (C)$$

$$(C) \xrightarrow[\text{Release of energy}]{\text{Deactivation}} 2H_2C=O \quad + \quad Heat \qquad (1.23)a^{**}$$

$$\underline{Overall\ equation:}\ 2(B) \longrightarrow \underset{--}{O_2} + 2H_2C=O + Energy \qquad (1.23)b^{**}$$

Stage 8: Same as Stage 3 of Equation 1.37a

$$\underline{Final\ Overall\ equation:}\ 2C_2H_6 + 6O_2 \longrightarrow 2CO_2 + 6H_2O + 2CO + Energy(From\ five\ Stages) \quad (1.39)$$

Hence methylene can indeed be produced as shown above, *depending on the operating conditions.* Stages 1 and 2b of Equations 1.32a and 1.34a respectively still remain as they are at lower operating conditions. Higher temperatures close to what is used for Cracking will be required to give methylene as will be seen downstream in the New Frontiers. ***Thus, the last case above compliments the two other cases of eleven and nine stages, because of the great importance and significance of OPERATING CONDITIONS and STAGE-WISE operations.***

For the ethane, the following takes place after Stage 1 of Equation 1.40a when oxidizing oxygen is readily available in the system.

Oxidation of Ethane continues:

(Oxidation of Ethene)

Stage 2:

$$\underset{--}{O_2} \underset{\text{Oxygen molecule}}{\overset{\text{Equilibruim State Existence of Oxidising}}{\rightleftharpoons}} 2en\cdot\ddot{\underset{\cdot\cdot}{O}}\cdot nn$$

$$2en\cdot O\cdot nn + \underset{(Stabilized)}{2H_2C=CH_2} \overset{\text{Oxidation}}{\rightleftharpoons} 2HO{\bullet}nn + \underset{(B)}{2e{\bullet}\overset{H}{\overset{|}{C}}=CH_2}$$

$$(B) \rightleftharpoons H{\bullet}e + e{\bullet}\overset{H}{\overset{|}{C}}=\underset{H}{\underset{|}{C}}{\bullet}n$$

$$2H{\bullet}e + 2nn{\bullet}OH \rightleftharpoons 2H_2O$$

$$(B) \xrightarrow[\text{Release of Energy}]{\text{Deactivation}} 2HC\equiv CH + Energy \qquad 1.41a$$

$$\underline{Overall\ Equation:}\ \underset{--}{O_2} + 2H_2C=CH_2 \longrightarrow 2H_2O + 2HC\equiv CH + Energy \quad 1.41b$$

In the second stage, it is shocking to find acetylene as one of the products, just as it was shocking to find ethene in Stage1. Water as vapor is found to be one of the products in almost all the stages. One can indeed begin to appreciate the great significance of the new concepts just introduced so far-

(i) Operating Conditions,

(ii) Stage-wise operations of everything in our world,

(iii) New Classifications for States of Existences,
(iv) Mechanisms for all systems and most importantly
(v) New classifications for radicals.

Looking very closely at the application of these new concepts in chemical reactions, one can begin to see that the concepts are not limited to chemistry alone, but to all disciplines. *Note that, in all the stages so far encountered, it is the electro-free or non-free-radical species (the males) that diffuses all the time in the presence or absence of the nucleo-free- or non-free- radicals (the females). The female does not diffuse under Equilibrium mechanism systems of operation.* As will be shown downstream, the female diffuses only when the male is not present in the system during Combination mechanism in polymerization systems and also under other conditions too early to be revealed. *Note also, that with oxidizing oxygen, it is the electro-non-free-radical end that diffuses all the time.* The same also applies when monomers or compounds are activated in the absence of initiators. ***The fact that it is only the males that move and work in all natural phenomena has a very deep meaning in humanity and many disciplines as will be shown in the New Frontiers and even herein- MOTHER NATURE.***

With the presence of more oxidizing oxygen in the system, the next stage follows.

(Oxidation of Acetylene)

<u>Stage 3:</u>

$$\underline{\underline{O_2}} \xrightleftharpoons[\text{Oxygen molecule}]{\text{Equilibruim State Existence of Oxidising}} 2en\cdot\ddot{\underset{\cdot\cdot}{O}}\cdot nn$$

$$\underset{(Stabilized)}{2en\cdot O\cdot nn + 2HC\equiv CH} \xrightleftharpoons{\text{Oxidation}} 2HO{\bullet}nn + \underset{(C)}{2\overset{H|}{C}\equiv C{\bullet}e}$$

$$(C) \rightleftharpoons 2H{\bullet}e + \underset{(D)}{2e{\bullet}C\equiv C{\bullet}n}$$

$$(D) \rightleftharpoons 4e{\bullet}\overset{\bullet n}{\underset{\bullet e}{C}}{\bullet}n$$

$$(E)\ \text{Activated Carbon Black}$$

$$2H{\bullet}e + 2nn{\bullet}OH \rightleftharpoons 2H_2O$$

$$(E) \xrightarrow[\text{Release of Energy}]{\text{Deactivation}} \underset{(Carbon\ black)}{4:\overset{\bullet nn}{\underset{\bullet en}{C}}} + Energy \qquad 1.42a$$

$$\underline{Overall\ Equation}: \underline{\underline{O_2}} + 2H_2C=CH_2 \longrightarrow 2H_2O + 4:\overset{\bullet nn}{\underset{\bullet en}{C}} + Energy\ 1.42b$$

In the stage above, the acetylene was oxidized to produce water, carbon black and energy. Activated carbon black was produced in the fourth step and this deactivated in the last step to give carbon black to complete the stage. ***Note that the Activated carbon black is different from Activated carbon*** as will shortly be shown. In the next and last stage, the carbon black was oxidized as follows. For the first time, one is beginning to show the structures of some carbon elements. ***From experimental observations, we know of the existences of Carbon black, Activated C, Coal, Coke, and so on; yet we don't know what they are! How can one continue to live in this kind of world dominated with illusions and fear of the unknown?*** Nevertheless, it is to be expected, because we don't fear what we know. ***In nature, it is the other way around. While in the natural world, it is fear of the known, in the physical world, it is fear of the unknown, noting that our world a complex one, is both physical and natural.***

Stages 4 &5:

$$\underline{\underline{O_2}} \xrightleftharpoons[\text{Oxygen molecule}]{\text{Equilibruim State Existence of Oxidising}} 2en\cdot\ddot{\underset{\cdot\cdot}{O}}\cdot nn$$

$$2en\cdot O\cdot nn + 2:\overset{\bullet n}{\underset{\bullet e}{C}} \xrightleftharpoons{\text{Oxidation}} 2nn\bullet O-\overset{\bullet e}{\underset{\bullet n}{C}}\bullet e$$

$$(Stabilized) \qquad\qquad (F)$$

$$\xrightarrow[\text{Release of Energy}]{\text{Deactivation}} 2:C=O + Energy \qquad 1.43a$$

$$\underline{Overall\ Equation}: 2\underline{\underline{O_2}} + 4\overset{\bullet n}{\underset{\bullet e}{C}}: \longrightarrow 4:C=O + Energy \qquad 1.43b$$

The stage above is a two-stage system in parallel, since four moles of Carbon black are involved. So there are indeed five stages.

FOR OXIDATION (PARTIAL)

$$\underline{Final\ Overall\ equation}: 2C_2H_6 + 5\underline{\underline{O_2}} \longrightarrow 4CO + 6H_2O + Energy(5)$$

$$(From\ five\ Stages) \qquad 1.43$$

$$\underline{Final\ Overall\ equation}: 2CH_4 + 3\underline{\underline{O_2}} \longrightarrow 2CO + 4H_2O + Energy(3)$$

$$(From\ three\ Stages) \qquad 1.43b$$

FROM COMBUSTION (FULL)

$$\underline{Final\ Overall\ equation}: 2C_2H_6 + 6O_2 \longrightarrow 2CO_2 + 6H_2O + 2CO + Energy(5)$$

$$(From\ eight\ or\ nine\ Stages) \qquad 1.44$$

While in Combustion, the formation of carbon monoxide cannot be prevented in the absence of oxidizing oxygen in the system or an oxidizing agent, in Oxidation, the formation can be prevented by not starving the system with oxidizing oxygen as already shown in Stage 4 of Equation 1.38a during the oxidation of methane. With presence of more oxidizing oxygen in the system, the fifth or sixth stage follows as recalled below.

Stages 6&7:

$$\underset{--}{O_2} \xrightleftharpoons[\text{Oxygen molecule}]{\text{Equilibruim State Existence of Oxidising}} 2en\cdot\ddot{\underset{..}{O}}\cdot nn$$

$$2en\cdot O\cdot nn + \underset{(Stabilized)}{2C=O} \xrightleftharpoons{\text{Oxidation}} \underset{(B)}{2e\bullet \overset{O}{\overset{\|}{C}} - O\bullet nn}$$

$$(B) \xrightarrow[\text{Release of Energy}]{\text{Deactivation}} 2CO_2 + Energy \qquad 1.45a$$

$$\underline{Overall\ Equation\ 2}: \underset{--}{O_2} + 4CO \longrightarrow 4CO_2 + Energy \qquad 1.45b$$

Stage 6 is also a two-stage system in parallel, bringing the total number of stages to seven.

FULL OXIDATION

$$\underline{Final\ Overall\ equation}: 2C_2H_6 + 7\underset{--}{O_2} \longrightarrow 4CO_2 + 6H_2O + Energy(7 or 8)$$

$$(From\ seven\ Stages) \qquad 1.46$$

During the oxidation of ethane, it was in Stages 4 and 5 that CO appeared as main product in place of CO_2, when in the system all the oxidizing oxygen has been used. With presence of more oxidizing oxygen, in Stages 6 and 7, no CO was produced. One can now begin to see one of the origins of so much CO_2 in our echo system, for when CO and oxidizing are present in the air, they combine to form CO_2. For full oxidation of ethane, seven stages are required, taking note of the molar ratios involved. While for methane, four moles of oxidizing oxygen are required, for ethane seven moles of oxidizing oxygen are required. One can begin to see the origin of $3\frac{1}{2}$ in Equation 1.31. That equation is strictly for Oxidation and not for Combustion. Note the stage where CO appears in the process.

1.3.5 **Oxidation of Propane (C_3H_8)**

This is the third member of the Alkane family in the Hydrocarbon family tree. With more carbon and hydrogen atoms in CH_2 series, one should expect to have more stages than the two cases considered so far. This is Mathematics. The exact molar ratios of the components involved to get different type of products, is all Mathematics.

Stage1:

$$\underset{--}{O_2} \xrightleftharpoons[\text{Oxygen molecule}]{\text{Equilibruim State Existence of Oxidising}} 2en\cdot\ddot{\underset{\cdot\cdot}{O}}\cdot nn$$

$$\underset{(Stabilized)}{2en\cdot O\cdot nn + 2C_3H_8} \xrightleftharpoons{\text{Oxidation}} \underset{(B)}{2e\bullet\overset{H}{\underset{H}{C}}-\overset{H}{\underset{H}{C}}-\overset{H}{\underset{H}{C}}-H} + 2nn\bullet OH$$

$$\longrightarrow \quad 2HO-C_3H_7 \qquad 1.47a$$

$$\underline{Overall\ Equation}: \underset{--}{O_2} + 2H_8C_3 \longrightarrow 2C_3H_7OH \qquad 1.47b$$

Worthy of note is that unlike methane and ethane, hydrogen atom (H•e) could not be released from (B) above, because when released the carbon center from where it was released cannot carry a nucleo-free-radical to form an alkene. CH_3 group cannot be released because it is more radical pushing than H. Notice how NATURE operates between the RICH and the POOR, whether of components or not. ***Here, unlike with the case of methane and ethane, propanol can readily be produced only from n-propane regardless the operating conditions, for as long as the exact molar ratios of the two components have been used.*** Above, the molar ratio is 1 to 1. With iso-propane, iso-propanol cannot be readily obtained at the high operating conditions. Instead, a stage similar to Stage 1 of Equation 1.40a will be obtained to produce propene.

$$\underline{Overall\ equation}: 2CH_4 + \underset{--}{O_2} \longrightarrow 2H_2O + Unactivated\ Methylene2(:CH_2)$$

$$OR\ 2METHANOL(CH_3OH) + Energy \qquad 1.48a$$

$$2C_2H_6 + \underset{--}{O_2} \longrightarrow 2H_2O + 2H_2C=CH_2 + Energy \qquad 1.48b$$

$$2C_3H_8 + \underset{--}{O_2} \longrightarrow 2C_3H_7OH \qquad 1.48c$$

After the first stage, methanol and ethanol can then be obtained as a second stage from the products at mild operating conditions as shown below for ethanol.

Stage 1:

$$H_2O \rightleftharpoons H\bullet e + nn\bullet OH$$

$$H\bullet e + \overset{H}{\underset{H}{C}}=\overset{H}{\underset{H}{C}} \xrightleftharpoons{\text{Activation}} H\bullet e + n\bullet\overset{H}{\underset{H}{C}}-\overset{H}{\underset{H}{C}}\bullet e$$

$$\underset{\longleftarrow}{\xrightarrow{Activation}} \quad H_3C - \overset{H}{\underset{H}{C}} \bullet e$$

$$(A)\,Ethylane$$

$$(A) \quad + \quad nn\bullet OH \longrightarrow H_5C_2OH \qquad 1.49a$$

$$\underline{Overall\ equation}: \ H_2O \ + \ H_2C = CH_2 \xrightarrow{Heat} H_5C_2OH \qquad 1.49b$$

Thus, methanol and ethanol can be obtained when the right operating conditions are used.

With propane, after the first stage of Equation 1.47a, the second stage follows in the presence of more oxidizing oxygen in the system.

Oxidation of C_3H_8 continues

Oxidation of Propanol

Stage 2:

$$\underset{--}{O_2} \underset{Oxygen\ molecule}{\overset{Equilibruim\ State\ Existence\ of\ Oxidising}{\rightleftharpoons}} 2en \cdot \ddot{\underset{..}{O}} \cdot nn$$

$$\underset{(Stabilized)}{2en \cdot O \cdot nn + \ 2C_3H_7OH} \quad \xrightleftharpoons{Oxidation} \underset{(C)}{2H - \overset{H}{\underset{H}{C}} - \overset{H}{\underset{H}{C}} - \overset{H}{\underset{H}{C}} - O\bullet en} + 2nn\bullet OH$$

$$(C) \quad \rightleftharpoons \quad 2(H_5C_2)HC = O \ + \ 2H\bullet e \ + \ Energy$$

$$2H \bullet e \ + \ 2nn \bullet OH \longrightarrow 2H_2O \qquad 1.50a$$

$$\underline{Overall\ Equation}: 2\underset{--}{O_2} + 2C_3H_8 \longrightarrow 2H_2O + 2(C_2H_5)HC = O + Energy \quad 1.50b$$

Note the hydrogen atom removed. It is from the center carrying OH group, that which is more radical-pushing than C_2H_5 group (See Equations 1.17a and 1.34a). When the H was removed, pentaldehyde was instantaneously formed with release of energy. This was then followed by the oxidation of the pentaldehyde in the presence of more oxidizing oxygen in the next stage to form ethanoic acid.

Oxidation of pentaldehyde

Stage 3:

$$\underline{\underline{O}}_2 \xrightleftharpoons[\text{Oxygen molecule}]{\text{Equilibruim State Existence of Oxidising}} 2en\cdot\ddot{\underset{..}{O}}\cdot nn$$

$$2en\cdot O\cdot nn + \underset{(Stabilized)}{2O = CH(C_2H_5)} \xrightleftharpoons{\text{Oxidation}} 2H - O\cdot nn + \underset{(B)}{2e\cdot \overset{O}{\overset{||}{C}} - C_2H_5}$$

$$\longrightarrow \underset{(C)}{2H_5C_2 - \overset{O}{\overset{||}{C}} - OH} \qquad 1.51a$$

$$\underline{Overall\ Equation}: \underline{\underline{O}}_2 + 2O = CH(C_2H_5) \longrightarrow 2H_5C_2COOH \qquad 1.51b$$

With the presence of more oxidizing oxygen in the system, oxidation of the acid continues as follows.

Oxidation of Ethanoic acid

Stage4:

$$\underline{\underline{O}}_2 \xrightleftharpoons[\text{Oxygen molecule}]{\text{Equilibruim State Existence of Oxidising}} 2en\cdot\ddot{\underset{..}{O}}\cdot nn$$

$$2en\cdot O\cdot nn + \underset{(Stabilized)}{2HO\overset{O}{\overset{||}{C}}C_2H_5} \xrightleftharpoons{\text{Oxidation}} 2HO\cdot nn + \underset{(B)}{2en\cdot O - \overset{O}{\overset{||}{C}}C_2H_5}$$

$$(B) \rightleftharpoons 2CO_2 + 2e\cdot C_2H_5 + Energy$$

$$2e\cdot C_2H_5 \rightleftharpoons 2H\cdot e + \underset{(C)}{2e\cdot \overset{H}{\underset{H}{C}} - \overset{H}{\underset{H}{C}}\cdot n}$$

$$2H\cdot e + 2nn\cdot OH \rightleftharpoons 2H_2O$$

$$(C) \xrightarrow{\text{Deactivation}} 2H_2C = CH_2 + Energy \qquad 1.52a$$

$$\underline{Overall\ Equation}: \underline{\underline{O}}_2 + 2C_2H_5COOH \longrightarrow 2CO_2 + 2H_2O + 2H_2C = CH_2 + Energy \qquad 1.52b$$

$$\underline{Overall\,Overall\ equation}: 2C_3H_8 + 4\underline{\underline{O}}_2 \longrightarrow 4H_2O + 2CO_2 + 2H_2C = CH_2 + Energy \qquad 1.52c$$

What today is called acetic acid (H_3CCOOH) should indeed be called methanoic acid or methanyloic acid, because what is carrying the COOH group is from methane or is from methanyl (not methyl) group. In the stage above, energy is released in the third and last steps, with formation of water and ethene. In the next stage, the ethene is next oxidized. A peroxide could not be formed above, otherwise it would have been a non-productive stage.

Stage 5:

$$\underset{--}{O_2} \xrightleftharpoons[\text{Oxygen molecule}]{\text{Equilibruim State Existence of Oxidising}} 2en\cdot\ddot{\underset{..}{O}}\cdot nn$$

$$2en\cdot O\cdot nn + \underset{(Stabilized)}{2H_2C=CH_2} \xrightleftharpoons{\text{Oxidation}} 2HO\bullet nn + \underset{(B)}{2e\bullet \overset{H|}{C}=CH_2}$$

$$(B) \rightleftharpoons 2H\bullet e + \underset{(C)}{2e\bullet \overset{H|}{C}=\underset{|H}{C}\bullet n}$$

$$2H\bullet e + 2nn\bullet OH \rightleftharpoons 2H_2O$$

$$(C) \xrightarrow[\text{Release of Energy}]{\text{Deactivation}} 2HC\equiv CH + Energy \quad 1.53a$$

$$\underline{Overall\ Equation}: \underset{--}{O_2} + 2H_2C=CH_2 \longrightarrow 2H_2O + 2HC\equiv CH + Energy \quad 1.53b$$

It is no longer shocking to find acetylene as one of the products, just as it was not shocking to find ethene in Stage 4. Water as vapor is found to be one of the products in almost all the stages.

With the presence of more oxidizing oxygen in the system, the next stage follows.

Stage 6:

$$\underset{--}{O_2} \xrightleftharpoons[\text{Oxygen molecule}]{\text{Equilibruim State Existence of Oxidising}} 2en\cdot\ddot{\underset{..}{O}}\cdot nn$$

$$2en\cdot O\cdot nn + \underset{(Stabilized)}{2HC\equiv CH} \xrightleftharpoons{\text{Oxidation}} 2HO\bullet nn + \underset{(C)}{2\overset{H|}{C}\equiv C\bullet e}$$

$$(C) \rightleftharpoons 2H\bullet e + \underset{(D)}{2e\bullet C\equiv C\bullet n}$$

$$(D) \rightleftharpoons 4e\bullet\overset{\bullet n}{\underset{\bullet e}{C}}\bullet n$$

$$(E)\ Activated\ Carbon\ Black$$

$$2H\bullet e + 2nn\bullet OH \rightleftharpoons 2H_2O$$

$$(E) \xrightarrow[\text{Release of Energy}]{\text{Deactivation}} 4:\overset{\bullet nn}{\underset{\bullet en}{C}} + Energy \quad 1.54a$$

$$(Carbon\ black)$$

$$\underline{Overall\ Equation}: \underset{--}{O_2} + 2H_2C=CH_2 \longrightarrow 2H_2O + 4:\overset{\bullet nn}{\underset{\bullet en}{C}} + Energy \quad 1.54b$$

In the stage above, the acetylene was oxidized to produce water, carbon black and energy. Activated carbon black was produced in the fourth step and this deactivated in the last step to give carbon black to complete the stage. In the last stage, the carbon black was oxidized as follows.

Stage 7&8:

$$\underset{--}{O_2} \xrightleftharpoons[\text{Oxygen molecule}]{\text{Equilibruim State Existence of Oxidising}} 2en\cdot\ddot{\underset{\cdot\cdot}{O}}$$

$$2en\cdot O\cdot nn + 2:\overset{\bullet n}{\underset{\bullet e}{C}} \xrightleftharpoons{\text{Oxidation}} 2\,nn\bullet O-\overset{\bullet e}{\underset{\bullet n}{C}}\bullet e$$

$$(Stabilized) \qquad\qquad (F)$$

$$\xrightarrow[\text{Release of Energy}]{\text{Deactivation}} 2:C=O \quad + \quad Energy \qquad 1.55a$$

$$\underline{Overall\ Equation}: 2\underset{--}{O_2} + 4\overset{\bullet n}{\underset{\bullet e}{C}}: \longrightarrow 4:C=O \quad + \quad Energy \qquad 1.55b$$

Stage 9&10:

$$\underset{--}{O_2} \xrightleftharpoons[\text{Oxygen molecule}]{\text{Equilibruim State Existence of Oxidising}} 2en\cdot\ddot{\underset{\cdot\cdot}{O}}\cdot nn$$

$$2en\cdot O\cdot nn + 2C=O \xrightleftharpoons{\text{Oxidation}} 2\,e\bullet\overset{\overset{O}{\|}}{C}-O\bullet nn$$

$$(Stabilized) \qquad\qquad (B)$$

$$(B) \xrightarrow[\text{Release of Energy}]{\text{Deactivation}} 2CO_2 \quad + \quad Energy \qquad 1.56a$$

$$\underline{Overall\ Equation}: \underset{--}{O_2} + 2CO \longrightarrow 2CO_2 \quad + \quad Energy \qquad 1.56b$$

Note that Stages 7 and 8, and 9, and 10 are taking place simultaneously in parallel and not in series when the system is full with oxidizing agents as shown schematically below in Figure 1.5.

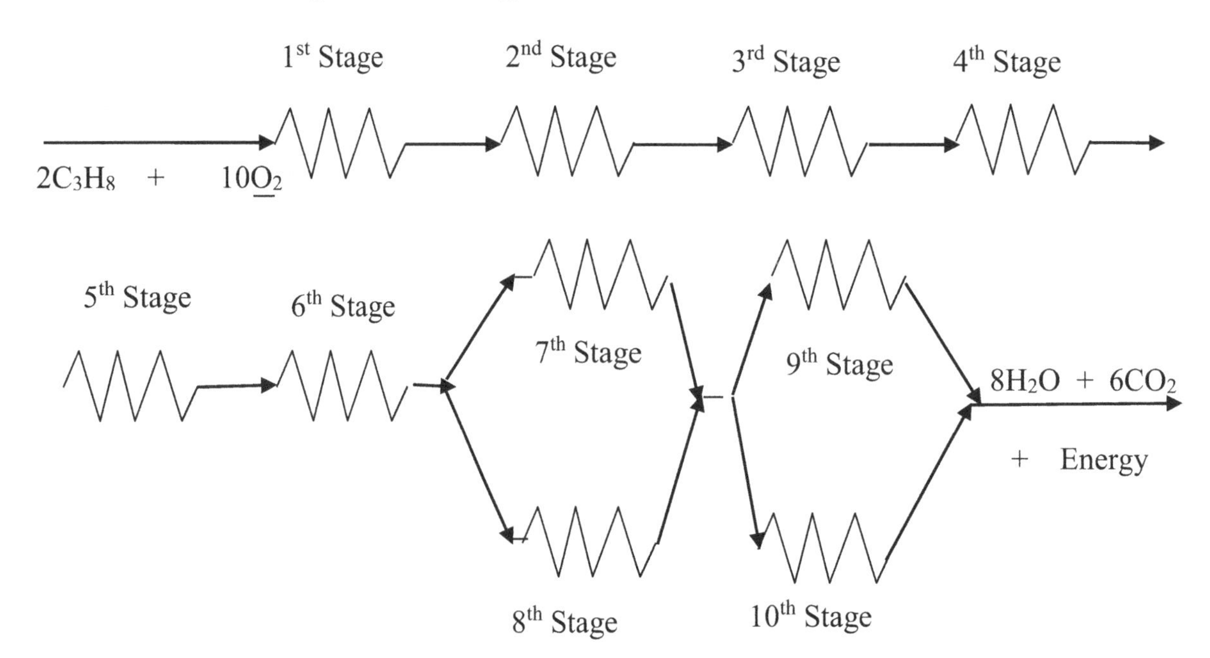

Figure 1.5 Series/Parallel configuration of oxidation of Propane

The final overall equation for the complete oxidation of propane, ethane and methane are therefore as follows.

$$\underline{Final\ Overall\ equation}:\ 2C_3H_8 + 10\underline{\underline{O_2}} \longrightarrow 6CO_2 + 8H_2O + Energy(9)$$

$$(From\ ten\ Stages) \quad 1.56c$$

$$\underline{Final\ Overall\ equation}:\ 2C_2H_6 + 7\underline{\underline{O_2}} \longrightarrow 4CO_2 + 6H_2O + Energy(7)$$

$$(From\ seven\ Stages) \quad (1.46)$$

$$\underline{Final\ overall\ equation}:\ 2CH_4 + 4\underline{\underline{O_2}} \xrightarrow{Oxidation} 4H_2O + 2CO_2 + Energy(4)$$

$$(From\ four\ Stages) \quad (1.36c)$$

In the oxidation of propane, ten stages were involved, in which the most important stages are the stages where energy is released, noting the types of products obtained in every stage.

Before completing this section on the combustion and oxidation of first gaseous members of hydrocarbons, there is need to recall what was shown in Equation 1.19 for States of existence of acetaldehyde ($HCOCH_3$), during the combustion of ethane. So far only the Equilibrium state of existence of Equation 1.19a was used at fairly mild operating conditions (See Stage 3 of Equation 1.18). That was what was used to complete provision of the mechanism for the combustion of the ethane. Can one use the same type of state of existence for the combustion of propane with $HCOC_2H_5$? Hence the need to consider the use of the Activated/Equilibrium state of existence of Equation 1.19b, a condition which is to be expected at the high operating conditions and in the presence of a passive catalyst such as an acid. Therefore. instead of Stage 3 of Equation 1.18, the followings are obtained for the continued combustion of ethane.

Stage 3:

$$CH_3CHO \underset{Existence\ of\ the\ Unactivated\ Aldehyde}{\overset{Equilibrium\ state\ of}{\rightleftharpoons}} H\cdot e + nn{\bullet}O-\underset{\underset{H}{|}}{\overset{\overset{n{\bullet}CH_2}{|}}{C}}\cdot e$$

$$H\cdot e + O=O \underset{Oxygen}{\overset{Activation\ of}{\rightleftharpoons}} H-O-O\cdot en$$

$$H-O-O\cdot en + e\cdot\underset{\underset{H}{|}}{\overset{\overset{n{\bullet}CH_2}{|}}{C}}-O{\bullet}nn \xrightarrow{Combination} H-O-O-\underset{\underset{H}{|}}{\overset{\overset{H}{|}}{C}}-\overset{\overset{H}{|}}{C}=O$$

$$(B)-A\ peroxide \quad 1.57$$

Stage4:

$$H-\overset{\overset{O}{\|}}{C}-\underset{\underset{H}{|}}{\overset{\overset{H}{|}}{C}}-O-O-H \xrightleftharpoons[\text{of Existence}]{\text{It's Equilibrium State}} H-O\cdot nn + en\bullet O-\underset{\underset{H}{|}}{\overset{\overset{H}{|}}{C}}-\overset{\overset{O}{\|}}{C}-H$$

$$en\cdot O-\underset{\underset{H}{|}}{\overset{\overset{H}{|}}{C}}-\overset{\overset{O}{\|}}{C}-H \rightleftharpoons \underset{(C)}{H-\overset{\overset{O}{\|}}{C}\cdot e} + H_2C=O + Energy$$

$$(C) \rightleftharpoons e\bullet\overset{\overset{O}{\|}}{C}\bullet n + H\bullet e$$

$$H\cdot e + nn\cdot OH \rightleftharpoons H_2O$$

$$e\cdot\underset{\underset{O}{\|}}{C}\bullet n \xrightarrow[\text{Energy Release}]{\text{Deactivation}} CO + Energy \qquad 1.58a$$

$$\underline{Overall\ Equation}: C_2H_6 + 2O_2 \longrightarrow CO + 2H_2O + H_2C=O + Energy \qquad 1.58b$$

Stage 5:

$$H_2C=O \xrightleftharpoons[\text{Existence of formaldehyde}]{\text{Activation State of}} H\cdot e + n\cdot\overset{\overset{H}{|}}{C}=O$$

$$H\cdot e + O=O \xrightleftharpoons[\text{Oxygen}]{\text{Activation of}} \underset{(A)}{H-O-O\cdot en}$$

$$A) + n\cdot\overset{\overset{H}{|}}{C}=O \xrightarrow{\text{Combination}} \underset{(Another\ peroxide)}{H-O-O-\overset{\overset{H}{|}}{C}=O} \qquad (1.14)$$

So far, note the absence of oxidizing oxygen as one of the products or as an intermediate.

Stage 6:

$$H-O-O-\overset{\overset{H}{|}}{C}=O \xrightleftharpoons[\text{Existence}]{\text{Equilibrium State of}} H-\ddot{\underset{..}{O}}\cdot nn + \underset{(C)}{en\cdot O-\overset{\overset{H}{|}}{C}=O}$$

$$(C) \quad \underset{Release}{\overset{Energy}{\rightleftharpoons}} CO_2 + H\cdot e + Energy$$

$$H\cdot e + nn\cdot OH \xrightarrow{Combination} H_2O \qquad (1.15a)$$

$$\underline{Overall\ Equation}: C_2H_6 + 3O_2 \longrightarrow CO + CO_2 + 2H_2O + Energy \qquad 1.59$$

With the use of Activated/Equilibrium state of existence of acetaldehyde of Equation 1.19b for the continued combustion of ethane, only six stages wherein energy is released four times in three stages were obtained. This is different from the case where the Equilibrium state of existence of acetaldehyde of Equation 1.19a was used. For this case, eight or nine stages with release of energy five times in four stages were obtained. The former required the use of one mole of ethane, while the latter required the use of two moles of ethane for their combustion to proceed. For both of them, the same products were obtained, but not the same intermediate products between stages. Methane, oxidizing oxygen, methanol, and more were not obtained here due to the different peroxides obtained after the second stage. Though the amount of energy generated may be the same for both cases, based on how oxidation takes place, that in which two moles of the hydrocarbon during oxidation are required after the oxidizing oxygen molecule produced inside a stage from the oxidizing agent is used in another stage are involved, the acetaldehyde under these operating conditions is therefore in EQUILIBRIUM state of existence as opposed to ACTIVATED/ EQUILIBRIUM state of existence. To keep the aldehydes and ketones, in the Activated/Equilibrium state of existence, presence of acids or alkalis are usually required as passive catalyst- a phenomenon called Enolization as will be explained downstream herein and in the New Frontiers. Vinyl alcohol which is very unstable molecularly rearranges to acetaldehyde under a particular operating condition (i.e., when removed from a very cold environment to our own Standard temperature and pressure (STP) environment), while acetaldehyde *enolizes* to give the vinyl alcohol under a different operating condition.

Based on the mechanisms which have been provided for Combustion and Oxidation of just hydrocarbons alone, one can begin to see how NATURE operates, noting that what has been shown so far is just a grain of sand in what will be seen in the NEW FRONTIERS. ***This text is just to lay the foundations for the launching of the first six Volumes of the New Frontiers.*** From what has been seen, though CO_2 still remains one of the products, how to make the use of hydrocarbon fuels environmentally friendly has been very deeply shown. The problem in part with respect to pollution in our environment is not that of CO_2, but those of CO, Oxidizing oxygen, NOs, SOs, deforestation and etc. Otherwise, we human-beings that emit CO_2 all the time for the plants to use, must be a curse to the environment if no plants were there. However, without the oxidizing oxygen, the ozone layer will not exist. It does not have to be used for full combustion, but for the production of different kinds of products. Only methane can be combusted without producing CO and therefore the only one that is environmentally friendly when combusted.

As has already been said based on the products obtained, the oxidizing oxygen produced cannot oxidize the carbon dioxide and water formed as shown below for the water. It can however oxidize CO.

1.3.6 Non-oxidation of Water

Stage 1:

$$\underset{--}{O_2} \xrightleftharpoons[\text{Existence of Oxidising oxygen}]{\text{Equilibrium State of}} 2en\cdot\ddot{\underset{\cdot\cdot}{O}}\cdot nn$$

$$2nn\cdot\ddot{\underset{\cdot\cdot}{O}}\cdot en + 2H_2O \xrightleftharpoons{\text{Abstraction}} 2H-O\cdot nn + 2en\cdot O-H$$

$$\xrightleftharpoons[\text{Existence of Hydrogen peroxide}]{\text{Equilibrium State of}} 2H-O-O-H$$

$$(Hydrogen\ peroxide)$$

$$(Stable,\ reactive\ and\ soluble) \qquad 1.60a$$

$$\underline{Overall\ Equation}:\ \underline{O_2} + 2H_2O \xrightleftharpoons[\text{Product}]{\text{No}} 2HOOH \qquad 1.60b$$

Stage 1:

$$\underset{--}{O_2} \xrightleftharpoons[\text{Existence of Oxidising oxygen}]{\text{Equilibrium State of}} 2en\cdot\ddot{\underset{\cdot\cdot}{O}}\cdot nn$$

$$2nn\cdot\ddot{\underset{\cdot\cdot}{O}}\cdot en + 2H_2O \xrightleftharpoons{\text{Abstraction}} 2H-O\cdot nn + 2en\cdot O-H$$

$$2HO\bullet en \rightleftharpoons 2H\bullet e + \underset{--}{O_2} + Energy$$

$$2H\bullet e + 2nn\bullet OH \xrightarrow[\underset{--}{O_2}]{\text{Suppressed by}} 2H_2O$$

$$(Stable,\ reactive\ and\ insoluble) \qquad 1.61a$$

$$\underline{Overall\ Equation}:\ \underline{O_2} + 2H_2O \longrightarrow 2H_2O + \underset{--}{O_2} + Energy \qquad 1.61b$$

See Equation 1.25a to ascertain the validity of the last equation above.

Based on the first equation, (Equation 1.60a), the oxidizing agent is stable, reactive, but SOLUBLE in water, since what is on the right of the symbols (HOOH), is different from what is on the left of the symbols (Oxidizing oxygen molecule and water). There is no single double headed arrow in the last step of the stage, clear indication of an unproductive stage. Such a stage with different compounds on the right-hand side (RHS) and left-hand side (LHS), is said to be **"STABLE, REACTIVE AND SOLUBLE"**. In the second equation (Equation 1.61a), due to presence of heat, HO•en was forced to release the atom (H•e) [which is the one that should carry the electro-radical, being more electropositive than oxygen] and energy, part of which is stored in the ring. Since in the last step of the stage, there is a single double headed arrow, the stage is productive, but with the same products as reactants on both sides of the symbols. Whether the stage is productive or not, once the RHS and LHS contain the same compounds, the stage is said to be **"STABLE, REACTIVE AND INSOLUBLE", with a difference.** The difference is release of energy, just as NaOH does in water. In Chapter 2, these will be clearly understood using what we use every day (e.g., NaCl).

Thus, the oxidizing oxygen molecule can be observed to be either soluble or insoluble in water depending on the operating conditions. For the first time, one is beginning to provide the clear definitions, the mechanisms of the concept of **SOLUBILITY and INSOLUBILTY,** two phenomena where it was wrongly believed that no chemical reactions take place. Hence, they have always been used synonymously with **MISCIBILITY/DISSOLUTION** and **IMMISCIBILITY/NON-DISSOLUTION**. Solubility and Dissolution are completely two different phenomena as will be explained herein downstream. One can see why both of them can exist side by side as phenomena in a system or one alone, i.e., that in which there is dissolution, but no solubilisation. ***The case where there is solubilisation, but no dissolution does not exist, because before a compound can be soluble in another compound, it must first dissolve in that compound.***

Sodium chloride dissolves and is soluble in water never to give hydrochloric acid and sodium hydroxide. But hydrochloric acid dissolves and reacts with sodium hydroxide to give sodium chloride (Salt) and water. Oxidizing oxygen molecule miscibilizes and is soluble or insoluble in water never to give hydrogen peroxide. But hydrogen peroxide decomposes when weakly heated to give water, oxidizing oxygen molecule and energy as has already been shown (See Stage 8 of Equation 1.25a) and recalled below.

Decomposition of Hydrogen peroxide

Stage 1:

$$2H-O-O-H \underset{Existence\ of\ Hyrogen\ peroxide}{\overset{Equilibrium\ State\ of}{\rightleftharpoons}} 2H-O\cdot en + 2nn\cdot O-H$$

$$2H-O\cdot en \underset{Release}{\overset{Energy}{\rightleftharpoons}} 2H\cdot e + \underline{O_2} + Energy$$

$$2H\cdot e + 2nn\cdot O-H \xrightarrow[presence\ of\ \underline{O_2}]{Suppressed\ by\ the} 2H_2O \qquad (1.25a)$$

$$\underline{Overall\ Equation}: 2HOOH \longrightarrow 2H_2O + \underline{O_2} + Energy \qquad (1.25b)$$

Hydrogen peroxide is herein an oxidizing agent wherein energy is produced. Worthy of note are Equations 1.25a and 1.61a. In the first, HOOH is decomposed to release energy, water and oxidizing oxygen, while in the second, water and oxidizing oxygen molecule products from decomposition, react again to give water, oxidizing oxygen and energy. The energy generated from the decomposition of two moles of hydrogen peroxide can therefore be used continuously under particular operating conditions to run a system. Two moles of hydrogen peroxide were required to produce one mole of oxidizing oxygen. If one mole of hydrogen peroxide had been used, the single element of oxidizing oxygen produced must be used within the single stage to give another product in addition to water. In such a case, energy cannot be produced continuously. From stage one above, Stage 2 follows and continues perpetually.

Semi-**Perpetual Oxidation of Water**

Stage 2:

$$\underline{O_2} \underset{Existence\ of\ \underline{O_2}}{\overset{Aquilibrium\ State\ of}{\rightleftharpoons}} 2en\cdot\ddot{\underset{\cdot\cdot}{O}}\cdot nn$$

$$2nn\cdot\ddot{\underset{\cdot\cdot}{O}}\cdot en + 2H_2O \overset{Abstraction}{\rightleftharpoons} 2H-O\cdot en + 2nn\cdot O-H$$

$$2H-O\cdot en \underset{Release}{\overset{Energy}{\rightleftharpoons}} 2H\cdot e + \underline{O_2} + Energy$$

$$2H\cdot e + 2nn\cdot O-H \xrightarrow[of\ \underline{O_2}]{Suppressed\ by\ presence} 2H_2O \qquad (1.61a)$$

$$\underline{Overall\ Equation}: \underset{(A)}{\underline{O_2}} + 2H_2O \xrightarrow[Water]{Oxidation\ of} \underset{(B)}{\underline{O_2}} + 2H_2O + Energy \qquad (1.61b)$$

*The Two Oxidizing Oxygen Molecules, (A) and (B) above, are very different in terms of the amount of Strain Energy in their rings, for which that of (A) is greater than that of (B). The first law with respect to Energy Conservation in Chemical Engineering must be obeyed. Even if energy was added from Stage*1, *both rings will still be different in energy content, because the reaction is exothermic.*

One expects that the oxidizing oxygen molecules produced in stages keep increasing in size as it is being used until the Strain in the ring becomes too small to make the increasing size of the ring exist as an imaginary ring. No energy can be provided for such a large ring to attain what is called the Minimum Required Strain energy (Min.RSE) if the ring is real, i.e., exists as will be shown in the New Frontiers ***in the Physics of Ringed compounds downstream.*** Therefore, a point is reached when the imaginary ring becomes just oxygen molecule as shown below, because in all systems living or non-living, energies are stored in visible Activation centers, invisible π-bonds carried by rings and σ-bonds and indeed all types of bonds. A ring contains plenty of stored energy depending on the family. These are called the STRAIN ENERGY (SE) which can only be released when the ring is opened either instantaneously if it has points of scission or via what are in general called FUNCTIONAL CENTERS. In general, there are three different kinds with different types, two of which we have barely seen so far- one with π-bonds and the other with vacant orbital/paired unbonded radicals such as in CO. Since the ring above is imaginary, and that stage is reactive, stable and insoluble, i.e., a non-productive stage, the energy seen in the Equation 1.61a is coming from what the oxygen center is carrying. Since, in this case no stable product is obtained in the process, this energy though small (Coming from full hetero-ring), is still enormous.

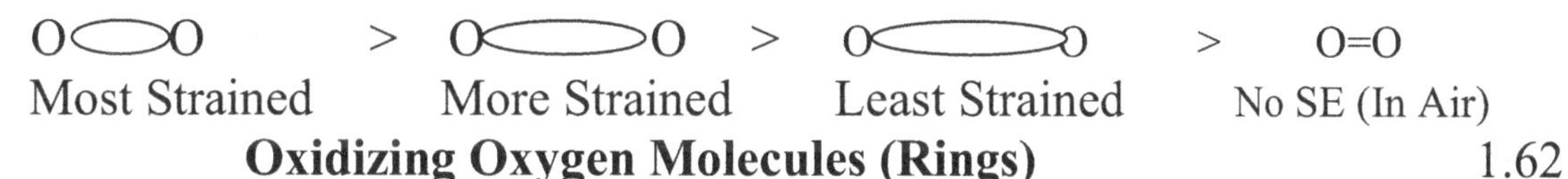

Oxidizing Oxygen Molecules (Rings) 1.62

Unknown is the fact that, when real rings belong to the same family, they carry the same SE, regardless the size of the ring. For example, all Cyclo-Alkane rings whether 3- or 4- or 5- or 6- etc. membered rings, contain the same amount of SE (about 25kcals/mole for the family). With increasing size, the ring becomes less strained and more difficult to open. All these will be fully explained in the New Frontiers, using the forces of Nature in Physics.

In general, the larger the size of the ring, the less strained is the ring. At the point where molecular oxygen is produced, fresh hydrogen molecules with hydrogen catalyst can be introduced into the system to begin a new cycle using molecular oxygen of the air from which HOOH is produced. Hence the process is semi-perpetual, noting that there is no perpetual one in this respect in our world, since they are all DEPENDENT variables. ***Only TIME an INDEPENDENT variable is perpetual in character not when it is used, but when it is not used. While only the real part is what we use, the imaginary part we cannot use.*** This more important part we neglect, because it cannot be comprehended being a symbol of the PAST. But it is there playing a major role. The Water now forms a reservoir for storage of energy without the need for a battery. All that is needed is a continuous source of HOOH. Indeed Table 1.2 summarizes the distinction between Molecular Oxygen and Oxidizing Oxygen molecule in terms of their structures, states of existence and other properties.

Table 1.2 Distinctions between Molecular Oxygen and Oxidizing Oxygen Molecule

Characters	Molecular Oxygen (In Air)	Oxidizing Oxygen Molecule
Structure	O=O (Linear) (A)	O⊂⊃O (B) A two-membered ring (∴Rings of different sizes)
Energy Carried	Inside the π-bond- An activation center. Unlike (B), heat is required.	Energy carried is different for the rings of different sizes. The larger, the less strained until (A) is obtained
Stability	Very stable	Very unstable; existing only transiently. (Always in Equilibrium)
Equilibrium State of existence	None	$(B) \underset{State}{\overset{Equilibrium}{\rightleftharpoons}} 2nn \cdot O \cdot en$
Activated State of Existence.	$(A) \rightleftharpoons nn \cdot O - O \cdot en$	None
Science	Of Chemistry	Of Chemistry and Physics (Springs)
Poison	Non-poisonous	Both, depending on the types of immediate neighbours (E.g. H_2)
Combination and Decomposition states	See Table I	None

The hydrogen peroxide which is going to be the continuous source of oxidizing oxygen molecule can be readily obtained from molecular hydrogen via forced combustion since a catalyst is required to keep H_2 in Equilibrium state of existence.

1.3.7 Forced Combustion of Hydrogen

Here, H_2 catalysts as opposed to the use of Flames in Space technology, which is also forced, must be used to keep the H_2 in Equilibrium state of Existence. This is almost like the case used in most hydrogen fuel cells where chemical means such as the use of Sodium hydride or sodium aluminium hydrides are involved.

Stage 1:

$$H_2 \xrightleftharpoons[H_2 \text{ in Equilibrium State of Existence}]{\text{Hyrogen catalyst used to keep}} H\cdot e + n\cdot H$$

$$H\cdot e + O=O \xrightleftharpoons[\text{Oxygen(Air)}]{\text{Activation of}} H-O-O\cdot en$$

$$H-O-O\cdot en + n\cdot H \xrightarrow[\text{Existence of HOOH}]{\text{Combination State of}} H-O-O-H \qquad 1.63a$$

$$\underline{Overall\ Equation}: H_2 + O_2 \xrightarrow{\text{Hydrogen Catalyst}} H-O-O-H \qquad 1.63b$$

Stage 2: Same as Stage 1 of Equation 1.25a.

Overall Equation: $2H_2 + 2O_2 \xrightarrow[\text{Catalyst}]{H_2} 2H_2O + O_2 + \text{Energy}$

(Air) (Oxidizing Oxygen)

1.64

In addition to the above, shown below is the most important stage in the reactions for generation of energy from H_2 and water.

Stage1:

$$H_2 \xrightleftharpoons[\text{CATALYST}]{H_2} H\cdot e + H\cdot n$$

$$H\cdot e + H_2O \rightleftharpoons H_2 + HO\cdot en + Energy$$

$$HO\cdot en \rightleftharpoons H\cdot e + en\cdot O\cdot nn + Energy$$

$$nn\cdot O\cdot en + H_2 \rightleftharpoons HO\cdot nn + H\cdot e$$

$$H\cdot e + n\cdot H \rightleftharpoons H_2$$

$$H\cdot e + nn\cdot OH \longrightarrow H_2O$$

$$\underline{Overall\ equation}: H_2 + H_2O \xrightarrow[CAT.]{H_2} H_2O + H_2 + Energy \qquad 1.65$$

Without the use of hydrogen catalyst, flames or some other chemical means, hydrogen molecule cannot be combusted, because it like F_2 is all the time in Stable state of Existence. With flames, either the Oxygen molecule is activated and attacks H_2 to give HOOH or the H_2 is broken into two parts to activate the O_2 to give HOOH in almost the same manner as first Stage 1 above. When chemical means are involved without the use of hydrogen catalysts, it is **not forced**. For example, one can use the following- sodium and ethanol. There are countless examples of such cases in different forms such as the use of sodium hydride and hydrogen. Sodium hydride is an intermediate product in the two cases since Sodium is an ionic metal far more electropositive than H which is also an ionic metal, but gaseous in character.

1.3.8 Non- Forced Combustion of Hydrogen

Using sodium and ethanol, the followings are obtained. This is almost typical of the cases used in most hydrogen fuel cells, except that here H_2 is produced in-situ.

Stage 1:

$$2Na \xrightleftharpoons{Excitation} 2Na\cdot e$$

$$2Na\cdot e + 2C_2H_5OH \xrightleftharpoons{Abstraction} 2C_2H_5ONa + 2H\cdot e$$

$$2H\cdot e \xrightarrow[State\ of\ existence\ of\ H_2]{Combination} H_2 \qquad 1.66a$$

$$\underline{Overall\ Equation}:\ 2Na + 2C_2H_5OH \longrightarrow 2C_2H_5ONa + H_2 \qquad 1.66b$$

Note how two hydrogen electro-free radicals or atoms have combined in the third step (Two like poles) to form stable hydrogen molecules, and the reason is because radicals do not repel and attract. This is very much unlike REAL charges- Covalent and Ionic as will be shown downstream. This makes it a productive stage (the Combination State of Existence of H_2).

Stage 2:

$$2Na \xrightleftharpoons{Excitation} 2Na\cdot e$$

$$2Na\cdot e + 2H_2 \xrightleftharpoons{Absrtaction} 2NaH + 2H\cdot e$$

$$2H\cdot e \xrightarrow[State]{Combination} H_2 \qquad 1.67a$$

$$\underline{Overall\ Equation}: 6Na + 4C_2H_5OH \longrightarrow 4C_2H_5ONa + 2NaH + H_2 \qquad 1.67b$$

Stage 3:

$$NaH \xrightleftharpoons[\text{of Exitence}]{\text{Its Equilibrium State}} Na\cdot e + H\cdot n$$

$$Na\cdot e + H_2 \xrightleftharpoons{\text{Abstraction}} NaH + H\cdot e$$

$$H\cdot e + O=O \xrightleftharpoons[\text{Oxygen molecule from Air}]{\text{Activation of}} H-O-O\cdot en$$

$$H-O-O\cdot en + H\cdot n \xrightarrow[\text{Existence of } H_2O_2]{\text{Combination State of}} H-O-O-H \quad 1.68a$$

$$\underline{Overall\ equation}: 12Na + 8C_2H_5OH + 2O_2(Air) \longrightarrow 8C_2H_5ONa + 4NaH + 2HOOH \quad 1.68b$$

Stage 4: Same as stage 1 of Equation 1.25a.

Overall Equation:

$$12Na + 8C_2H_5OH + \underset{\text{(Air)}}{2O_2} \longrightarrow 8C_2H_5ONa + 4NaH + \underset{\text{(Oxidizing oxygen molecule)}}{O_2} + 2H_2O + \text{Energy} \quad 1.69$$

Note the quantity of sodium and ethanol required for hydrogen combustion produced in situ. However, when sodium hydride is directly involved, it becomes an Active catalyst with hydrogen molecules externally supplied to the fuel cell or system continuously. As it seems, when energy is released using molecular oxygen of the air, oxidizing oxygen and water vapor are released to the environment. This is not environmentally friendly. Without water being involved, a battery for storage of energy will still be required.

When methane combusts, no oxidizing oxygen molecule is produced. But when ethane and other higher normal aliphatic members in their gaseous state combust, carbon monoxide in addition to the known products of water and carbon dioxide are produced. Almost the same applies to the combustion of hydrogen molecule, however with oxidizing oxygen molecule as product. During combustion of hydrogen, the oxidizing oxygen produced can be a pollutant if there is no hydrogen or "water" molecule around. If there is H_2, it can be removed to give for example water vapor as follows.

1.3.9 Oxidation of Hydrogen

Stage 1:

$$\underline{O_2} \xrightleftharpoons[\text{Existence of } \underline{O_2}]{\text{Equilibruim State of}} 2en\cdot\ddot{\underset{..}{O}}\cdot nn$$

$$2nn\cdot\ddot{\underset{..}{O}}\cdot en + 2H_2 \xrightleftharpoons{\text{Abstraction}} 2H-O\cdot nn + 2H\cdot e$$

$$\xrightarrow[\text{Operating conditions of the System}]{\text{Suppressed by the}} 2H_2O \quad 1.70a$$

$$\underline{Overall\ Equation}: \underline{O_2} + \quad 2H_2 \xrightarrow[Operating\ Conditions]{Depends\ on\ the} 2H_2O \qquad 1.70b$$

$$(E.g\ from\ Air)$$

$$\underline{Overall\ Equation}: 2O_2 + 4H_2 \xrightarrow[For\ Combustion\ of\ H_2]{H_2\ Catalyst} 4H_2O + Energy$$

$$(Three\ Stages) \quad (Air) \qquad (Water\ Vapor) \qquad 1.70c$$

Temperature, Pressure, presence of excess oxidizing oxygen elements or molecules, and other unique types of neighbours are the types of operating conditions that will suppress the Equilibrium State of Existence of water in the third step above. Oxidizing oxygen elements can be used differently from some oxidizing agents to give a productive or non-productive stage.

1.3.10 Combustion of Benzene (An Aromatic Compound)

Benzene is the first member of Aromatic hydrocarbons and as such, it must be unique.

Stage 1:

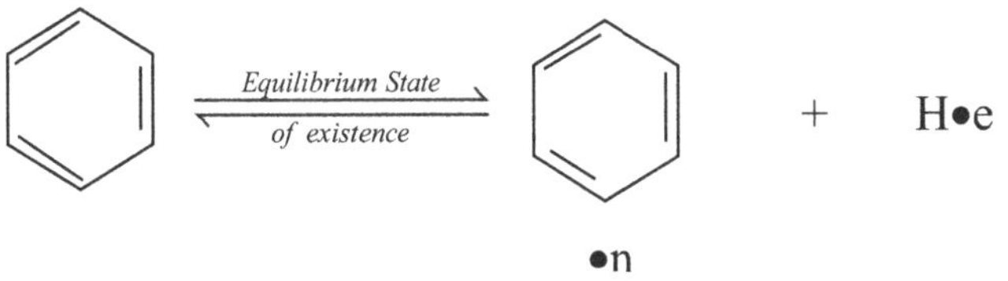

[Benzene] ⇌ (Equilibrium State of existence) [Benzene ring] •n + H•e
(A)

H•e + O =O ⇌ H – O – O •en
(B)

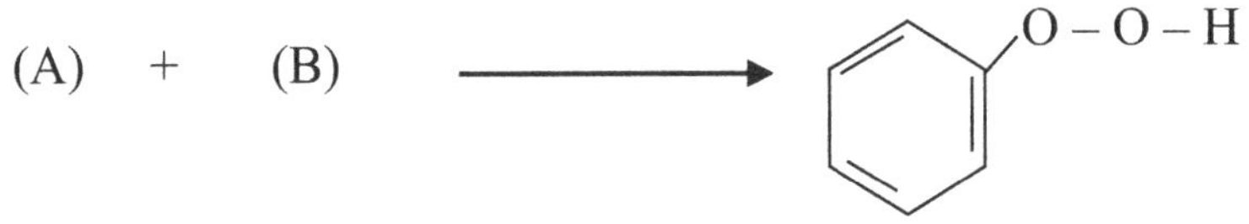

(A) + (B) ⟶ [Benzene ring]–O – O – H
(C)

1.71

Stage 2:

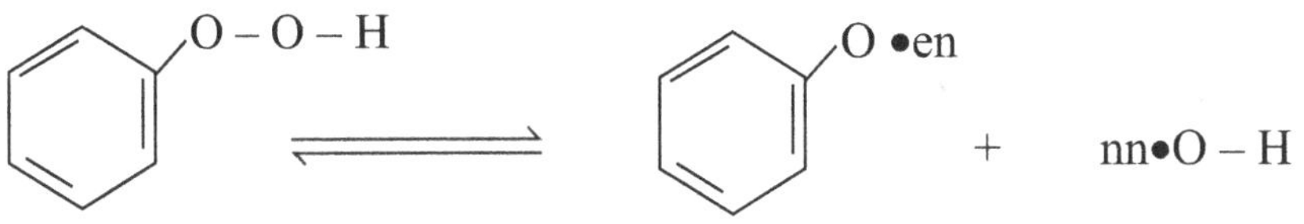

[Benzene ring]–O – O – H ⇌ [Benzene ring]–O •en + nn•O – H
(C) (D)

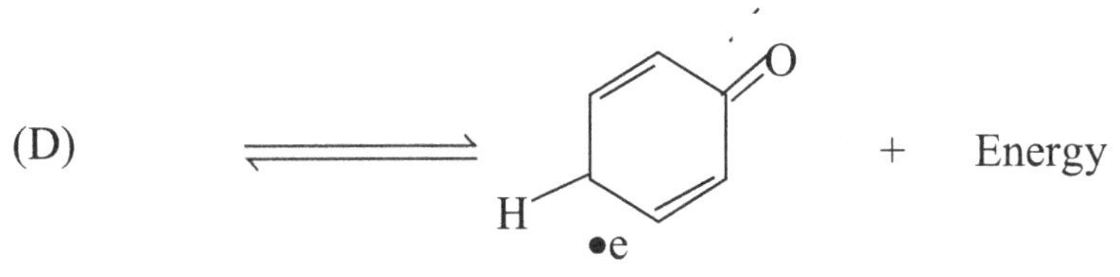

(D) ⇌ [Ring]=O, H, •e + Energy
(E)

(E) + nn• OH ⟶ [structure: O, H, OH]

(F) An Acid

1.72a

Overall equation: $H_6C_6 + O_2 \longrightarrow$ (F) + Energy 1.72b

The Equilibrium state of existence of (F) above is shown in the next stage wherein as an acid it is hydrogen that is held. Based on the structure, it is an acid. All the mechanisms so far used and shown are Equilibrium mechanisms, wherein in the last step of a stage, it is either equilibrium step (non-productive) or combination step (productive). This indeed is a New Science.

Stage 3:

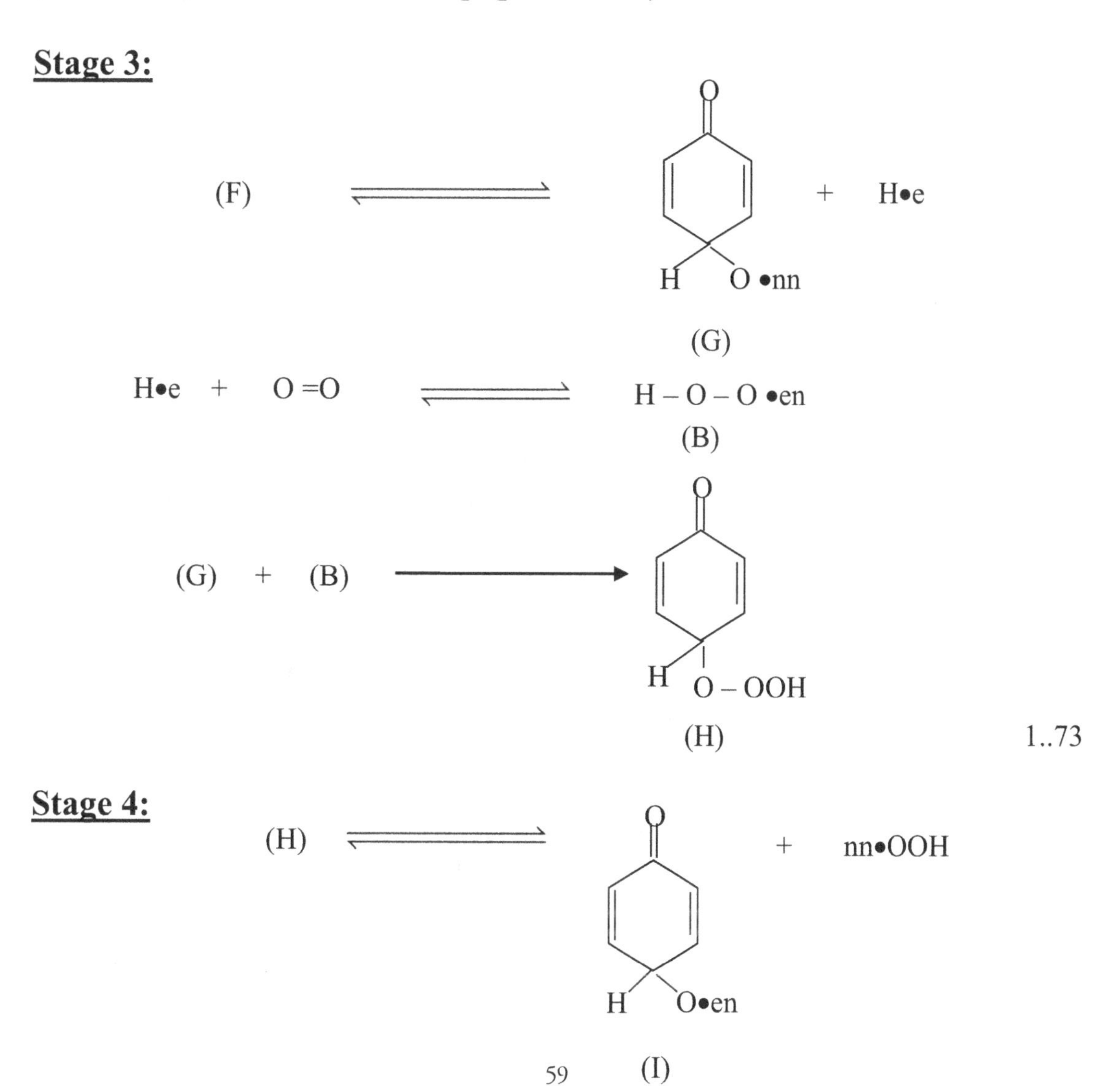

$$(I) \rightleftharpoons \text{[para-benzoquinone structure]} + H{\bullet}e + Energy$$

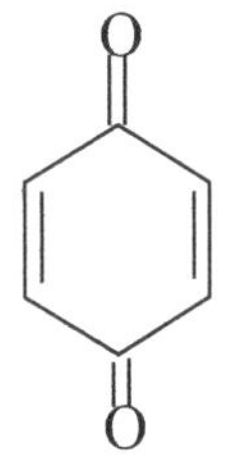

(I)
Aromatic Carbon dioxide

$$H{\bullet}e + nn{\bullet}OOH \longrightarrow HOOH \qquad 1.74$$

Stage 5:

$$2H-O-O-H \underset{\text{Existence of Hyrogen peroxide}}{\overset{\text{Equilibrium State of}}{\rightleftharpoons}} 2H-O\cdot en + 2nn\cdot O-H$$

$$2H-O\cdot en \underset{\text{Release}}{\overset{\text{Energy}}{\rightleftharpoons}} 2H\cdot e + \underline{O_2} + Energy$$

$$2H\cdot e + 2nn\cdot O-H \xrightarrow[\text{presence of } \underline{O_2}]{\text{Suppressed by the}} 2H_2O \qquad (1.25a)$$

$$\underline{Overall\ Equation}: 2HOOH \longrightarrow 2H_2O + \underline{O_2} + Energy \qquad (1.25b)$$

Overall equation: $2H_6C_6 + 4O_2 \longrightarrow 2\text{Benzoquinone} + \text{Energy} + 2H_2O + \underline{O_2}$ 1.75a

Worthy of note are the followings-

(i) The combustion of methane the first member of the aliphatic hydrocarbons is uniquely different from the combustion of benzene the first member of the Aromatic hydrocarbons. With benzene, oxidizing oxygen is obtained as one of the products. There are five stages as opposed to four stages for methane with energy release coming from three stages as opposed to two for methane. Obviously, the combustion of benzene provides more energy than the combustion of methane.

(ii) While two moles of benzene are required to complete its combustion, only one mole of methane is required.

(iii) In place of carbon dioxide obtained during combustion of methane, is an aromatic equivalent called para- benzoquinone, (I) above. These are shown below.

$O = C = O$ Versus

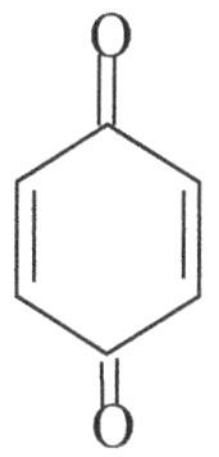

Aliphatic Carbon dioxide

Aromatic Carbon dioxide 1.75b

(iii) Like the aliphatic CO_2, the aromatic CO_2 can be activated as follows-

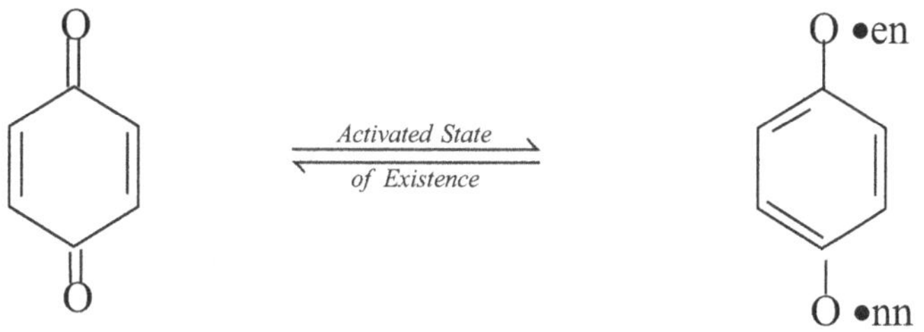

Activated State of Existence

O •en

O •nn

Aromatic Carbon dioxide

(J) Activated Aromatic Carbon dioxide 1.75c

$$O = C = O \underset{Existence(\mathbf{Heat})}{\overset{Activated\ State\ of}{\rightleftharpoons}} O = \underset{\cdot e}{C} - O \cdot nn \quad 1.75d$$

(Aliphatic CO_2)

In CO_2, the electro-free-radical could not move to the O center, because one of laws of Boundary as will be shown in the New Frontiers (where the O center cannot carry more than eight radicals in its last shell) will be broken. This unique character of benzoquinone is what makes it carcinogenic of the external type, that is, that created from the pollution resulting from use of aromatic fuels in our environment. Many moles of (J) above after release of oxidizing oxygen molecules and energy can add to each other in chains to form deadly aromatic cyclic ethers cells, a polymeric living compound as well as dead cells as shown in Figure 1.6.

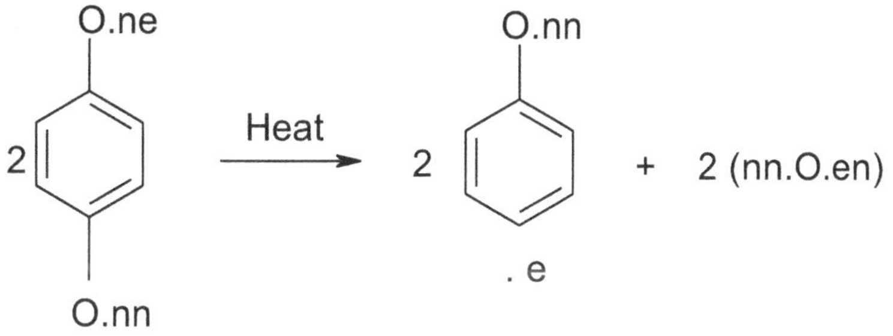

O.ne

2

O.nn

Heat

O.nn

2

. e

+ 2 (nn.O.en)

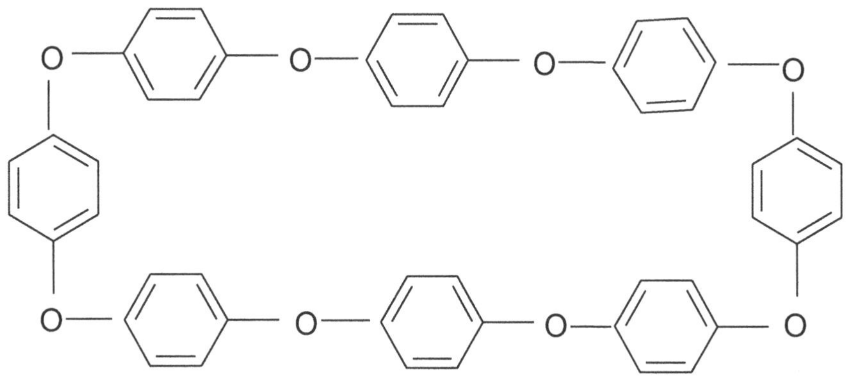

(A) Eight-membered Dead Cell (A Tumoil)

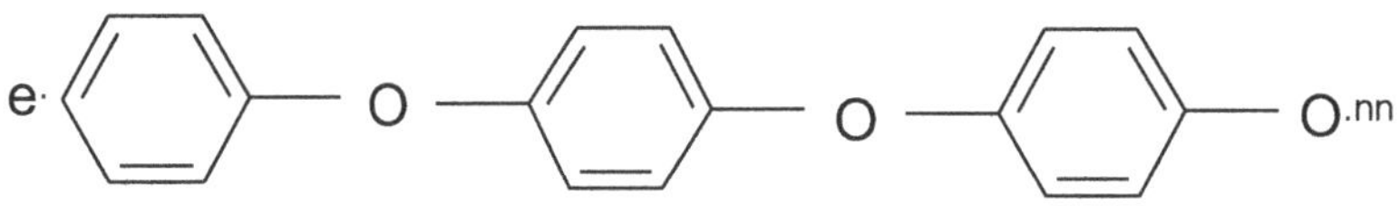

(B) Living Three- membered Cell (A Poison)

Figure 1.6 Types of Dead and Living Cells Obtained from Combustion of Aromatic Hydrocarbons

(iv) In the combustion of benzene, the ring cannot yet be broken or opened. To be broken or opened requires either saturation of the ring and therefore different operating conditions or some other means as will be shown downstream. Benzene is well known to combust with high Octane number.[27] What it does to the environ-ment is well known, but how it does it is unknown. ***This is the CO_2 that is the poison, not aliphatic CO_2.***

1.3.11 Combustion of Toluene (An Alipharomatic Compound)

Toluene is herein said to be alipharomatic, because it a hybrid of a group from methane (Aliphatic) and a group from benzene (Aromatic). Why benzene is aromatic and cyclooctatetraene is not aromatic will be explained downstream in the New Frontier.

Stage 1:

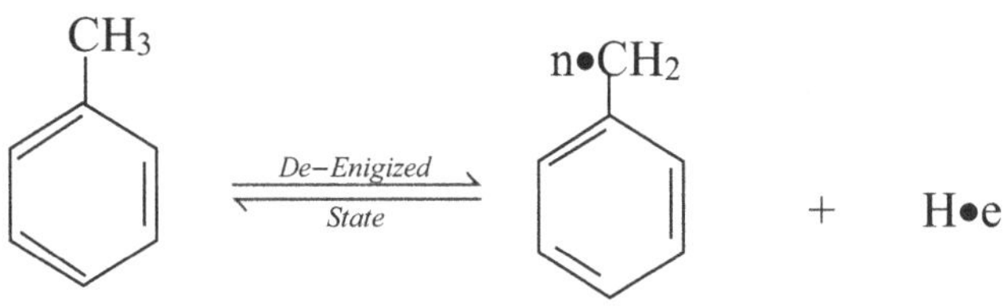

CH_3 — De–Enigized State ⇌ n•CH_2 + H•e

(A)

H•e + O =O ⇌ H – O – O •en
(B)

HOOCH$_2$

.(A) + (B) ⟶ [benzene ring]

(C) A peroxide 1.76

Stage 2:

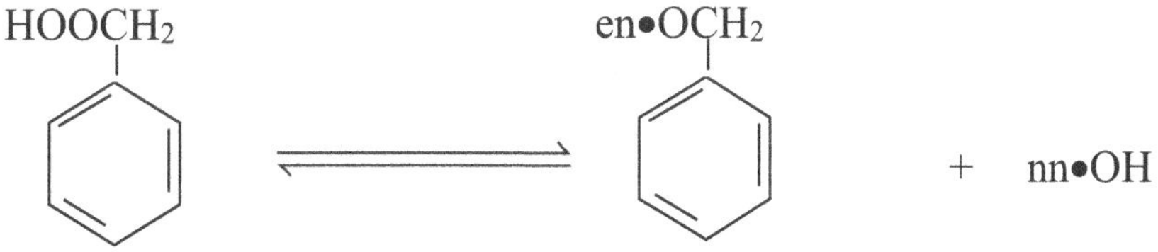

HOOCH$_2$ en•OCH$_2$

[benzene ring] ⇌ [benzene ring] + nn•OH

(C) A peroxide (D)

O = CH

(D) ⇌ [benzene ring] + H•e + Energy

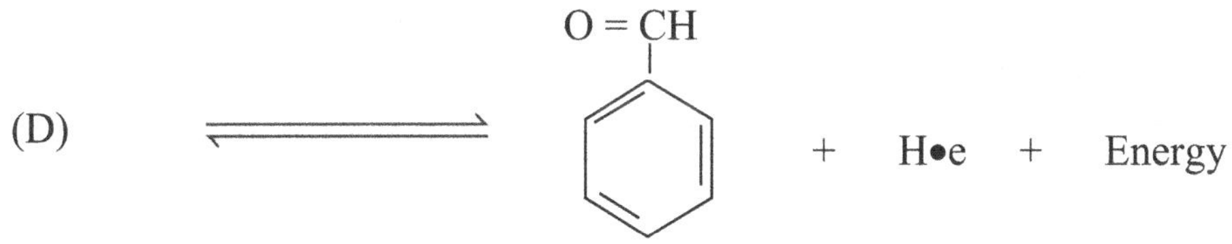

(E)

H•e + nn•OH ⟶ H_2O 1.77a

Overall equation: $H_3CC_6H_5$ + O_2 ⟶ H_2O + (E) + Energy
1.77b

With the presence of more O_2 in the system, combustion of the benzaldehyde continues as follows in stage 3. However, note so far how CHEMICAL ENERGIES in form of HEAT are released in these systems. There are many but limited ways by which these forms of energies are released. How these energies manifest their presence are based on the applications of the forces of nature in Physics with respect to Light, Sound and Feelings.

Stage3:

(E) ⇌ H·e + O=C·n (attached to benzene ring) (F)

H·e + O=O ⇌ H—O—O·en

HOO·en + (F) ⟶ HOOC=O (attached to benzene ring) (G) 1.78a

Overall Equation: $2C_6H_5CH_3 + 4O_2 \longrightarrow 2H_2O + 2(G)$ 1.78b

Another peroxide partly aromatic in character is thus formed. This decomposes via Equilibrium mechanisms as follows in the next stage, noting first and foremost the Equilibrium states of existences of compounds encountered so far. These are the finger-prints for the compounds. For every compound, there is one and only one finger print where possible. Nature does not operate indiscriminately, does not differentiate, but only INTERGRATES. Yet in Mathematics, there is Differentiation and Integration, because, in the physical side of our world, before Integration can exist, it must co-exist with Differentiation, just as before "Sweet" can exist, it must co-exist with "Bitter", otherwise none will exist. *Nature also abhors non-linearity, vacuum and more. Yet they exist in our physical world. Non-linearity itself is a form of inherent disturbance. It is based on what Nature abhors, that one can begin to see the world we live in and indeed how to solve problems, since Life is problem solving.* We cannot escape from the fact that EVERYTHING in our world are both for GOOD and BAD, all depending on how applied (THE OPERATING CONDITIONS). Otherwise, how can one know what is good or sweet or friendly or kind or much more when they do not co-exist with the opposite side of the spectrum of Life? *These were seen from the new added definition of the ATOM, based on RADICALS; something one thought at the beginning to be very minor, but now found to be the ESSENCE OF THE NEW FRONTIERS, and indeed "THE BEGINNING OF A NEW DAWN FOR HUMANITY.*

Stage 4:

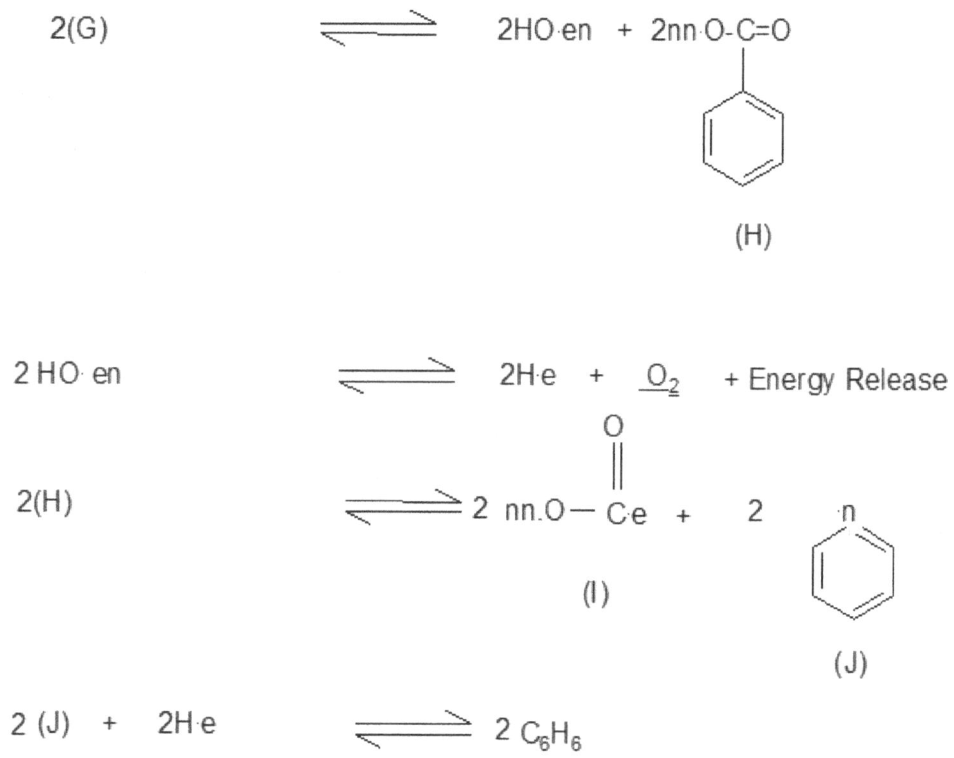

$$2(G) \rightleftharpoons 2HO{\cdot}en + 2nn{\cdot}O\text{-}C{=}O$$

(attached to a benzene ring)

(H)

$$2\,HO{\cdot}\,en \rightleftharpoons 2H{\cdot}e + \underline{O_2} + \text{Energy Release}$$

$$2(H) \rightleftharpoons 2\;nn.O-\overset{\displaystyle O}{\overset{\|}{C}}{\cdot}e + 2\;{\cdot}n\text{(benzene ring)}$$

(I) (J)

$$2\,(J) + 2H{\cdot}e \rightleftharpoons 2\,C_6H_6$$

$$2\,(I) \longrightarrow 2CO_2 + \text{Energy Release} \qquad 1.79a$$

Overall Equation: $2C_6H_5CH_3 + 4O_2 \longrightarrow 2C_6H_6 + 2CO_2 + 2H_2O + \underline{O_2} + \text{Energy}$

(Air) (Oxidizing Oxygen) 1.79b

Since toluene like other higher aromatic hydrocarbons is partly aliphatic and partly aromatic, hence on combustion, presence of water, aliphatic CO_2 and oxidizing oxygen can be observed as part of the products. With presence of more O_2 in the system, combustion of the benzene formed does not indeed continue, like the case with ethane and higher members. The presence of oxidizing oxygen will not allow it to continue. If it has continued, the following would have been obtained as the overall equation.

Overall Equation: $2C_6H_5CH_3 + 8O_2 \longrightarrow 2\underline{O_2} + 4H_2O + 2CO_2 +$

(Nine Stages) (Air) (Oxidizing

Oxygen)

$+ 2\ C_6H_4O_2$ + More Energy 1.80

In view of the fact that oxidizing oxygen is non-productive with water, it oxidizes benzene as will shortly be shown to give the final overall equation.

Overall Equation: $2C_6H_5CH_3 + 6O_2 \longrightarrow 4H_2O + 2CO_2 + 2C_6H_4O_2 +$ Energy 1.81
(Seven Stages)

The oxidizing oxygen is far more unstable than benzene, though benzene as will be shown in the New Frontiers, has a dual character in terms of States of existence- ***the ENERGIZED and DE-ENERGIZED states***. This is not an exception to the laws of Nature in Chemistry. These states start manifesting their presence strongly only when groups are carried. When it is in the ENERGIZED state, the ring is ***resonance stabilized*** and the group carried cannot be tempered with, i.e., the group is shielded. When it is in the DE-ENERGIZED state, the ring is not resonance stabilized and the group carried by it can be tempered with. In the first step of Equation 1.71, benzene was in the ENERGIZED state. In the first step of Equation 1.76, the benzene in toluene was kept in the DE-ENERGIZED state. The ENERGIZED Equilibrium state of existence of toluene is as follows.

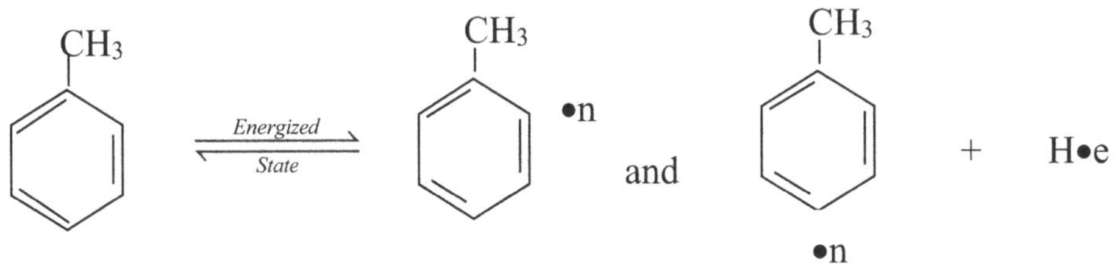

CH3 (Energized State ⇌) CH3 •n and CH3 •n + H•e

(A) 2Ortho- (B) Para- 1.82a

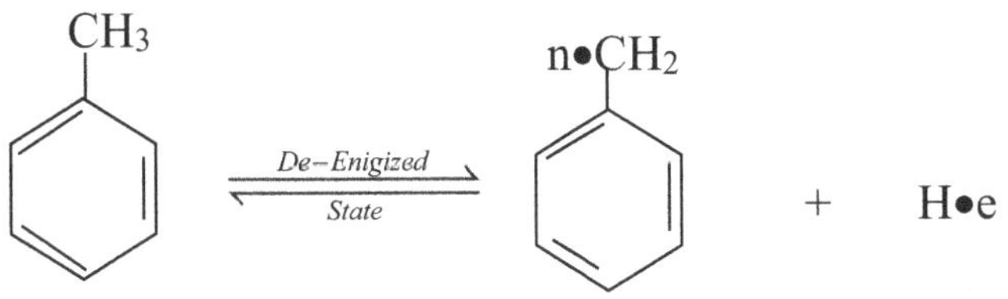

CH3 (De-Enigized State ⇌) n•CH2 + H•e

(A) 1.82b

With the presence of oxidizing oxygen, the benzene is kept in the DE-ENERGIZED state, i.e., stable state of existence, in which nothing is held. The benzene is then oxidized as follows in the next stage.

(a) Oxidation of Benzene

Stage 5:

$$\underline{\underline{O_2}} \rightleftharpoons 2nn{\bullet}O{\bullet}en$$

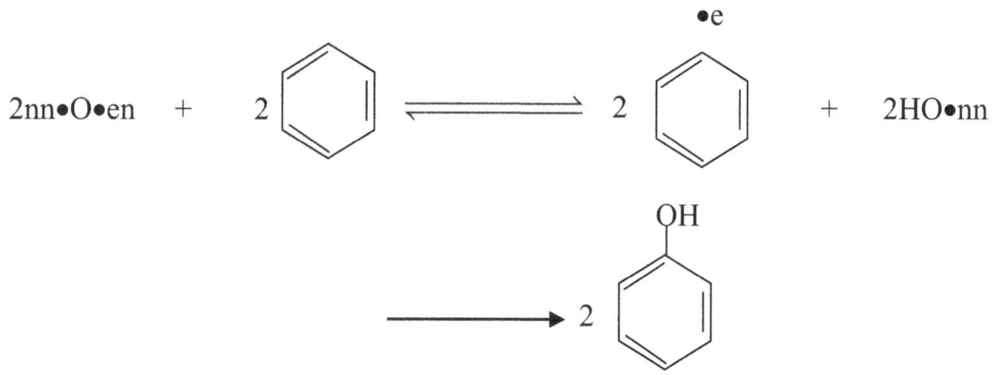

2nn•O•en + 2 [benzene] ⇌ 2 [benzene •e] + 2HO•nn

⟶ 2 [phenol]

(K) Phenol

1.83a

Overall equation: $O_2 + 2C_6H_6 \longrightarrow 2HOC_6H_5$ 1.83*b*

(b) Combustion of phenol

Stage 6:

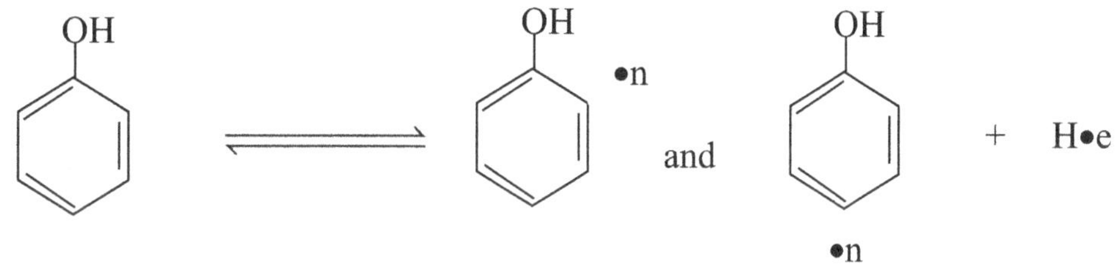

[phenol] ⇌ [phenol •n (A)] and [phenol •n (B)] + H•e

(A) (B)

H•e + O=O ⇌ H – O – O •en
(C)

(A) and (B) + (C) ⟶ [phenol with O – OH ortho (D)] and [phenol with O – OH para (E)]

(D) (E)

1.84

The phenol in the absence of oxidizing oxygen is found to have its Equilibrium state of existence in the ENERGIZED state, in which the H atoms in the ortho- and para-positions are held, because as will be shown downstream in the New Frontiers, the OH group is a strong ***radical-pushing group*** of greater capacity than alkylane (or alkanyl) groups such CH_3 which in turn is of greater capacity than H. However, whether in the energized or de-energized state, the same products will finally be obtained. Note that in the combustion of benzene, only p-benzoquinone was shown. Indeed, there must be some fractions (more) of ortho-benzoquinone if not blocked. Using the para-version above, i.e., (E) which indeed is less, the followings are obtained in the next stage which will obviously be in parallel with the large fraction of ortho- version in this stage.

Stage 7:

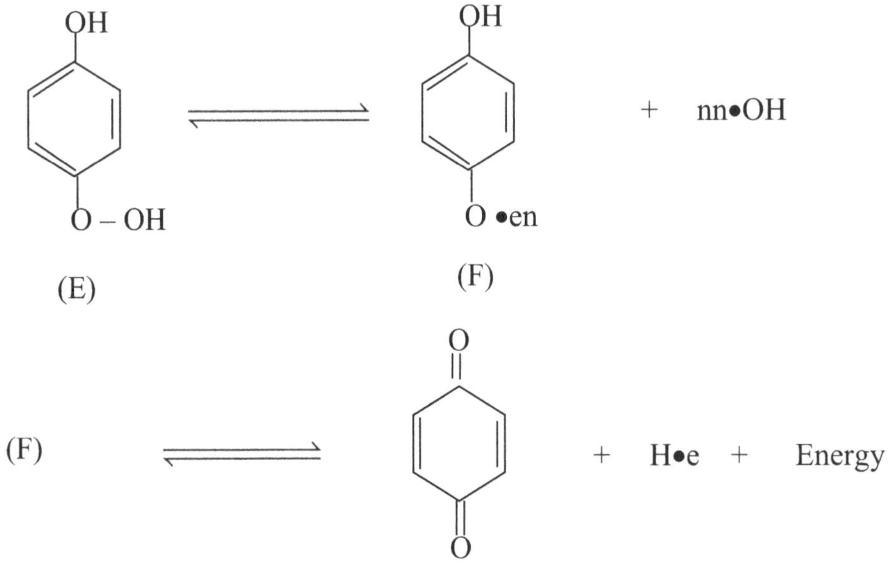

OH

O – OH

(E)

$\rightleftharpoons$

OH

O •en

(F)

+ nn•OH

(F) $\rightleftharpoons$

O

O

+ H•e + Energy

H•e + nn• OH $\longrightarrow$ H_2O 1.85

Final Overall equation: $2C_6H_5CH_3 + 6O_2 \;(AIR) \longrightarrow 2CO_2 + 4H_2O + 2C_6H_4O_2\;(G) + Energy(5)$ (1.81)

On the whole, there are seven stages, in which energy was released five times in four stages. After the presence of benzene in the fourth stage, this was followed by the presence of phenol via oxidation of the benzene in the fifth stage. In the sixth stage, combustion of the phenol followed and ended in the seventh stage with release of the quinones (both ortho- and para-), water and more energy. The only product that is not environmentally friendly is the para-benzoquinone in particular, in view of its linear and non-steric character. Indeed, one has seen the oxidation of benzene during the combustion of toluene. One has also seen the combustion of phenol in two stages-Stages 6 and 7 during the combustion of toluene. The product obtained for the combustion of phenol is the same as the product obtained from the oxidation of phenol as shown below.

1.3.12 **Oxidation of Phenol**

With the presence of oxidizing oxygen, the phenol will remain in a stable state of existence.

Stage 1: $\underline{O_2} \rightleftharpoons 2nn{\bullet}O{\bullet}en$

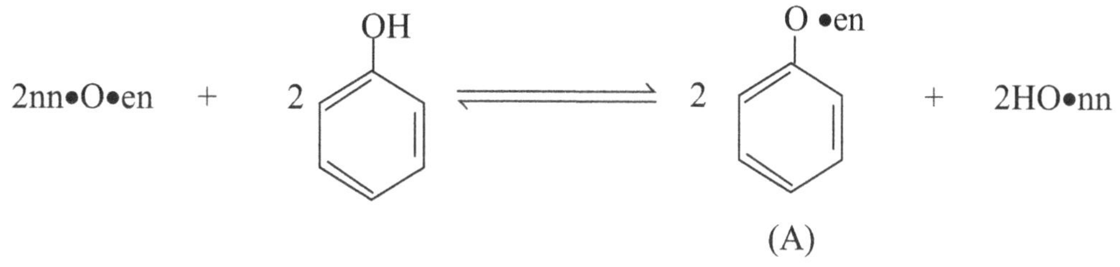

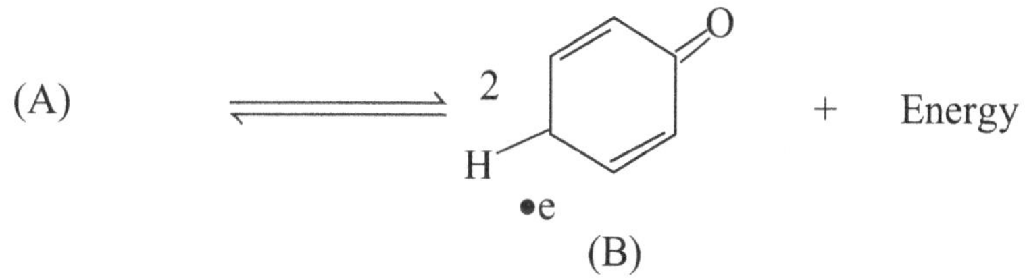

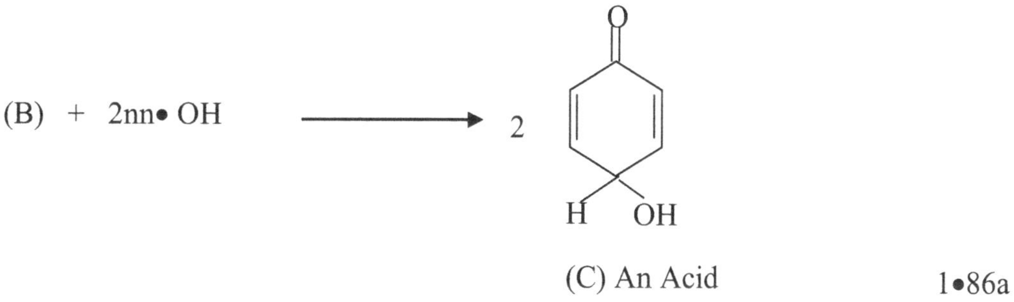

2nn•O•en + 2 (phenol, OH) ⇌ 2 (O •en) + 2HO•nn

(A)

(A) ⇌ 2 (H, •e, O) + Energy

(B)

(B) + 2nn• OH ⟶ 2 (O; H OH)

(C) An Acid 1•86a

Overall equation: $2H_5C_6OH + \underline{O_2} \longrightarrow$ (C) + Energy 1.86b

Stage 2: $\underline{O_2} \rightleftharpoons 2nn{\bullet}O{\bullet}en$

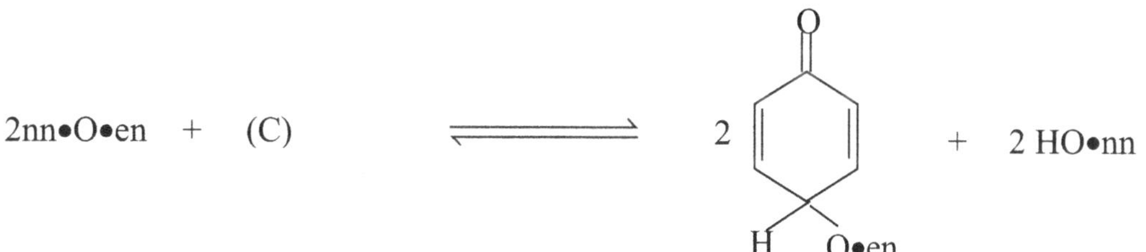

2nn•O•en + (C) ⇌ 2 (O; H O•en) + 2 HO•nn

(D)

(D) $\rightleftharpoons$ 2 [p-benzoquinone] + 2H•e + Energy

2H•e + 2nn•OH ⟶ $2H_2O$ 1.87a

<u>Overall equation</u>: $2C_6H_5OH + 2\underline{\underline{O_2}} \longrightarrow 2C_6H_4O_2 + 2H_2O + Energy$ 1.87*b*

Only two stages are required for the oxidation of phenol, just like for its combustion, both giving the same products- water, benzoquinones. But unlike with combustion, energy is released here in the two stages. Note that the amount of energy released in all the steps where released are not equal.

1.3.13 **Oxidation of Toluene**

Unlike with phenol where the same products were obtained via both oxidation and combustion, this has not been the case with toluene and other higher alipharomatic hydrocarbons. Why?

Stage 1: $\underline{\underline{O_2}} \rightleftharpoons 2nn{\bullet}O{\bullet}en$

2nn•O•en + 2 [toluene, $C_6H_5CH_3$] $\rightleftharpoons$ 2 [C_6H_5–e•CH_2] (A) + 2nn•OH

⟶ 2 [$C_6H_5CH_2OH$] (B) 1.88

(B) is an alipharomatic alcohol. Whether (A), decomposes or not to give a carbene, the same alcohol (B) will remain obtained. In the next stage, in the absence of carbene, the following will still be obtained.

Stage 2:

$$O_2 \rightleftharpoons 2nn{\bullet}O{\bullet}en$$

$$2nn{\bullet}O{\bullet}en + (B) \rightleftharpoons 2\ C_6H_5\text{-}H_2CO{\bullet}en\ (C) + 2nn{\bullet}OH$$

$$(C) \rightleftharpoons 2\ C_6H_5\text{-}CH{=}O\ (D) + 2H{\bullet}e + Energy$$

$$2H{\bullet}e + 2HO{\bullet}nn \longrightarrow 2H_2O \qquad 1.89a$$

In this stage, benzaldehyde, an aromatic aldehyde is obtained. In the process, energy was released. Note that the energy in question is chemical- Heat. In the presence of more oxidizing oxygen molecules, the next stage follows under the same operating conditions.

Stage 3:

$$O_2 \rightleftharpoons 2nn{\bullet}O{\bullet}en$$

$$2nn{\bullet}O{\bullet}en + (D) \rightleftharpoons 2\ C_6H_5\text{-}C(=O){\bullet}e\ (E) + 2nn{\bullet}OH$$

$$(E) + 2nn{\bullet}OH \longrightarrow 2\ C_6H_5\text{-}C(=O)OH\ (F) \qquad 1.90a$$

$$\underline{Overall\ equation}: 2C_6H_5CH_3 + 3O_2 \longrightarrow 2C_6H_5COOH + 2H_2O + Energy(1)$$

$$(Small)\ 1.90b$$

In general, it is believed that with alipharomatic hydrocarbons, *"the carbon atom adjacently placed to the ring is more susceptible to oxidation than the remaining carbon atoms of the alkyl (Real name-alkylane) group and is the initial point of attack. For chains having two or more carbon atoms, the probable course of reaction is by way of the* ***alcohol, ketone, and enol.*** *Since, all of the intermediates are oxidized more readily than the hydrocarbon, the only isolable aromatic compound is the carboxylic acid"*[50].

$$ArCH_2CH_2R \longrightarrow \underset{(A)}{ArCHOHCH_2R} \longrightarrow \underset{(B)}{Ar\underset{\displaystyle O}{\overset{\|}{C}}CH_2R} \longrightarrow \underset{(C)}{Ar\underset{\displaystyle OH}{\overset{|}{C}}=CHR} \longrightarrow$$

$$\longrightarrow ArCOOH + HOOCR \qquad 1.91^{50}$$

Where $Ar \equiv C_6H_5$

Based on how the reactions take place in stages, with proper control, some of the products can be isolated. One has seen from above how (A) above which is (B) in Stage 1 of Equation 1.88, (B) above which is (D) in Stage 2, and the acid (F) in Stage 3 can be isolated. ***If the only isolable aromatic compound is carboxylic acid, then what this implies is that the oxidizing oxygen cannot oxidize the carboxylic acid.*** The only possible reason is because the acid may be in Energized or De-energized Equilibrium state of existence in the presence of oxidizing oxygen which is also all the time in Equilibrium state of existence.

O
‖
COH (benzene ring) ⇌ [De-Energized State] O ‖ CO•nn (benzene ring) + H•e ; O ‖ COH (benzene ring) ⇌ [Energized State] O ‖ COH (benzene ring with n• at meta position) + H•e

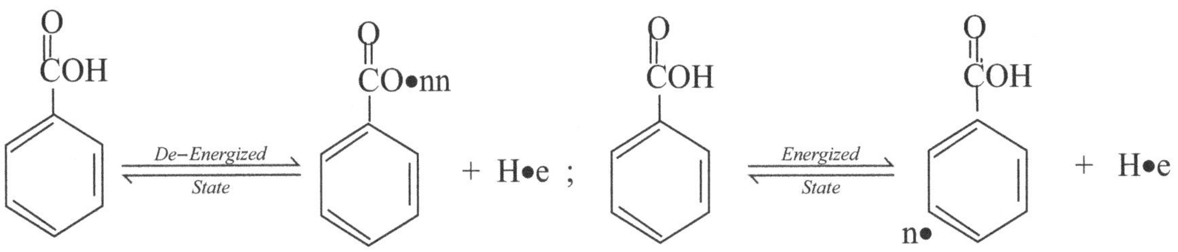

1.92

Note that it is the H atom in the meta-position that is held in Equilibrium state of existence, because the COOH group is ***a radical-pulling group*** as will be shown downstream in the New Frontiers. When both are in Equilibrium state of existence, the aromatic carboxylic acid is *"stable, reactive and insoluble"* in the oxidizing oxygen molecule. But as it seems, based on how NATURE operates, though almost identical types of products cannot be obtained via COMBUSTION and OXIDATION for the same family of compounds, it is believed that oxidation beyond carboxylic acid does not continue, because at the operating conditions, the acid would have decomposed in the De-energized state to release CO_2 and benzene. ***This could not take place because of the presence of oxidizing oxygen which has kept it in the Energized Equilibrium state of existence.*** However, if the carboxylic acid can be stabilized, oxidation of the acid can still continue in the presence of more oxidizing oxygen in the system as shown below.

Stage 4:

$$\underline{\underline{O_2}} \rightleftharpoons 2nn \bullet O \bullet en$$

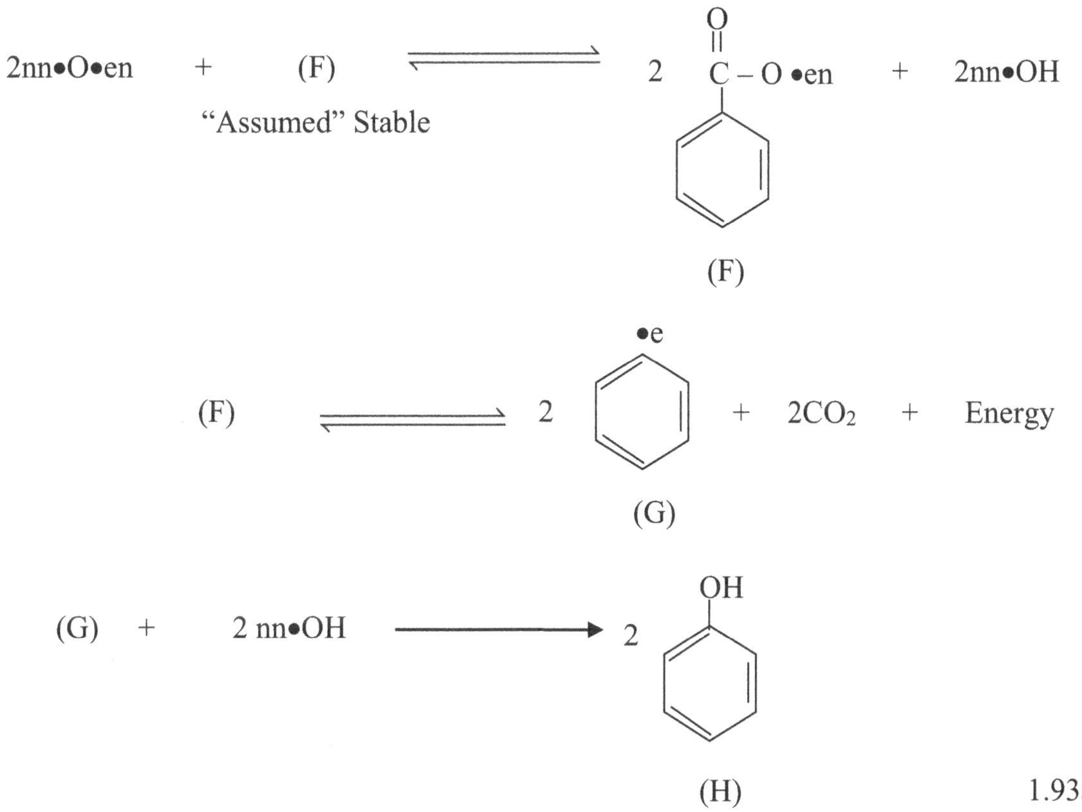

2nn•O•en + (F) ⇌ 2 C₆H₅–C(=O)–O •en + 2nn•OH

"Assumed" Stable

(F)

(F) ⇌ 2 •e-C₆H₅ + $2CO_2$ + Energy

(G)

(G) + 2 nn•OH ⟶ 2 C₆H₅OH

(H) 1.93a

$$\underline{\text{Overall equation}}: C_6H_5COOH + \underline{\underline{O_2}} \longrightarrow C_6H_5OH + CO_2 + Energy \quad 1.93b$$

With the presence of oxidizing oxygen, the phenol also an acid, but far weaker than the carboxylic acid, will remain in a stable state of existence as has already been shown, but recalled below for the purpose of continuity and understanding.

Stage 5:

$$\underline{\underline{O_2}} \rightleftharpoons 2nn \bullet O \bullet en$$

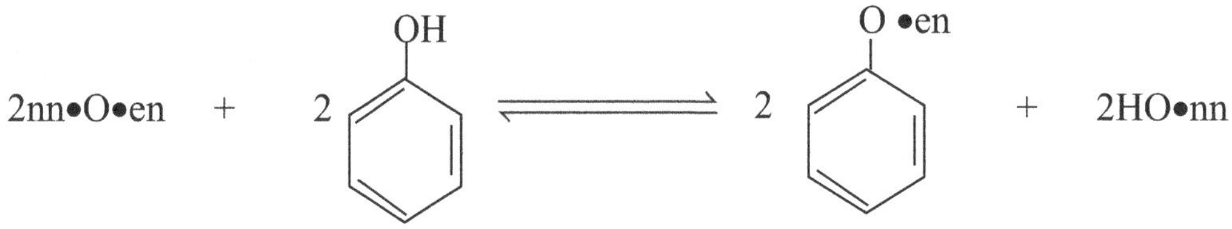

$$2nn{\bullet}O{\bullet}en \quad + \quad 2\ (\text{phenol, } C_6H_5OH) \rightleftharpoons 2\ (\text{phenoxy, } C_6H_5O{\bullet}en) \quad + \quad 2HO{\bullet}nn$$

(A)

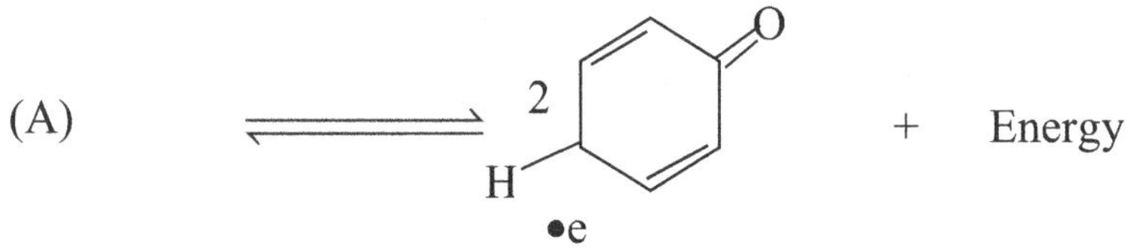

$$(A) \rightleftharpoons 2\ (\text{cyclohexadienone with } H^{\bullet e}) \quad + \quad \text{Energy}$$

(B)

$$(B) \quad + \quad 2nn{\bullet}\ OH \longrightarrow 2\ (\text{4-H, 4-OH cyclohexadienone})$$

(C) An Acid (1.86a)

Overall equation: $2H_5C_6OH \quad + \quad \underline{O_2} \longrightarrow (C) \quad + \quad \text{Energy}$ (1.86b)

Stage 6: $\underline{O_2} \rightleftharpoons 2nn{\bullet}O{\bullet}en$

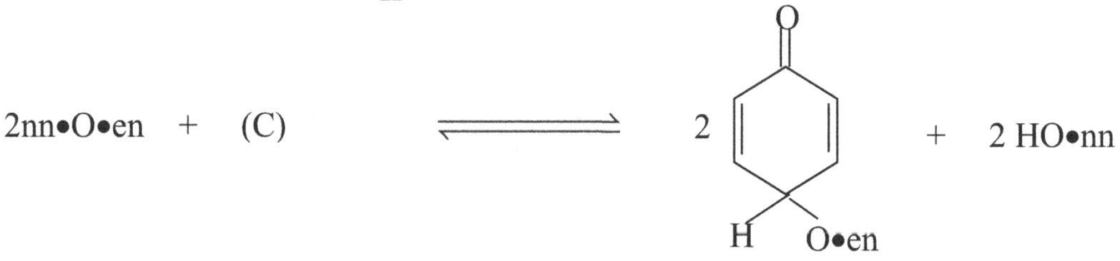

$$2nn{\bullet}O{\bullet}en \quad + \quad (C) \rightleftharpoons 2\ (\text{4-H, 4-O}{\bullet}\text{en cyclohexadienone}) \quad + \quad 2\ HO{\bullet}nn$$

(D)

$$(D) \rightleftharpoons 2\ (\text{p-benzoquinone}) \quad + \quad 2H{\bullet}e \quad + \quad \text{Energy}$$

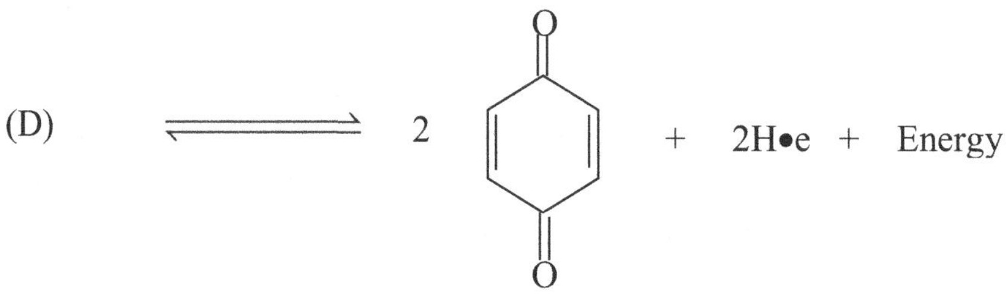

$$2H{\bullet}e \quad + \quad 2nn{\bullet}OH \longrightarrow 2H_2O \qquad (1.87a)$$

$$\underline{Overall\ equation}: 2C_6H_5OH + 2\underset{--}{O_2} \longrightarrow 2C_6H_4O_2 + 2H_2O + Energy \quad (1.87b)$$

$$\underline{Overall\ equation}: 2C_6H_5COOH + 3\underset{--}{O_2} \longrightarrow 2C_6H_4O_2 + 2H_2O + Energy(3) + 2CO_2 \quad 1.94$$

$$\underline{Overall\ equation}: 2C_6H_5CH_3 + 6\underset{--}{O_2} \longrightarrow 2C_6H_4O_2 + 4H_2O + Energy(4) + 2CO_2 \quad 1.95$$

Worthy of note is that when the carboxylic acid was oxidized, so much energy was generated. In the impossible three-stage system above, energy was generated in every stage none of which is via de-activation. One has now provided the answer to the "Why?" at the beginning of the sub-section. Indeed the carboxylic acid cannot be oxidized. But under harsher operating conditions, it may be oxidized. Because of the presence of enol, (C) in Equation 1.91, and its absence so far, one will next consider the oxidation of Ethyl benzene.

1.3.14 **Oxidation of Ethyl Benzene.**

We started with H in benzene, then CH_3 group in toluene and now with C_2H_5 group, all ***radical-pushing groups*** with increasing capacity.

Stage 1:

$$\underset{--}{O_2} \rightleftharpoons 2nn{\bullet}O{\bullet}en$$

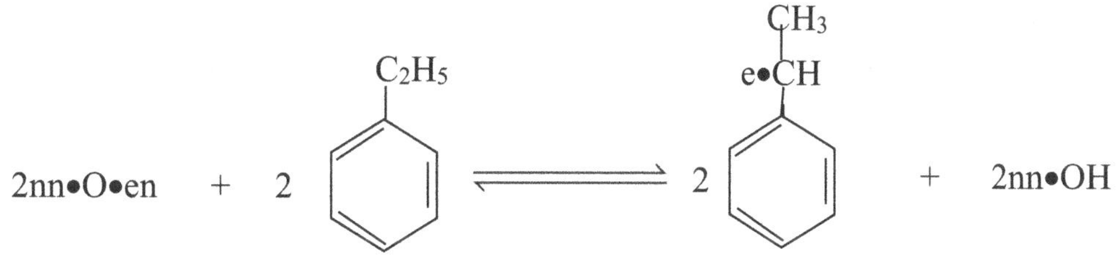

$$2nn{\bullet}O{\bullet}en + 2\ C_6H_5C_2H_5 \rightleftharpoons 2\ C_6H_5(e{\bullet}CH{-}CH_3) + 2nn{\bullet}OH$$

(A)

$$(A) \rightleftharpoons 2\ C_6H_5(e{\bullet}CH{-}CH_2{\bullet}n) + 2H{\bullet}e$$

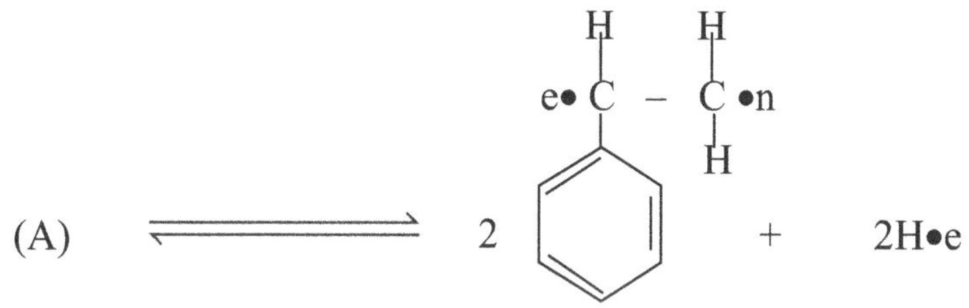

(B)

$$2H{\bullet}e + 2HO{\bullet}nn \rightleftharpoons 2H_2O$$

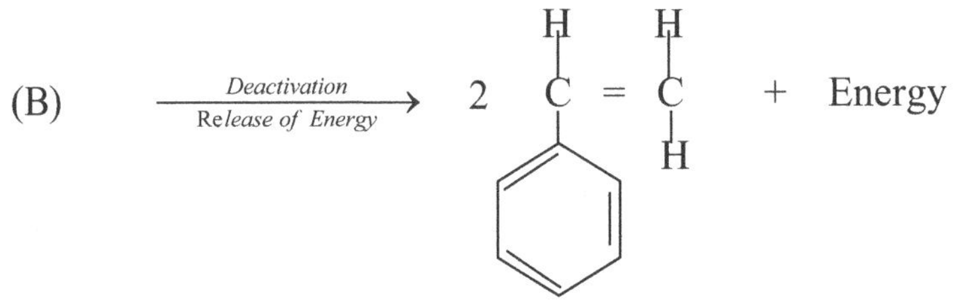

(B) $\xrightarrow[\text{Release of Energy}]{\text{Deactivation}}$ 2 C_6H_5–CH = CH_2 + Energy

(C) 1.96a

Overall equation: $2C_6H_5C_2H_5 + \underline{\underline{O_2}} \longrightarrow 2H_2C = CHC_6H_5 + 2H_2O$

$+ \ Energy$ 1.96*b*

Note that, in the second step, it was the H on the C center next to the ring that was abstracted, because if it was the one externally located that was abstracted, H would not have been released in the third step to form (B). That carbon center where the H was removed is richer in H than any C center, because it is the one directly carrying the "non-resonance stabilized ring" which has five H atoms. If it was a ring that was resonance stabilized, the situation would have been different. It is important to note at this point in time that the world or the academia know little or nothing about RESONANCE STABILIZATION phenomena, until they see what is contained in the NEW FRONTIERS. Like all things, they think they do. ***How can any compound that can undergo resonance stabilization do it chargedly, when charges cannot be removed or moved from their carriers?*** That is what can be seen in all textbooks on the subject universally!

The (C) formed an alipharomatic alkene is styrene. It is shocking to find styrene as an intermediate product in the oxidation of ethyl benzene using only one mole of oxidizing oxygen molecule. With more oxidizing oxygen molecules, Stage 2 follows.

Stage 2: $\underline{\underline{O_2}} \rightleftharpoons 2nn{\bullet}O{\bullet}en$

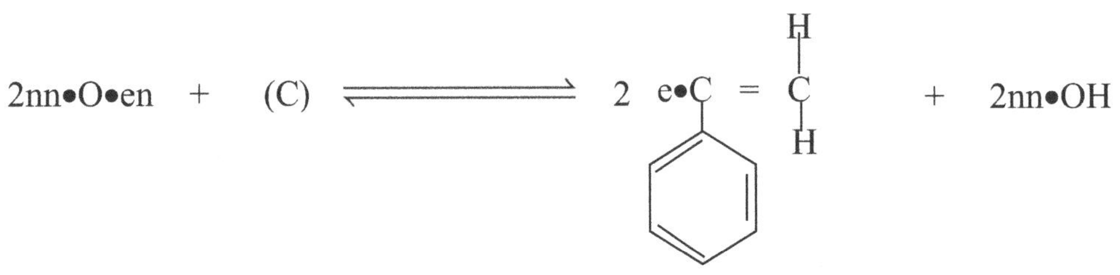

2nn•O•en + (C) $\rightleftharpoons$ 2 e•C(C_6H_5) = CH_2 + 2nn•OH

(D)

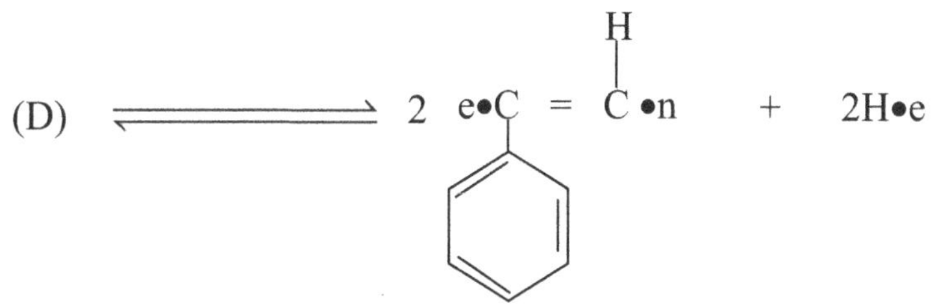

(D) ⇌ 2 e•C(C₆H₅) = C(H) •n + 2H•e

(E)

2H•e + 2HO •nn ⇌ $2H_2O$

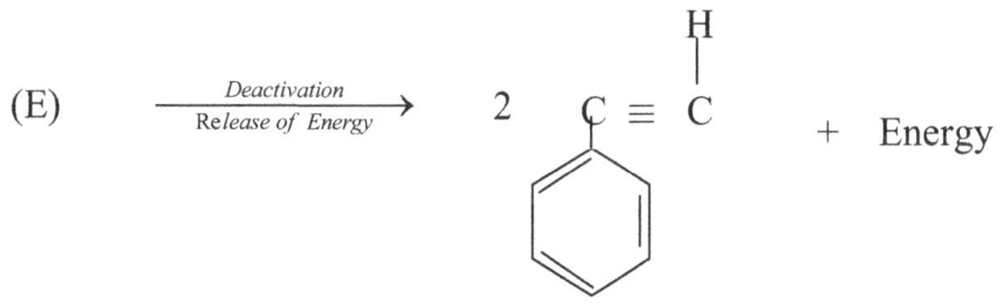

(E) $\xrightarrow[\text{Release of Energy}]{\text{Deactivation}}$ 2 C(C₆H₅) ≡ C–H + Energy

(F) 1.97a

Overall equation: $2C_6H_5C_2H_5 + 2O_2 \longrightarrow 2HC \equiv CC_6H_5 + 4H_2O + Energy$ 1.97*b*

A phenyl acetylene is observed to be the last product in this stage of five steps. Worthy of note is that little or nothing is known about this compound. The reason is probably because it is very unstable. It decomposes instantaneously as shown below in the next stage even in the presence of oxidizing oxygen molecule.

Stage 3:

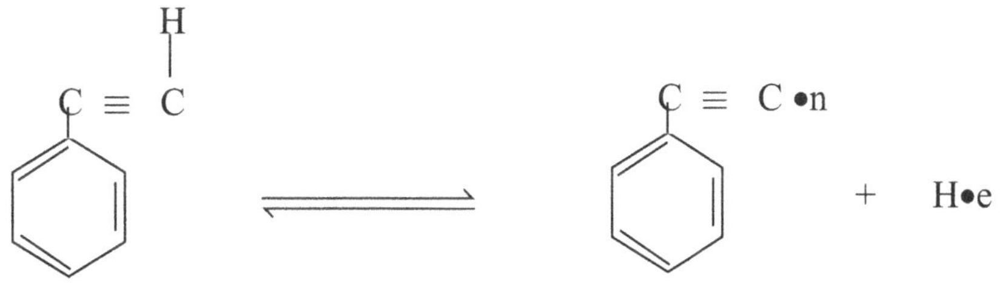

C(C₆H₅) ≡ C–H ⇌ C(C₆H₅) ≡ C •n + H•e

(F) (G)

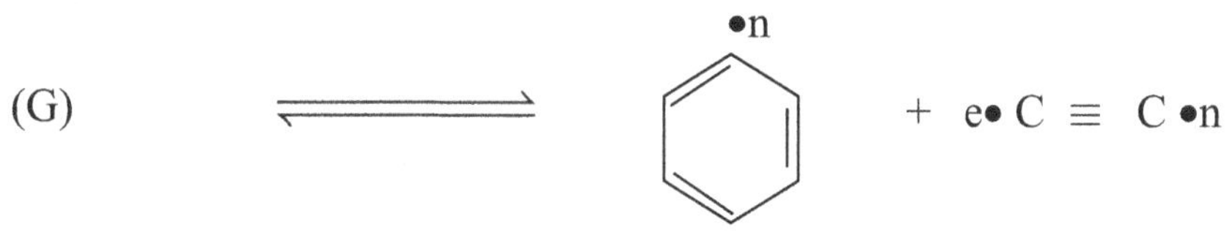

(G) ⇌ •n(C₆H₅) + e• C ≡ C •n

(H)

e• C ≡ C •n ⇌ 2 e•C(•n)(•e)•n

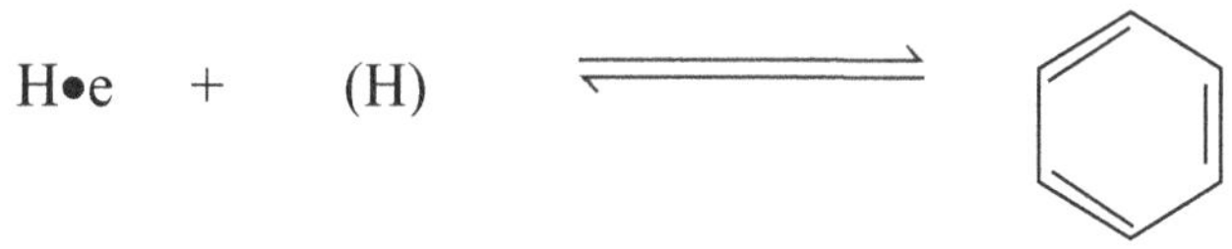

$$H{\bullet}e \quad + \quad (H) \quad \rightleftharpoons \quad \text{[benzene ring]}$$

$$2 \quad e{\bullet}\overset{\bullet n}{\underset{\bullet e}{C}}{\bullet}n \xrightarrow[\text{Release of Heat}]{\text{Deactivation}} 2 \quad e{\bullet}\overset{\bullet\bullet}{C}{\bullet}n \quad + \quad \text{Energy}$$

Activated Carbon Carbon Black

1.98a

$$\underline{Overall\ equation}: 2Phenyl\ Benzene \longrightarrow 2Benzene + 4Carbon\ Black + Energy(3) \qquad 1.98b$$

The phenyl acetylene decomposes instantaneously to benzene, carbon black and release of energy. (F) in Stage 2 is the product which could not be isolated or even oxidized in Stage 3.

Stage 4:

$$\underline{O_2} \rightleftharpoons 2nn{\bullet}O{\bullet}en$$

$$2nn{\bullet}O{\bullet}en \quad + \quad 2 \quad e{\bullet}\overset{\bullet\bullet}{C}{\bullet}n \quad \rightleftharpoons \quad 2nn{\bullet}O-\overset{\bullet e}{\underset{\bullet n}{C}}{\bullet}e$$

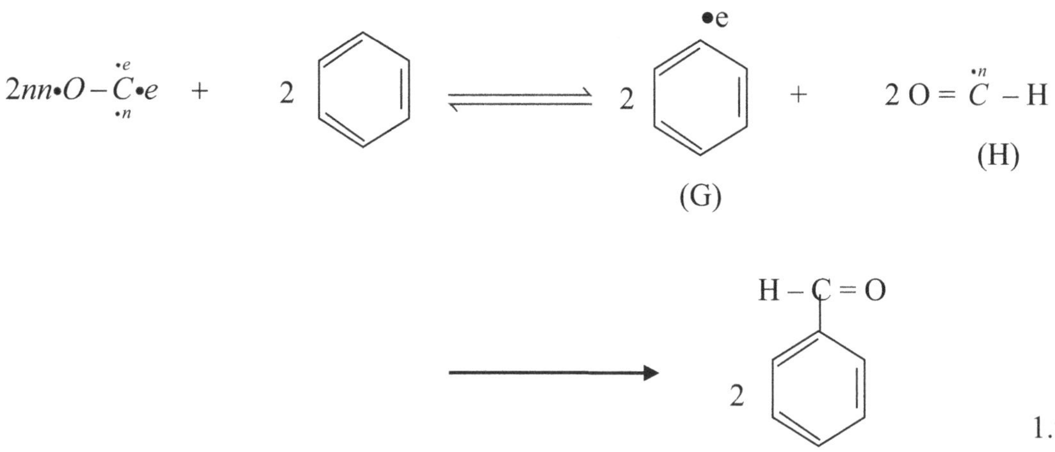

$$2nn{\bullet}O-\overset{\bullet e}{\underset{\bullet n}{C}}{\bullet}e \quad + \quad 2\ \text{[benzene ring]} \quad \rightleftharpoons \quad 2\ \text{[benzene ring with } {\bullet}e\text{]} \quad + \quad 2\,O=\overset{\bullet n}{C}-H$$

(G) (H)

$$\longrightarrow 2\ \text{[benzene ring with } H-C=O\text{]} \qquad 1.99a$$

(I)

$$\underline{Overall\ equation}: 2C_6H_5C_2H_5 + 3\underline{O_2} \longrightarrow 2C_6H_5CHO + 2C\ (Carbon\ Black) + 4H_2O + Energy(3) \qquad 1.99b$$

In this stage, only two moles of the carbon black were consumed leaving two moles behind. After being activated by the oxidizing oxygen in the second step via the paired unbonded radicals as has already been encountered in Stage 4 of Equation 1.43a, the electro-free-radical end of the carbon monoxide comes to immediately abstract H atom from the de-energized benzene to give (G) and (H) to form benzaldehyde in the last step. Between the carbon black and the benzaldehyde formed, the more stable is the benzaldehyde. Therefore, the carbon black is first oxidized as already shown and recalled below for the purpose of continuity.

Stage 5:

$$\underline{\underline{O_2}} \underset{\text{Oxygen molecule}}{\overset{\text{Equilibruim State Existence of Oxidising}}{\rightleftharpoons}} 2en\cdot\ddot{\underset{..}{O}}\cdot nn$$

$$2en\cdot O\cdot nn + 2:\overset{\bullet n}{\underset{\bullet e}{C}} \overset{\text{Oxidation}}{\rightleftharpoons} 2\,nn\bullet O-\overset{\bullet e}{\underset{\bullet n}{C}}\bullet e$$

(Stabilized) (F)

$$\xrightarrow[\text{Release of Energy}]{\text{Deactivation}} 2:C=O \quad + \quad Energy \qquad (1.43a)$$

$$\underline{Overall\ Equation}:\ \underline{\underline{O_2}} + 2\overset{\bullet n}{\underset{\bullet e}{C}}: \longrightarrow 2:C=O \quad + \ Energy \qquad (1.43b)$$

Stage 6:

$$\underline{\underline{O_2}} \underset{\text{Oxygen molecule}}{\overset{\text{Equilibruim State Existence of Oxidising}}{\rightleftharpoons}} 2en\cdot\ddot{\underset{..}{O}}\cdot nn$$

$$2en\cdot O\cdot nn + \ 2C=O \overset{\text{Oxidation}}{\rightleftharpoons} 2\,e\bullet\overset{\overset{O}{\|}}{C}-O\bullet nn$$

(Stabilized) (B)

$$(B) \xrightarrow[\text{Release of Energy}]{\text{Deactivation}} 2CO_2 \ + \ Energy \qquad (1.45a)$$

$$Overall\ equation:\ \underline{\underline{O_2}} + 2CO \longrightarrow 2CO_2 \ + \ Energy \qquad (1.45b)$$

$$\underline{Overall\ equation}:\ 2C_6H_5C_2H_5 + 5\underline{\underline{O_2}} \longrightarrow 2C_6H_5CHO + 2CO_2 + 4H_2O + Energy(5) \quad 1.100$$

With the removal of CO in the system, this was next followed by the oxidation of the benzaldehyde to the acid as again recalled below.

Stage 7:

$$\underline{\underline{O_2}} \rightleftharpoons 2nn\bullet O\bullet en$$

$$2nn\bullet O\bullet en \ + \ (I) \rightleftharpoons 2\ \overset{\overset{O}{\|}}{C}\bullet e\ (\text{attached to phenyl ring}) \ + \ 2nn\bullet OH$$

(J)

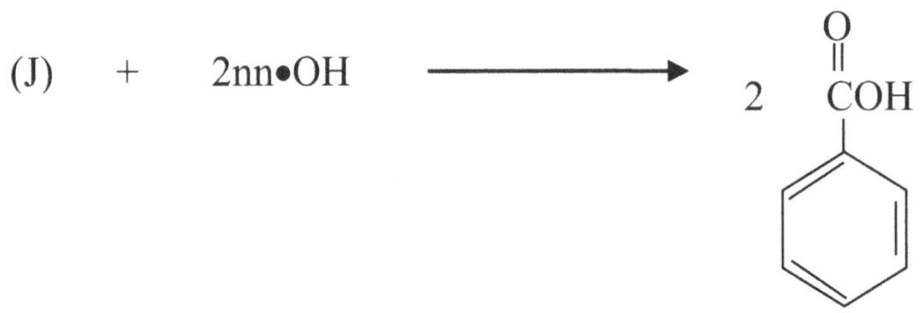

(J) + 2nn•OH ⟶ 2 C_6H_5COOH

(K) (1.90a)

Overall equation: $2C_6H_5C_2H_5 + 6O_2 \longrightarrow 2C_6H_5COOH + 4H_2O + 2CO_2 + Energy(5)$ 1.101

On the whole, seven stages were involved for the full oxidation of ethyl benzene, while for toluene, three stages were involved.

As has been noted, because the carboxylic acid would decompose under the operating conditions to give benzene and CO_2 as shown below in the absence of oxidizing oxygen in the system, oxidation of the carboxylic acid does not indeed take place.

Stage 1: {In the absence of Oxidizing Oxygen}

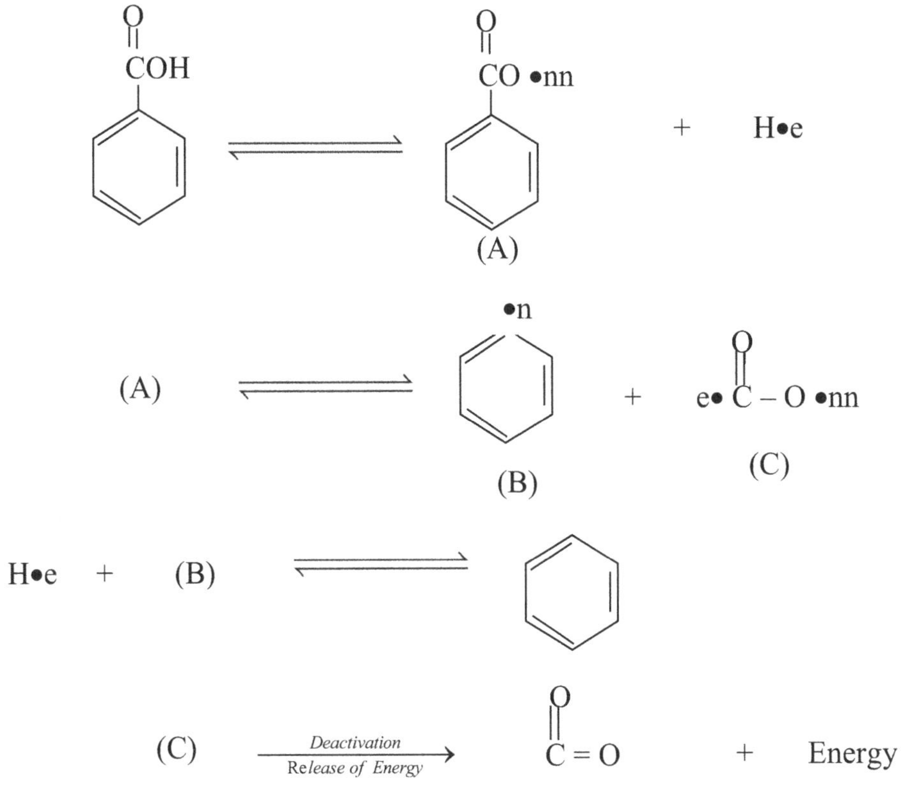

C_6H_5COOH ⇌ C_6H_5CO–O•nn (A) + H•e

(A) ⇌ •nC_6H_5 (B) + e•C(=O)–O•nn (C)

H•e + (B) ⇌ C_6H_6

(C) $\xrightarrow[\text{Release of Energy}]{\text{Deactivation}}$ O=C=O + Energy 1.102a

Overall equation: $C_6H_5COOH \longrightarrow C_6H_6 + CO_2 + Energy$ 1.102*b*

Unlike methyl benzene (Toluene), where for the types of products identified as intermediates in Equation 1.91- the (A)s - alcohols, the (B)s – ketones, the (C)s – Enols, only the alcohol and aldehyde were identified, while with ethyl benzene, only an alkene (Styrene) and aldehyde were observed. No alcohol was identified with ethyl benzene. We are yet to identify the Enols.

1.3.15 Oxidation of Propyl Benzene

Propyl benzene (Real name is propylane benzene), the third member of alipharomatic hydrocarbons, looks as if it is becoming more aliphatic in character, which indeed is not the case, because of the strong unique character of the benzene ring.

Stage 1: $O_2 \rightleftharpoons 2nn{\bullet}O{\bullet}en$

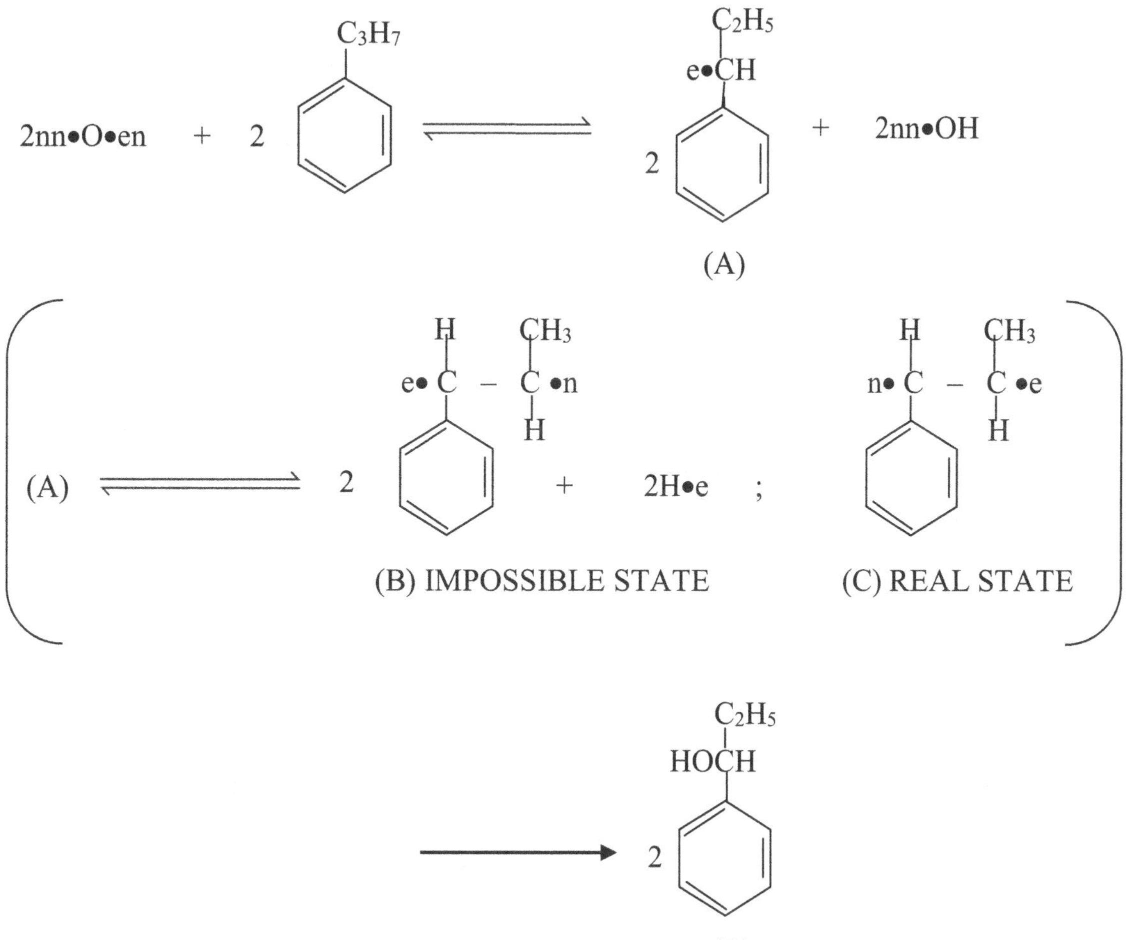

2nn•O•en + 2 [benzene ring with C_3H_7] ⇌ 2 [benzene ring with e•CH–C_2H_5] + 2nn•OH

(A)

(A) ⇌ 2 [benzene ring with e•C(H) – C(CH3)(H)•n] + 2H•e ; [benzene ring with n•C(H) – C(CH3)(H)•e]

(B) IMPOSSIBLE STATE (C) REAL STATE

⟶ 2 [benzene ring with HOCH–C_2H_5]

(D) 1.103a

$$\underline{Overall\ equation}: 2C_6H_5C_3H_7 + \underset{--}{O_2} \longrightarrow 2C_6H_5CH(OH)C_2H_5 \qquad 1.103b$$

Note that H atom could not be rejected after the second step as indicated above, because the radical-pushing capacity of the phenyl group is just a little larger than H, but less than that of CH_3 group. (D) an alcohol is obtained here and oxidized as follows.

Stage 2: $\underset{--}{O_2} \rightleftharpoons 2nn{\bullet}O{\bullet}en$

2nn•O•en + (D) ⇌ 2 [en• O – CH(C₂H₅)–C₆H₅] (E) + 2 nn• OH

(E) ⇌ 2 [O = C(C₂H₅)–C₆H₅] (F) + 2H •e + Energy

$$2H{\bullet}e + 2HO{\bullet}nn \longrightarrow 2H_2O \qquad 1.104a$$

$$\underline{Overall\ equation}: 2C_6H_5C_3H_7 + 2\underset{--}{O_2} \longrightarrow 2C_6H_5COC_2H_5 + 2H_2O + Energy \quad 1.104b$$

Here, the real ketone as opposed to aldehydes identified so far can be seen. One would have expected the corresponding case of (F) to appear for ethyl benzene with CH_3 group in place of C_2H_5. Instead, it was the appearance of styrene that prevented its formation.

Stage 3: $\underset{--}{O_2} \rightleftharpoons 2nn{\bullet}O{\bullet}en$

2nn•O•en + (F) ⇌ 2 [e• CH₂–CH₂–C(=O)–C₆H₅] (G) + 2nn •OH

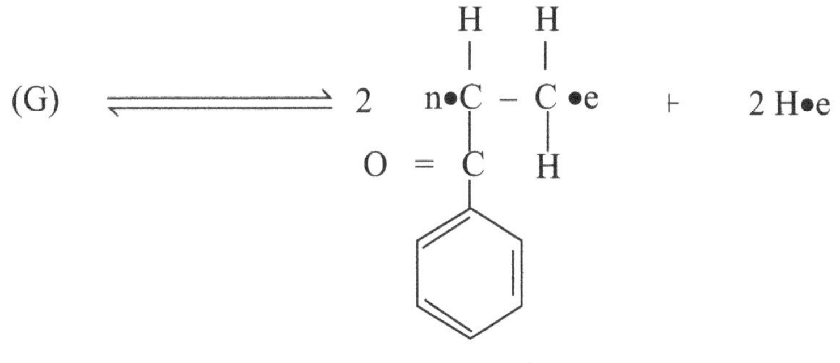

$$(G) \rightleftharpoons 2\ n{\bullet}\overset{H}{\underset{O=C(C_6H_5)}{C}} - \overset{H}{\underset{H}{C}}{\bullet}e + 2\,H{\bullet}e$$

(H)

$$2H{\bullet}e + 2HO{\bullet}nn \rightleftharpoons 2H_2O$$

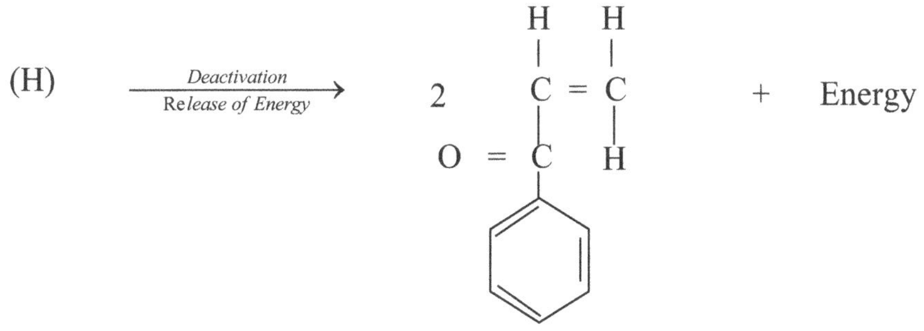

$$(H) \xrightarrow[\text{Release of Energy}]{\text{Deactivation}} 2\ \overset{H}{\underset{O=C(C_6H_5)}{C}} = \overset{H}{\underset{H}{C}} + \text{Energy}$$

(I) 1.105a

$$\underline{\textit{Overall equation}}: 2C_6H_5C_3H_7 + 3O_2 \longrightarrow 2C_6H_5(CO)HC = CH_2 + 4H_2O + Energy(2) \qquad 1.105b$$

Notice the hydrogen atom that has been abstracted in the second step. It is from the C center carrying more H atoms, that which is externally located. The carbon center adjacently located to the benzene ring no longer has H on it. The C_6H_5CO group is ***a strong radical-pulling group*** as will be shown in the New Frontiers. Hence, the C center carrying it is carrying a nucleo-free-radical after H is released (because of the heat) in the third step of the stage. In the last step where the stage becomes productive, energy is released after deactivation of (H) from the third step. The (I) formed as will be shown in the next chapter is an Electrophile (Male) with X and Y centers as exist in humans and with other animals' chromosomes. All the monomers we have encountered so far such as ethylene (Real name Ethene), styrene, and more have been ***Nucleophiles (Females)*** carrying only X centers. With presence of more oxidizing oxygen molecules in the system, oxidation of (I) continues as follows.

Stage 4:

$$\underset{--}{O_2} \rightleftharpoons 2nn{\bullet}O{\bullet}en$$

$$2nn{\bullet}O{\bullet}en + (I) \rightleftharpoons 2\,\overset{-}{O}=C(C_6H_5)-C(H)=C(H){\bullet}e + 2nn{\bullet}OH$$

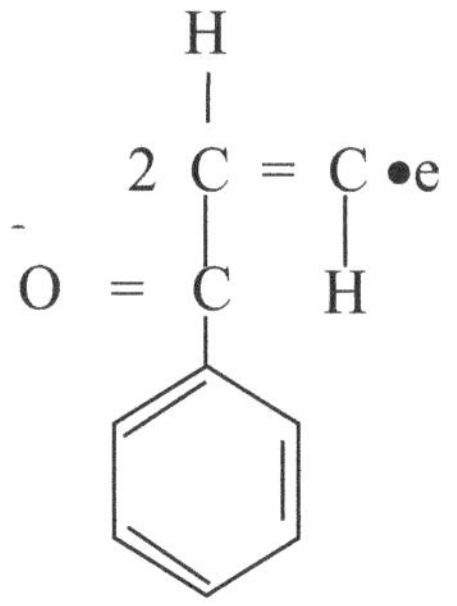

(J)

$$(J) \rightleftharpoons 2\,O=C(C_6H_5)-C({\bullet}n)=C(H){\bullet}e + 2H{\bullet}e$$

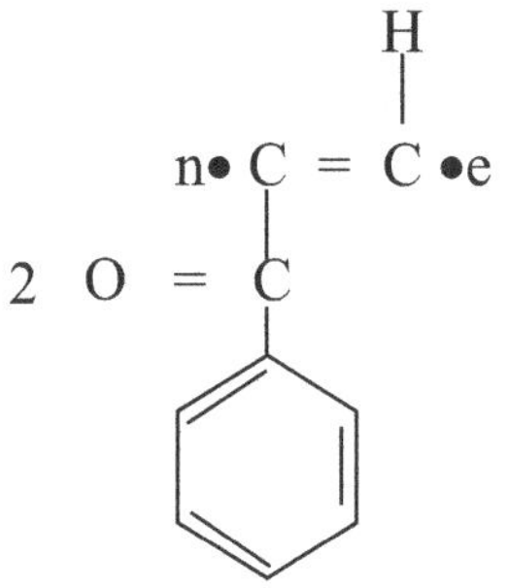

(K)

$$2H{\bullet}e + 2HO{\bullet}nn \rightleftharpoons 2H_2O$$

$$(K) \xrightarrow[\text{Release of Energy}]{\text{Deactivation}} 2\,O=C(C_6H_5)-C\equiv C-H + \text{Energy}$$

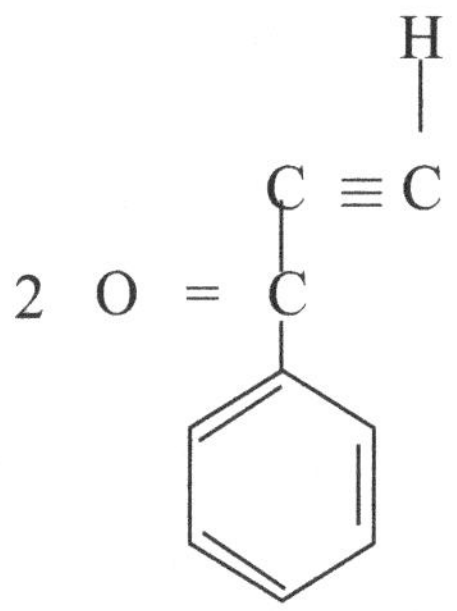

(L) 1.106a

$$\underline{\text{Overall equation}}: 2C_6H_5C_3H_7 + 4\underset{--}{O_2} \longrightarrow 2C_6H_5CO(C\equiv CH) + 6H_2O + Energy(3) \quad 1.106b$$

Just like the case of phenyl acetylene a Nucleophile which was found to be unstable, so also is the phenyl carbonyl acetylene an Electrophile noted to be very unstable, both in the presence or absence of oxidizing oxygen molecule. (L) breaks down immediately as follows.

Stage 5:

$$(L) \rightleftharpoons 2\ O = C(-C \equiv C\bullet n)(-C_6H_5) + 2H\bullet e$$

(M)

$$(M) \rightleftharpoons 2\ O = C\bullet n(-C_6H_5) + 2\,e\bullet C \equiv C\bullet n$$

(N)

$$2e\bullet C \equiv C\bullet n \rightleftharpoons 4\ e\bullet \overset{\bullet n}{\underset{\bullet e}{C}}\bullet n$$

$$2H\bullet e + 2(N) \rightleftharpoons 2\ O = C(-H)(-C_6H_5)$$

(O)

$$4\ e\cdot \overset{\bullet n}{\underset{\bullet e}{C}}\cdot n \xrightarrow[\text{Release of Heat}]{\text{Deactivation}} 4\ e\bullet \overset{\bullet\bullet}{C}\bullet n + \text{Energy}$$

Activated Carbon Black — Carbon Black

1.107

<u>*Overall Eqyation*</u>: $2C_6H_5C_3H_7 + 4O_2 \longrightarrow 2C_6H_5CHO + 4C(Carbon\ Black)$

$+ 6H_2O + Energy(4)$ 1.108

When (L) decomposed, benzaldehyde and carbon black were obtained, just like the case with ethyl benzene. As has been said and repeated again, between the carbon black and the benzaldehyde formed the more stable is the benzaldehyde. Therefore, the carbon black is first oxidized as already shown and recalled below for the purpose of continuity.

Stage 6 &7:

$$\underset{--}{O_2} \xrightleftharpoons[\text{Oxygen molecule}]{\text{Equilibruim State Existence of Oxidising}} 2en\cdot\ddot{O}$$

$$2en\cdot O\cdot nn + \underset{(Stabilized)}{2:\overset{\bullet n}{\underset{\bullet e}{C}}} \xrightleftharpoons{\text{Oxidation}} \underset{(F)}{2\,nn\bullet O-\overset{\bullet e}{\underset{\bullet n}{C}}\bullet e}$$

$$\xrightarrow[\text{Release of Energy}]{\text{Deactivation}} 2:C=O \quad + \quad Energy \qquad (1.43a)$$

$$\underline{Overall\ Equation}: \underset{--}{O_2} + 2\overset{\bullet n}{\underset{\bullet e}{C}}: \longrightarrow 2:C=O \quad + \quad Energy \qquad (1.43b)$$

Stage 8&9:

$$\underset{--}{O_2} \xrightleftharpoons[\text{Oxygen molecule}]{\text{Equilibruim State Existence of Oxidising}} 2en\cdot\ddot{O}\cdot nn$$

$$2en\cdot O\cdot nn + \underset{(Stabilized)}{2C=O} \xrightleftharpoons{\text{Oxidation}} \underset{(B)}{2\,e\bullet\overset{O}{\overset{\|}{C}}-O\bullet nn}$$

$$(B) \xrightarrow[\text{Release of Energy}]{\text{Deactivation}} 2CO_2 \quad + \quad Energy \qquad (1.45a)$$

$$\underline{Overall\ Equation}\ : \underset{--}{O_2} + 2CO \longrightarrow 2CO_2 \quad + \quad Energy \qquad (1.45b)$$

$$\underline{Overall\ equation}: 2C_6H_5C_3H_7 \quad + \quad 8\underset{--}{O_2} \longrightarrow 2C_6H_5CHO \quad + \quad 4CO_2 \quad + \quad 6H_2O$$
$$+ \quad Energy(6) \quad 1.109$$

With the removal of CO in the system, this was next followed by the oxidation of the benzaldehyde to the acid as again recalled below.

Stage 10:

$$\underset{--}{O_2} \rightleftharpoons 2nn\bullet O\bullet en$$

$$2nn\bullet O\bullet en \quad + \quad (O) \rightleftharpoons 2\ \underset{(M)}{C_6H_5\overset{O}{\overset{\|}{C}}\bullet e} \quad + \quad 2nn\bullet OH$$

$$(D) \quad + \quad 2nn\bullet OH \longrightarrow 2\ \underset{(P)}{C_6H_5\overset{O}{\overset{\|}{C}}OH} \qquad (1.90a)$$

$$\underline{Overall\ equation}: 2C_6H_5C_3H_7 + 9\underline{O}_2 \longrightarrow 2C_6H_5COOH + 6H_2O + 4CO_2 + Energy(8) \qquad 1.110$$

As has been noted, because the carboxylic acid would decompose under the operating conditions to give benzene and CO_2 in the absence of oxidizing oxygen in the system, oxidation of the carboxylic acid does not indeed take place. The carboxylic acid seems to be kept in the ENERGIZED Equilibrium state of existence in the presence of oxidizing oxygen. On the whole, ten stages were involved for the full oxidation of propyl benzene. Nevertheless, one has observed that Enols are not parts of the products based on the operating conditions. If enol is formed, it can only come from ethyl benzene, propyl benzene under different operating conditions such as ***presence of acid or alkali catalysts and under mild operating conditions.*** Just as acetaldehyde for example enolizes to vinyl alcohol as shown below, so also the ketones here enolizes to give the alcohols (Enols).

1.3.16 <u>Enolization of Acetaldehyde</u>

While aldehydes, ketones and aromatic ketones can undergo this phenomenon, benzaldehyde cannot.

<u>Stage 1:</u>

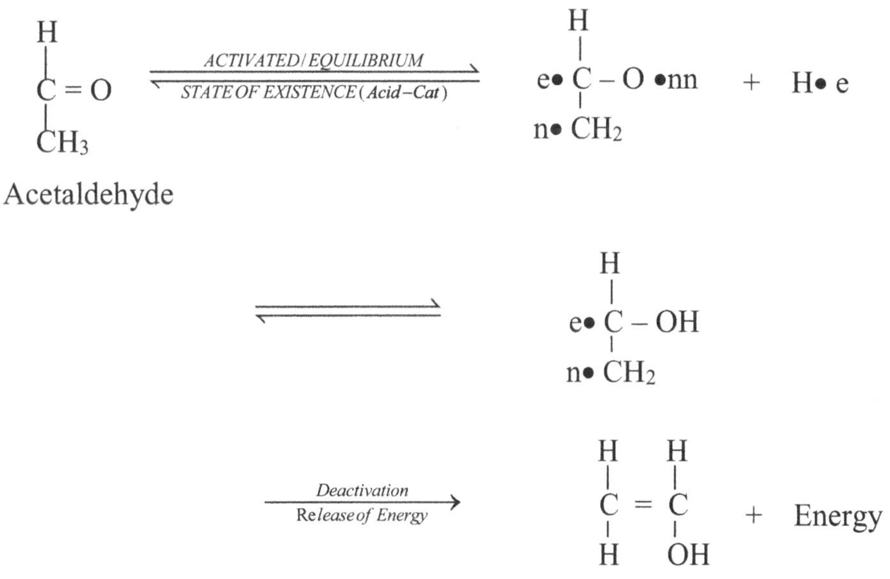

$$\underset{\text{Acetaldehyde}}{H-\overset{|}{C}(=O)-CH_3} \xrightleftharpoons[STATE\ OF\ EXISTENCE(Acid-Cat)]{ACTIVATED/EQUILIBRIUM} e\bullet \overset{H}{\underset{n\bullet CH_2}{C}} - O\ \bullet nn + H\bullet e$$

$$\rightleftharpoons e\bullet \overset{H}{\underset{n\bullet CH_2}{C}} - OH$$

$$\xrightarrow[Release\ of\ Energy]{Deactivation} \underset{\text{Vinyl alcohol}}{\overset{H}{\underset{H}{C}} = \overset{H}{\underset{OH}{C}}} + Energy \qquad 1.111a$$

$$\underline{Overall\ equation}: H_3CCHO \xrightarrow[Base\ catalyzed]{Acid\ or} H_2C = CHOH + Energy \qquad 1.111b$$

The acetaldehyde has been enolized to vinyl alcohol which itself is very unstable at STP. In the presence of dilute acids or alkalis which are passive in character (like "Policemen"), it is forced to exist in ACTIVATED/EQUILIBRIUM state of existence as shown in the first step of the stage. O being more electronegative than C accepts the H to form the – ol (Enolization). At STP or a little higher, it molecularly rearranges to acetaldehyde which is far more stable than the vinyl alcohol. What makes the alcohol unstable is the H on the OH group (Polar/Ionic) adjacently located to a double bond; for if it was CH_3 group (Polar/Non-ionic), the rearrangement will not take place as will be fully explained in the New Frontiers under another operating condition - heat as shown below. [It is all taking from the RICH and giving to the POOR, and never the other way around and accepting what is exactly due to you, and not more than what is not due to you].

Stage 1:

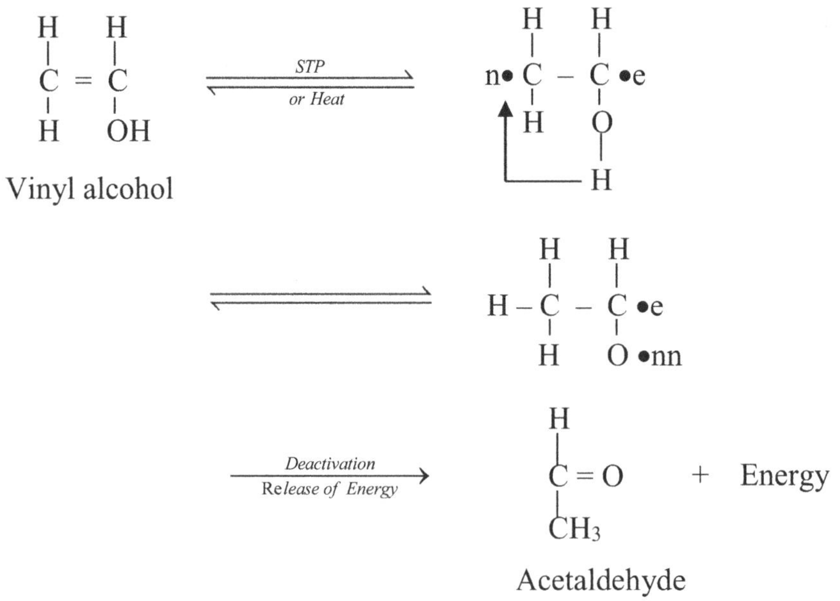

```
H   H                            H   H
|   |                            |   |
C = C      ---STP--->        n• C - C •e
|   |      <--or Heat--          |   |
H   OH                         ↑ H   O
                               |     |
Vinyl alcohol                  └─────H

                                 H   H
                                 |   |
           ===========>       H- C - C •e
                                 |   |
                                 H   O •nn

                                 H
                                 |
      --Deactivation-->          C = O     +   Energy
      Release of Energy          |
                                 CH3

                              Acetaldehyde             1.112a
```

MOLECULAR REARRANGEMENT OF FIRST KIND OF FIRST TYPE

$$\underline{Overall\ equation}: H_2C = CH(OH) \xrightarrow[or\ Heat]{STP} CH_3CHO \ + \ Energy \qquad 1.112b$$

Note that the two reactions above are only reversible under two different operating conditions, not as wrongly represented in present-day Science for reversible reactions. As will be seen in the New Frontiers, there are different kinds of molecular rearrangements phenomena. These are just some of the few methods of rearrangements of molecules called **Tautomerism**. Enolization is one of them. So also, are Imidization, Chloridization and more to be seen downstream. Each kind has different types. The case above is ***molecular rearrangement of the first kind of the first type.*** The transfer species here which is H is very important, because it the same that is involved in ***Activated/Equilibrium state of existence*** and is the one that will make the monomer ***not undergo the route which is not natural to it during INITIATION STEP in polymerization systems*** and is also the one which will ***kill the growing polymer chain by starvation of the system with monomer during PROPAGATION*** in polymerization systems in the route natural to the monomer. In the New Frontiers, this was called ***THE LAWS OF CONSERVATION OF TRANSFER OF TRANSFER SPECIES.*** It must be emphasized at this point in time, that what we have been seeing so far mark the beginning of a NEW SCIENCE and indeed the beginning of a NEW DAWN FOR HUMANITY.

The enols said to be observed during oxidation of alipharomatic hydrocarbons[51], can be explained as follows. During the oxidation of ethyl benzene, in stage 1 of Equation 1.96, the formation of methyl phenyl ketone was prevented due to the operating condition and ready ease of rejection of H atom to form activated styrene which deactivated to release energy. At a milder operating condition, the following would have taken place in place of that stage and succeeding stages.

Stage 1:

$$\underline{\underline{O_2}} \rightleftharpoons 2nn{\bullet}O{\bullet}en$$

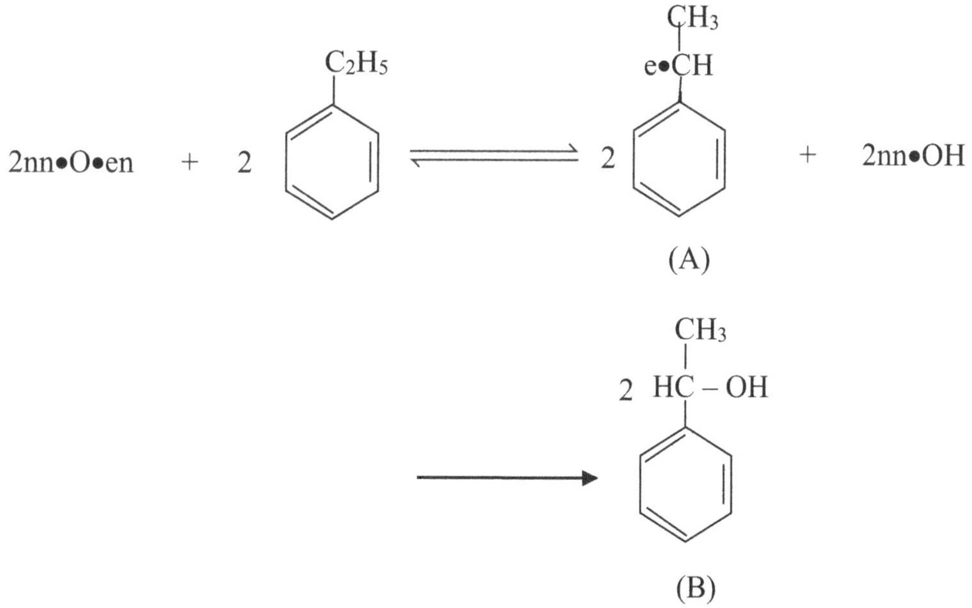

$$2nn{\bullet}O{\bullet}en + 2\ C_6H_5C_2H_5 \rightleftharpoons 2\ C_6H_5\text{–}CH({\bullet}e)CH_3\ (A) + 2nn{\bullet}OH$$

$$\longrightarrow 2\ C_6H_5\text{–}HC(OH)CH_3\ (B) \qquad [1.96a]$$

Overall equation: $2C_6H_5C_2H_5 + \underline{\underline{O_2}} \longrightarrow 2C_6H_5CH(OH)CH_3$ [1.96*b*]

Stage 2:

$$\underline{\underline{O_2}} \rightleftharpoons 2nn{\bullet}O{\bullet}en$$

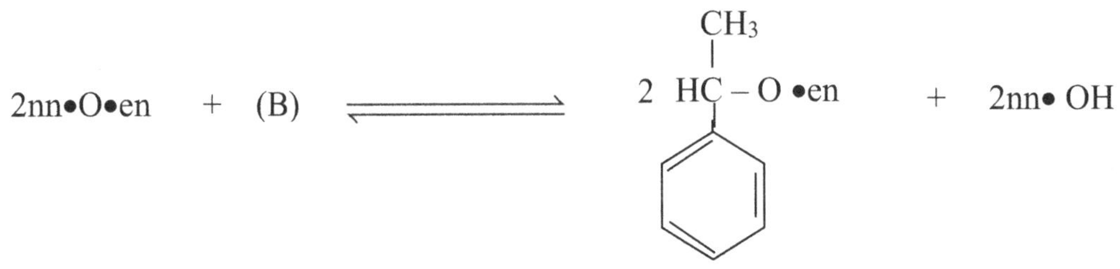

CH_3
|
$2nn\bullet O\bullet en$ + (B) $\rightleftharpoons$ 2 HC – O •en + 2nn• OH
(benzene ring)

(C)

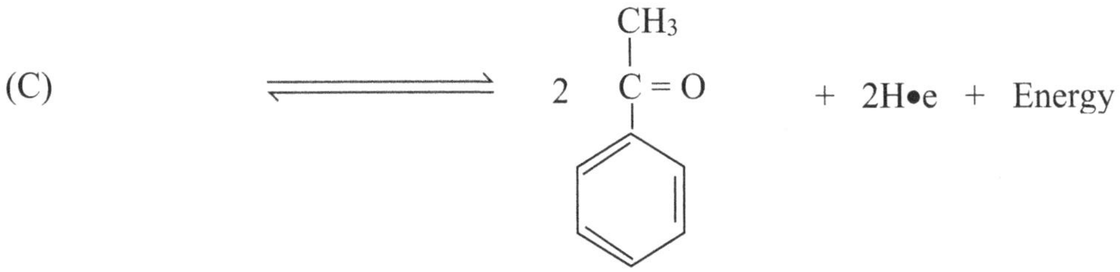

CH_3
|
(C) $\rightleftharpoons$ 2 C = O + 2H•e + Energy
(benzene ring)

(D)

$$2H\bullet e + 2HO\bullet nn \longrightarrow 2H_2O \qquad 1.113a$$

$$\underline{Overall\ equation}: 2C_6H_5C_2H_5 + 2\underline{\underline{O_2}} \longrightarrow 2C_6H_5COCH_3 + 2H_2O + Energy \qquad 1.113b$$

Whether there is no oxidizing oxygen left in the system or not, it seems that enolization follows as shown below.

Stage 3:

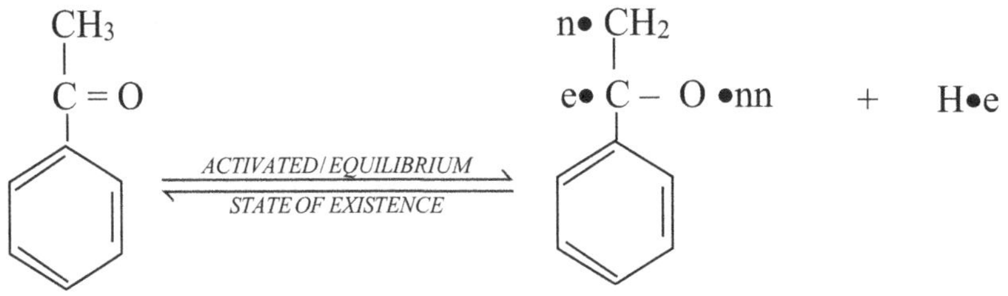

CH_3 n• CH_2
| |
C = O $\underset{STATE\ OF\ EXISTENCE}{\overset{ACTIVATED/EQUILIBRIUM}{\rightleftharpoons}}$ e• C – O •nn + H•e
(benzene ring) (benzene ring)

(D) (E)

n• CH_2
|
$\rightleftharpoons$ e• C – OH
(benzene ring)

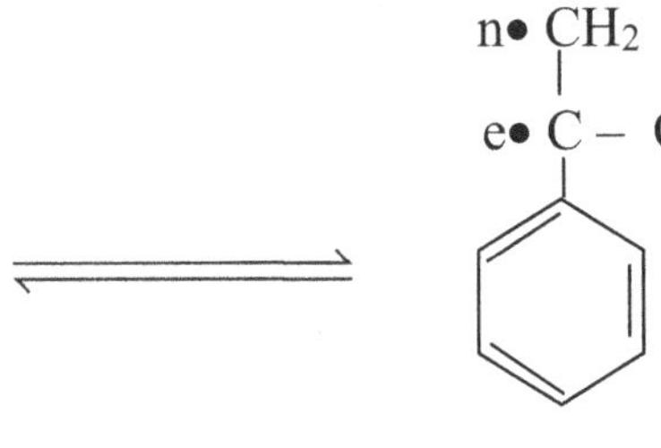

(F[1])

$$\xrightarrow[\text{Release of Energy}]{\text{Deactivation}} \quad C_6H_5(OH)C = CH_2 \quad + \quad \text{Energy}$$

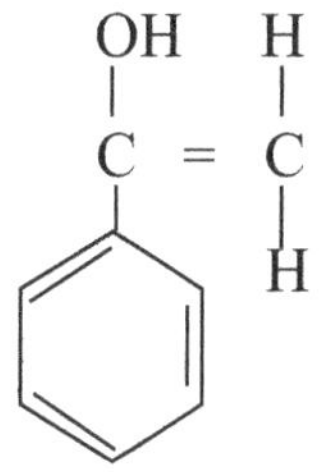

(F) An Enol 1.114a

$$\underline{\text{Overall equation}}: C_6H_5COCH_3 \xrightarrow[\text{or Base catalyst}]{\text{Acid}} C_6H_5(OH)C = CH_2 \qquad 1.114b$$

$$\underline{\text{Overall equation}}: 2C_6H_5C_2H_5 + 2\underline{O_2} \xrightarrow[\text{or Based catalyzed}]{\text{Acid}} 2C_6H_5(OH)C = CH_2 + 2H_2O + \text{Energy}(2) \qquad 1.114c$$

If there was oxidizing oxygen left in the system after Stage 2, either the enol (F) will not be present in the system or in the presence of oxidizing oxygen, the ketone formed (D) was kept in Activated/Equilibrium state of existence (E). If the latter is not the case, the oxidizing oxygen molecules from the oxidizing agent *was then* added into the system after formation of (F) to continue the oxidation, wherein the aromatic acid will be obtained as shown below. It is believed that if the latter took place, ***then the ketone will not be identified as shown in Equation 1.91.***

Stage 4:

$$\underline{O_2} \rightleftharpoons 2nn{\bullet}O{\bullet}en$$

$$2nn{\bullet}O{\bullet}en + (F) \rightleftharpoons 2\ C_6H_5(OH)C = C{\bullet}e(H) + 2\ nn{\bullet}OH$$

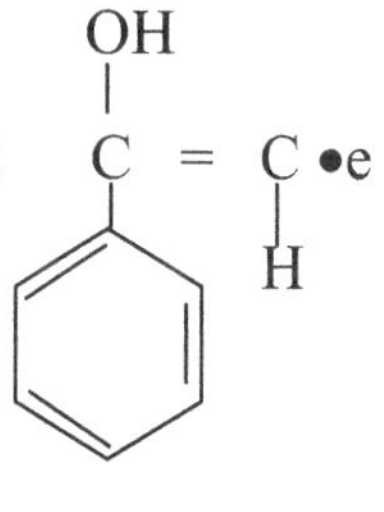

(G)

$$\longrightarrow 2\ C_6H_5(OH)C = C(OH)H$$

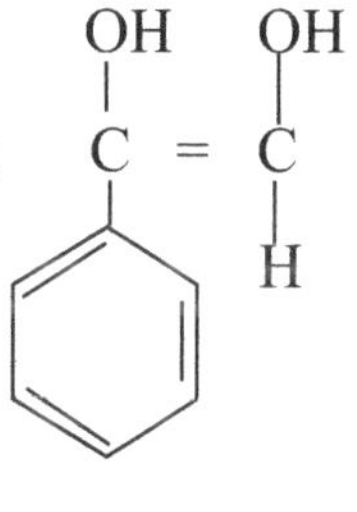

(H) 1.115a

$$\underline{Overall\ equation}: 2C_6H_5C_2H_5 + 3\underline{\underline{O_2}} \longrightarrow 2C_6H_5(OH)C = CH(OH) + 2H_2O$$
$$+ \ Energy(2) \qquad 1.115b$$

Worthy of note is that the H atom on the OH group could not be tempered with by the oxidizing oxygen, because as will be shown in the New Frontiers, when a group is internally located in dienes or trienes or the likes and the case above, the group is resonance stabilized. This is the case with for example the CH_3 group in isoprene, but not with that of pentadiene where the CH_3 group is externally located. Secondly, note that (G) above cannot release any transfer species (e• C_6H_5 or en• OH) against one of the laws of Nature wherein the C center cannot carry a nucleo-free-radical when any of the groups (OH or phenyl group) is released. Hence (H) was formed making the stage productive, since the product formed has a finger-print where H is the component held when made to exist in Equilibrium state of existence. This is why the last step showing a double-headed single arrow symbol is what it is (Combination state of Existence).

Stage 5: $\underline{\underline{O_2}} \rightleftharpoons 2nn{\bullet}O{\bullet}en$

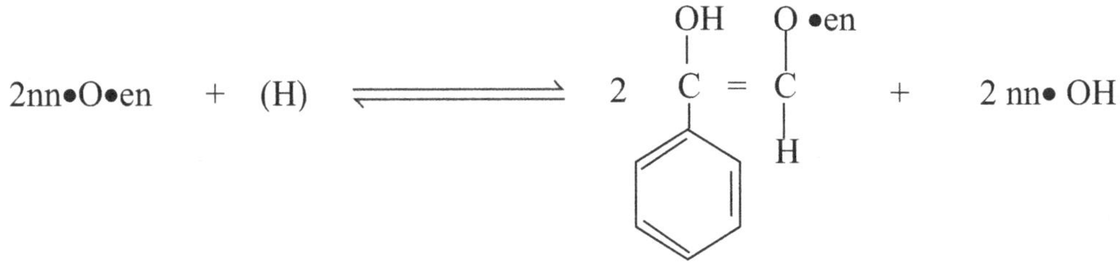

2nn•O•en + (H) ⇌ 2 C_6H_5–C(OH) = C(O•en)(H) + 2 nn• OH

(I)

⇌ 2 C_6H_5–C(OH) = C = O + H• e + Energy

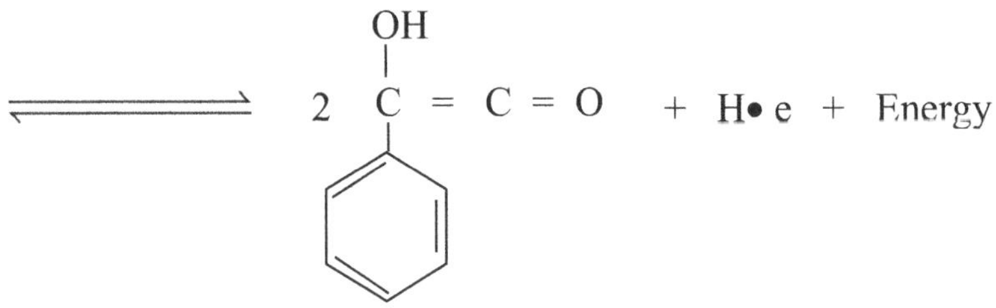

(J)

2H•e + 2HO •nn ⟶ $2H_2O$ 1.116a

$$\underline{Overall\ equation}: 2C_6H_5C_2H_5 + 4\underline{\underline{O_2}} \longrightarrow 2C_6H_5(OH)C = C = O + 4H_2O$$
$$+ \ Energy(3) \qquad 1.116b$$

An Electrophile, an aromatic resonance stabilized ketene (J) is obtained as product above in this stage. How it is resonance stabilized is shown below.

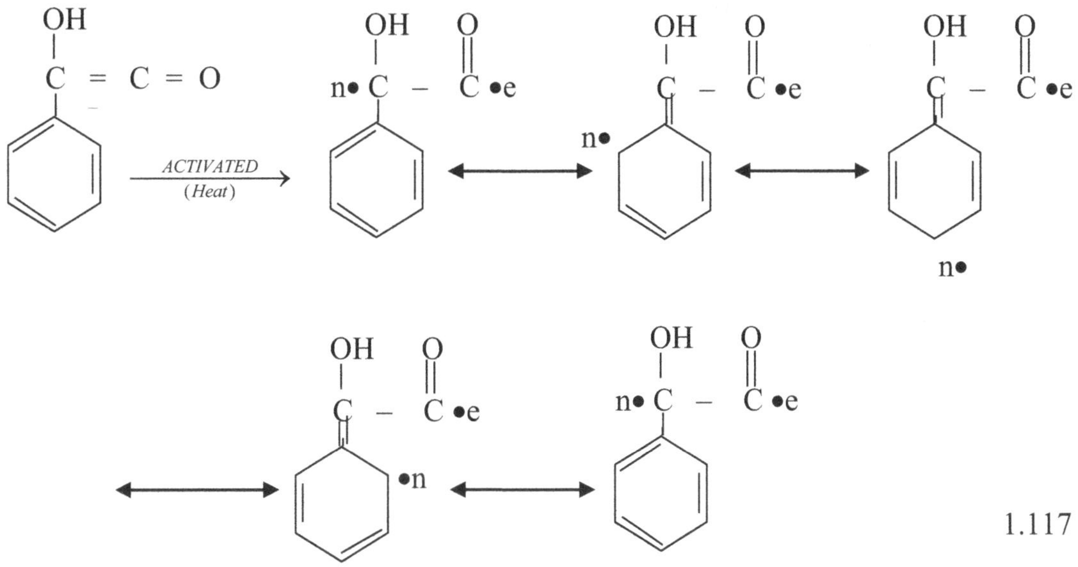

ACTIVATED (Heat)

1.117

RESONANCE STABILIZATION OF THE FIRST KIND OF THE SECOND TYPE

[The C = C center is Y and the C = O center is X]

Note that all the aromatic aldehydes and Ketones are Electrophiles (Males) with X and Y centers. While all ACTIVATION CENTERS ARE NUCLEOPHILIC (X≡ FEMALE) IN CHARACTER, the Y center appears when an hetero center such C = O, C ≡N, etc. center is conjugatedly or cumulatively placed to a C = C, C ≡ C center and more to be seen in the New Frontiers.

The resonance stabilization above is said to be of the second type, because the electro-free-radical moved from the π-bond to grab an adjacently places nucleo-free-radical to form another π-bond, leaving a nucleo-free-radical in front. In the first type, a visible electro-free-radical comes to grab a nucleo-free-radical from an adjacently located π- bond to also form another π- bond and leave an electro-free-radical in front. As will been seen in the New Frontiers, there are different kinds and types of Resonance stabilization phenomena. So also, there are different kinds and types of Electrophiles (Males) and Nucleophiles (Females). As a (the Ketene) different type of Electrophile (Male), the resonance stabilization cannot go beyond the Carbon- Carbon double bond, i.e., the C = O center cannot be involved. It is for this reason methyl acrylate which will be seen in the next chapter, is not known to be resonance stabilized. There are lots to be seen downstream about how NATURE operates, because ***NATURE as defined in the New Frontiers is nothing else other than ORDERLINESS, whether GODLY or SATANIC in the real and imaginary domains, that which takes no non-senses and abhors Vacuum or Non-linearity or differentiation and more as will be seen down-stream.*** The Benzoquinone which we have encountered during combustion of aromatic hydrocarbons, already said to be an Electrophile is also resonance stabilized, as shown below.

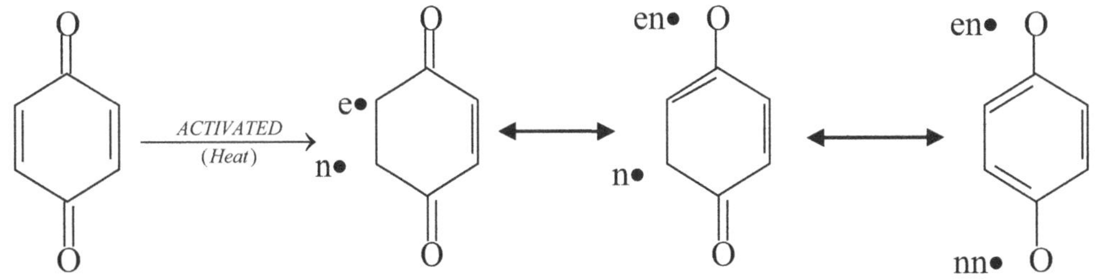

```
 O                        O           en• O             en• O
 ||                       ||              |                 |
(ring)  --ACTIVATED-->  e•(ring)  <-->   (ring)   <-->    (ring)
          (Heat)        n•               n•
 ||                       ||              ||                |
 O                        O               O             nn• O
```

1.118

RESONANCE STABILIZATION OF FIRST KIND OF THE FIRST AND SECOND TYPES CHARACTER

Shown below is resonance stabilization in pentadiene a Nucleophile (i.e. Female)

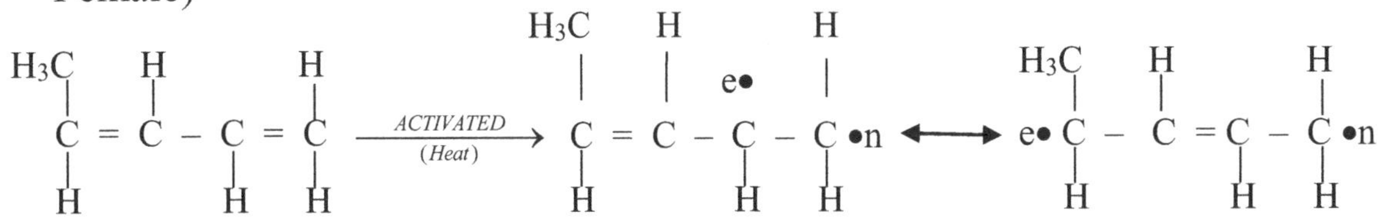

```
H3C  H     H                H3C  H       H                H3C  H      H
 |   |     |                 |   |   e•  |                 |   |      |
 C = C - C = C  --ACTIVATED-->  C = C - C - C •n  <-->  e• C - C = C - C •n
 |       |   |     (Heat)    |       |   |                 |      |   |
 H       H   H               H       H   H                 H      H   H
```

1.119

RESONANCE STABILIZATION OF THE FIRST KIND OF THE FIRST TYPE

Note that in all of them, the activation center first activated is the least Nucleophilic center. That is how Nature operates, that which is to be expected without the need of making necessary and unnecessary assumptions as done in present-day Science. When the wrong center is not first activated, the exercise is nonsense and that is what we have been doing in general. Downstream in the New Frontiers, one will show the order of Nucleophilicities of all ACTIVATION centers that exist in our world because of applications to other things. One will give the order of all RADICAL-PUSHING and RADICAL-PULLING groups that also exist in our world. One will see the order of so many things. It is only in CHEMISTRY that that which we call LOGICS exists and not in MATHEMATICS. For the last case above (Pentadiene), it was the visible electro-free-radical that grabbed the nucleo-free-radical from the adjacently located π-bond to form another π- bond and give the 1,4-mono-form above, the more stable mono-form. It started with 3,4- mono-form to 1,4- or 4,1- mono-form. Note that when double or triple bonds are conjugatedly (resonance stabilized) or cumulatively (resonance or non-resonance stabilized) placed, two ACTIVATION centers cannot be activated at the same time as done in present-day Science indiscriminately. One center can be activated one at a time. It is only when the two or more double or triple bonds are isolatedly placed, that two activation centers can be activated at the same time provided they are of the same nucleophilic capacity, since NATURE is ORDERLINESS.

In ones attempt to provide the mechanisms of COMBUSTION and OXIDATION, one has been forced to introduce only a grain of the new concepts in the New Frontiers. The ketene (J) of Equation 1.116a obtained from the oxidation of the enol (F) formed in Stage 3 of Equation 1.114a, cannot be oxidized, since the OH group carried by it as shown in Equation 1.117 is resonance stabilized, i.e., heavily shield from any form of attack. The OH group cannot be used to provide Equilibrium state of existence for the ketene. That the Enol cannot further be oxidized while the ketone from which it was formed oxidized, is impossible. Hence, it is believed and clearly be seen that the aromatic ketene (F) is very unstable and therefore decomposes instantaneously as shown below.

Stage 6:

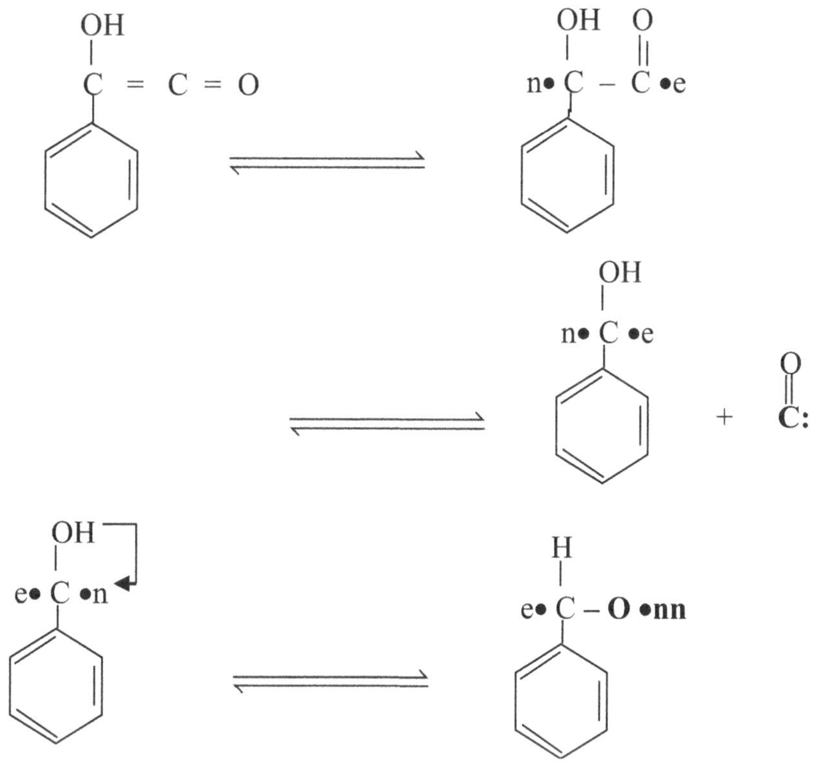

OH
C = C = O

⇌

OH O
n• C – C •e

⇌

OH
n• C •e + O=C:

OH
e• C •n

⇌

H
e• C – **O •nn**

(J) A Carbene (K)

MOLECULAR REARRANGEMENT OF THE SECOND KIND OF THE FIRST TYPE

$$\xrightarrow[\text{Release of Energy}]{\text{Deactivation}} \begin{matrix} H \\ | \\ C=O \\ | \\ C_6H_5 \end{matrix} \quad + \quad \text{Energy}$$

(L) 1.120a

$$\underline{Overall\ equation}: 2C_6H_5C_2H_5 + 4\underset{--}{O_2} \longrightarrow 2C_6H_5CHO + 2CO + 4H_2O + Energy(4) \qquad 1.120b$$

The aromatic ketene decomposes instantaneously to give a carbene and CO, both of which are not environmentally friendly. Imagine what damage this would do to the environment if there was no oxidizing oxygen molecule left in the system. However, in that stage, the aromatic carbene was made to undergo what is called ***Molecular rearrangement of the Second kind of the first type*** in the New Frontiers to form benzaldehyde as shown above. Of the two compounds formed, the more stable is the benzaldehyde. Therefore, the CO is first oxidized.

Stage 7:

$$\underset{--}{O_2} \xrightleftharpoons[\text{Oxygen molecule}]{\text{Equilibruim State Existence of Oxidising}} 2en\cdot\ddot{\underset{..}{O}}\cdot nn$$

$$2en\cdot O\cdot nn + \underset{(Stabilized)}{2C=O} \xrightleftharpoons{\text{Oxidation}} \underset{(B)}{2e\bullet\overset{O}{\overset{||}{C}}-O\bullet nn}$$

$$(B) \xrightarrow[\text{Release of Energy}]{\text{Deactivation}} 2CO_2 + Energy \qquad (1.45a)$$

$$\underline{Overall\ Equation}: \underset{--}{O_2} + 2CO \longrightarrow 2CO_2 + Energy \qquad (1.45b)$$

Stage 8:

$$\underset{--}{O_2} \rightleftharpoons 2nn\bullet O\bullet en$$

$$2nn\bullet O\bullet en + (L) \rightleftharpoons 2\ \begin{matrix} O \\ || \\ C\bullet e \\ | \\ C_6H_5 \end{matrix} + 2nn\bullet OH$$

$$\longrightarrow 2\ \begin{matrix} OH \\ | \\ C=O \\ | \\ C_6H_5 \end{matrix} \qquad 1.121$$

Via Enolization

Overall equation: $2C_6H_5C_2H_5 + 6\underline{\underline{O_2}} \longrightarrow 2C_6H_5COOH + 2CO_2 + 4H_2O + Energy(5)$ 1.122

Via No Enolization

Overall equation: $2C_6H_5C_2H_5 + 6\underline{\underline{O_2}} \longrightarrow 2C_6H_5COOH + 4H_2O + 2CO_2 + Energy(5)$ (1.101)

It is not shocking to note that via Enolization the number of stages have been increased from seven to eight stages with same energy generated. Based on what we have seen so far, it is obvious that when a benzene ring carries triple or cumulenic bonds adjacently or conjugatedly placed as shown below, the compound is very unstable.

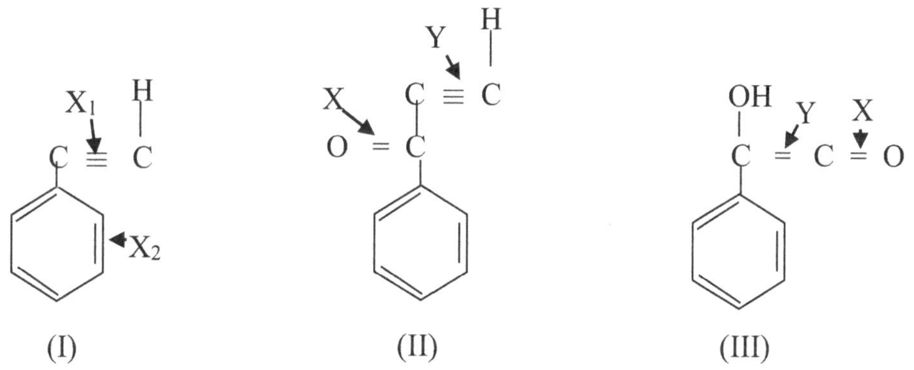

(I) (II) (III) 1.123

(I) is a Nucleophile, while (II) and (III) are Electrophiles. At mild operating conditions, they readily decompose.

For the propyl benzene, unlike the case of ethyl benzene, the formation of ketone could not be prevented as shown in Equation 1.104a. The ketone therein was immediately oxidized in the next stage, not to give an enol as shown in Equation 1.91, but to produce an equivalent of acrolein (Phenyl acrolein) which continued further oxidation. For this particular case, it seems that if the ketone was formed, a fraction was immediately oxidized, while the remaining fraction of the ketone was then allowed to immediately rearrange to enol before oxidation continued. Thus, two parallel stages are therefore formed- one for the ketone and the other for the enol. From stage 2 therefore, the followings are obtained for the fraction allowed to enolize.

Stage 3:

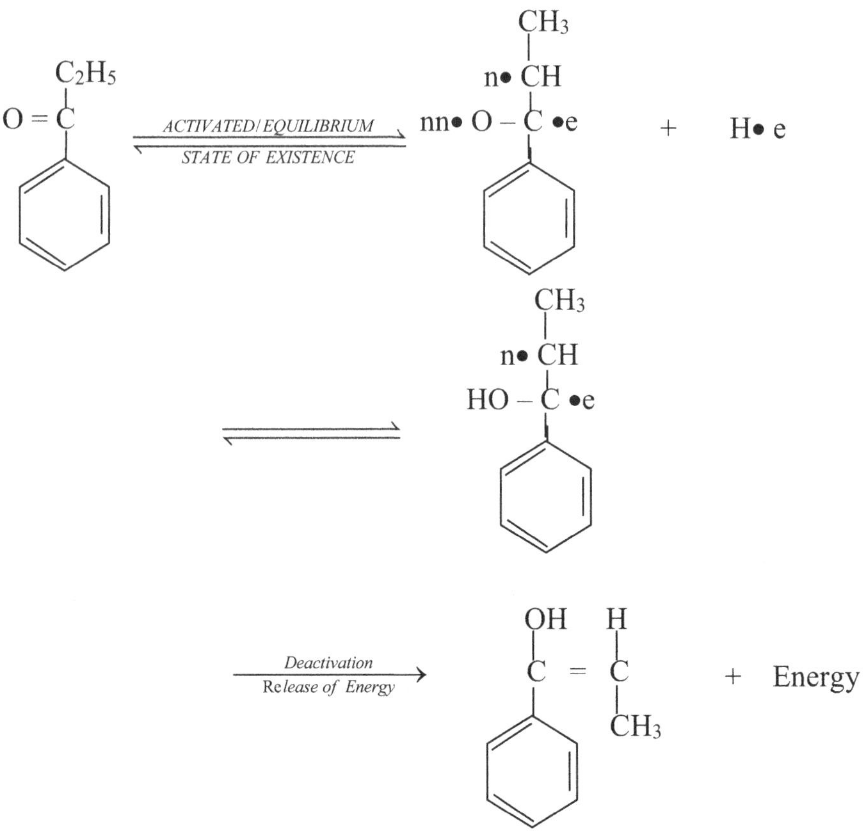

$$O=C(C_2H_5)(C_6H_5) \underset{\text{STATE OF EXISTENCE}}{\overset{\text{ACTIVATED/EQUILIBRIUM}}{\rightleftharpoons}} nn\bullet O - C(\bullet e)(C_6H_5) - CH(n\bullet) - CH_3 \quad + \quad H\bullet e$$

$$\rightleftharpoons HO - C(\bullet e)(C_6H_5) - CH(n\bullet) - CH_3$$

$$\underset{\text{Release of Energy}}{\overset{\text{Deactivation}}{\longrightarrow}} C_6H_5(OH)C = C(H)CH_3 \quad + \quad \text{Energy}$$

(A)

1.124a

$$\underline{\text{Overall equation}}: 2C_6H_5C_3H_7 + 2\underline{\underline{O_2}} \longrightarrow 2C_6H_5(OH)C = CHCH_3 + 2H_2O + Energy(2) \qquad 1.124b$$

Stage 4:

$$\underline{\underline{O_2}} \rightleftharpoons 2nn\bullet O\bullet en$$

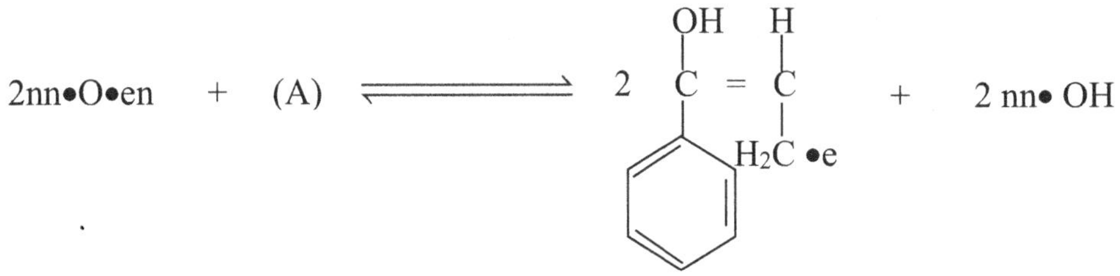

$$2nn\bullet O\bullet en \quad + \quad (A) \rightleftharpoons 2\; C_6H_5(OH)C = C(H)H_2C\bullet e \quad + \quad 2\;nn\bullet OH$$

(B)

$$\longrightarrow 2\; C_6H_5(OH)C = C(H)H_2COH$$

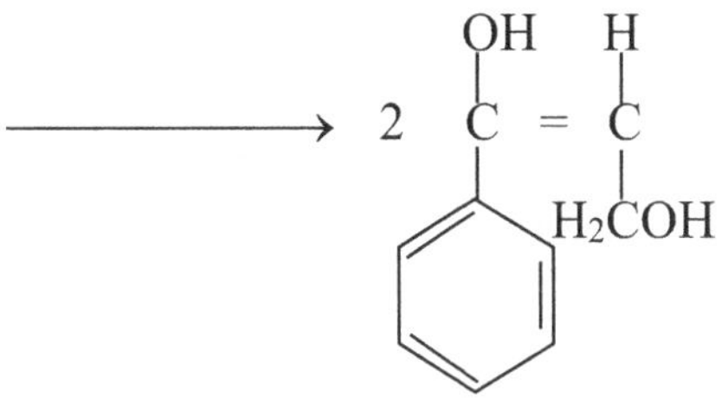

(C)

1.125a

$$\underline{Overall\ equation}: 2C_6H_5C_3H_7 + 3\underset{--}{O_2} \longrightarrow 2C_6H_5(OH)C = CH(CH_2OH) + 2H_2O + Energy(2) \quad 1.125b$$

(B) above could not reject H, because the C center carrying a $R_2C=$ group cannot carry a nucleo-free-radical, since the radical-pushing capacity of $H_2C=$ group is very large compared to all alkyl (alkylane) groups which are all of greater capacity than that of H. Hence the OH group added in the last step to complete the stage to give (C).

Stage 5: $\underset{--}{O_2} \rightleftharpoons 2nn\bullet O\bullet en$

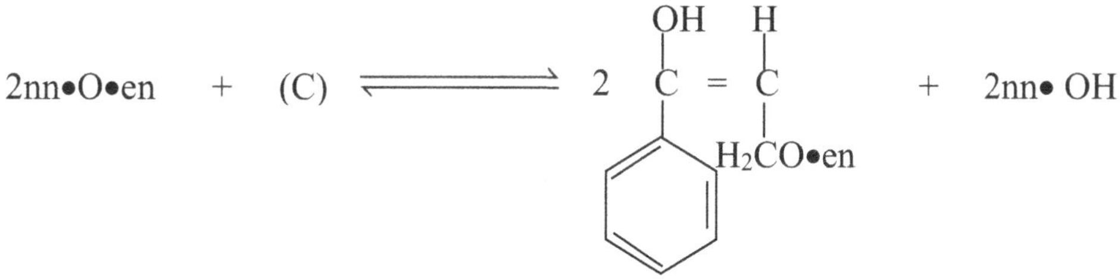

2nn•O•en + (C) ⇌ 2 [C(OH)(C₆H₅)=C(H)(H₂CO•en)] + 2nn• OH

(D)

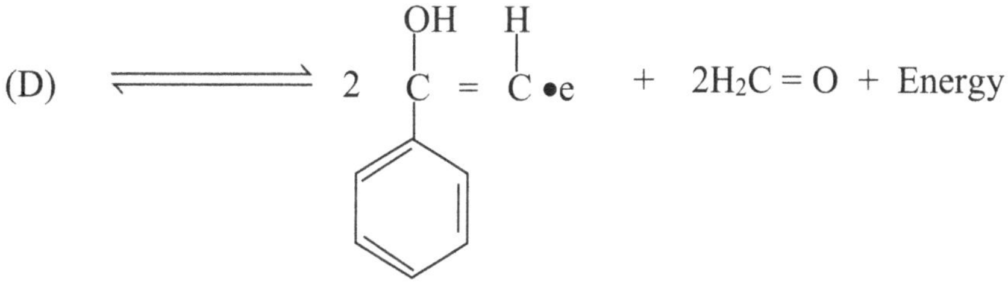

(D) ⇌ 2 [C(OH)(C₆H₅)=C(H)•e] + $2H_2C = O$ + Energy

(E)

(E) + 2nn• OH ⟶ 2 [C(OH)(C₆H₅)=C(H)(OH)]

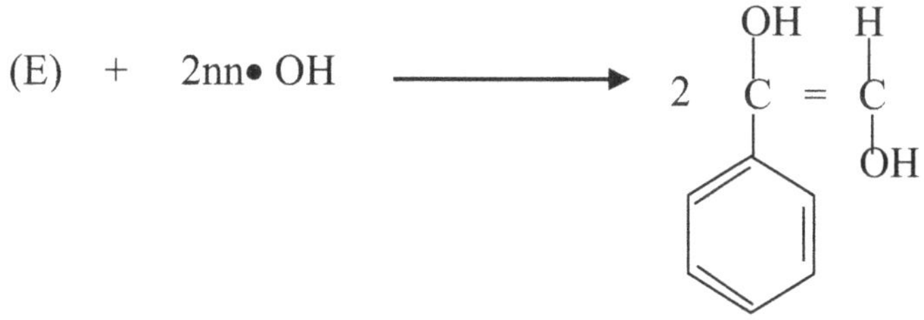

(F) 1.126a

$$\underline{Overall\ equation}: 2C_6H_5C_2H_5 + 4\underset{--}{O_2} \longrightarrow 2C_6H_5(OH)C = CH(OH) + 2H_2O + 2H_2C = O + Energy(3) \quad 1.126b$$

The (F) obtained above was also obtained during the oxidation of ethyl benzene via enolization. For the first time, formaldehyde has appeared as one of the products from the oxidation of alipharomatic hydrocarbons. This started with propyl benzene and it will be so for higher ones. Between the diol (F) and formaldehyde, the diol is less stable. Therefore, the next stage follows with oxidation of the diol as re-called below.

Stage 6: $\underline{O_2} \rightleftharpoons 2nn{\bullet}O{\bullet}en$

2nn•O•en + (F) ⇌ 2 [$OH-C(C_6H_5)=C(H)-O$ •en] + 2 nn• OH

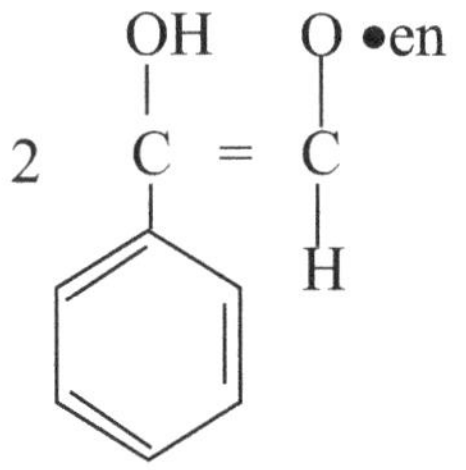

(G)

⇌ 2 [$OH-C(C_6H_5)=C=O$] + 2H• e + Energy

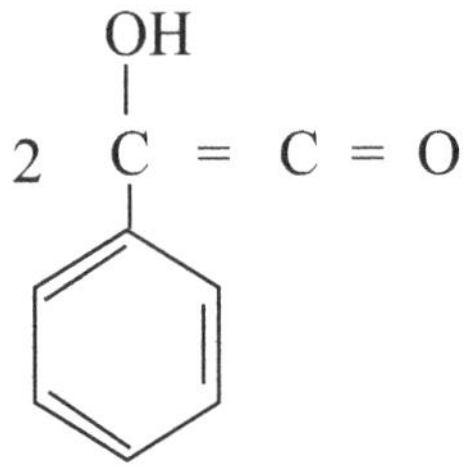

(H)

2H•e + 2HO •nn ⟶ $2H_2O$ (1.116a)

$$\underline{\text{Overall equation}}: 2C_6H_5C_3H_7 + 5\underline{O_2} \longrightarrow 2C_6H_5(OH)C=C=O + 4H_2O$$
$$+ \; 2H_2C=O \; + \; Energy(4) \qquad 1.127$$

Stage 7:

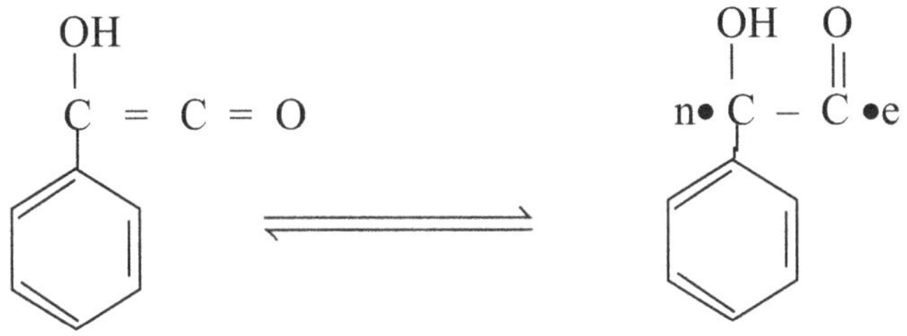

$OH-C(C_6H_5)=C=O$ ⇌ n•$C(OH)(C_6H_5)-C(=O)$ •e

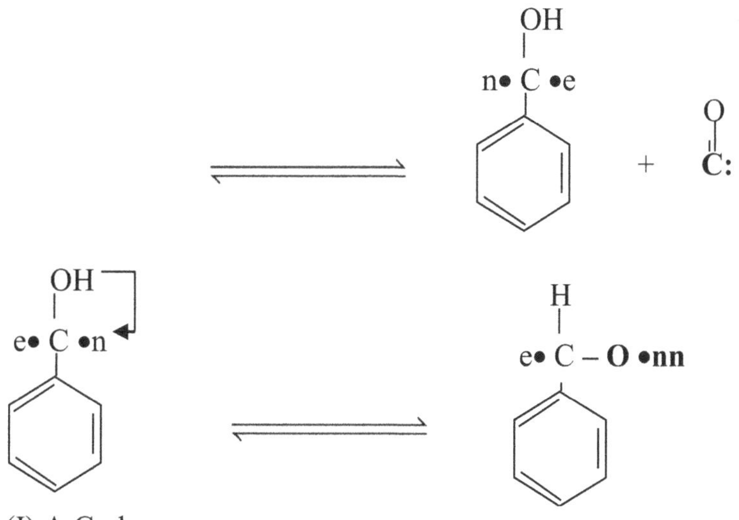

OH
|
n• C •e

$\rightleftharpoons$ [benzene ring] + O=C:

OH
|
e• C •n

H
|
e• C – O •nn

$\rightleftharpoons$

(I) A Carbene

(J)

MOLECULAR REARRANGEMENT OF THE SECOND KIND OF THE FIRST TYPE

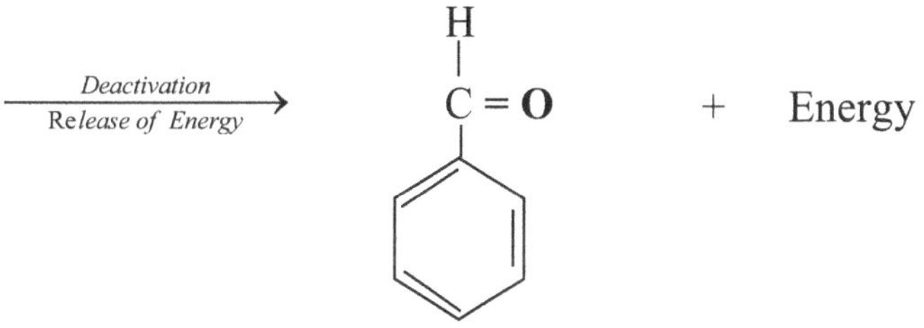

H
|
$\xrightarrow[\text{Release of Energy}]{\text{Deactivation}}$ C = O + Energy

(K (1.120a)

$$\underline{Overall\ equation}: 2C_6H_5C_3H_7 + 5\underset{--}{O_2} \longrightarrow 2C_6H_5CHO + 2CO + 4H_2O$$

$$+ 2H_2C=O + Energy(5) \quad 1.128$$

Between the benzaldehyde, formaldehyde and carbon monoxide, the most stable is benzaldehyde, followed by the formaldehyde and lastly CO. Therefore, oxidation begins with CO, followed by formaldehyde and finally the benzaldehyde.

Stage 8:

$$\underset{--}{O_2} \xrightleftharpoons[\text{Oxygen molecule}]{\text{Equilibruim State Existence of Oxidising}} 2en\cdot\ddot{\underset{\cdot\cdot}{O}}\cdot nn$$

$$2en\cdot O\cdot nn + 2C=O \xrightleftharpoons{\text{Oxidation}} 2e\bullet\overset{O}{\overset{\|}{C}}-O\bullet nn$$

(*Stabilized*) (*B*)

$$(B) \xrightarrow[\text{Release of Energy}]{\text{Deactivation}} 2CO_2 + Energy \quad (1.45a)$$

$$\underline{Overall\ Equation}: \underset{--}{O_2} + 2CO \longrightarrow 2CO_2 + Energy \quad (1.45b)$$

$$\underline{Overall\ equation}: 2C_6H_5C_3H_7 + 6\underset{--}{O_2} \longrightarrow 2C_6H_5CHO + 2CO_2 + 4H_2O + 2H_2C=O + Energy(6) \quad 1.129$$

Stage 9:

$$\underset{--}{O_2} \xrightleftharpoons[\text{Oxygen molecule}]{\text{Equilibruim State Existence of Oxidising}} 2en\cdot\ddot{\underset{..}{O}}\cdot nn$$

$$\underset{(Stabilized)}{2en\cdot O\cdot nn + 2O=CH_2} \xrightleftharpoons{\text{Oxidation}} 2H-O\cdot nn + \underset{(B)}{2e\cdot\overset{O}{\overset{||}{C}}-H}$$

$$(B) \rightleftharpoons 2H{\bullet}e + \underset{|(A)}{2n{\bullet}\overset{O}{\overset{||}{C}}{\bullet}e}$$

$$2H{\bullet}e + 2nn{\bullet}OH \rightleftharpoons 2H_2O$$

$$(A) \xrightarrow[\text{release of Energy}]{\text{Deactivation}} 2:C=O + Energy \quad 1.51a$$

$$\underline{Overall\ Equation}: \underset{--}{O_2} + 2O=CH_2 \longrightarrow 2H_2O + 2CO + Energy \quad 1.51b$$

Stage 10:

$$\underset{--}{O_2} \xrightleftharpoons[\text{Oxygen molecule}]{\text{Equilibruim State Existence of Oxidising}} 2en\cdot\ddot{\underset{..}{O}}\cdot nn$$

$$\underset{(Stabilized)}{2en\cdot O\cdot nn + 2C=O} \xrightleftharpoons{\text{Oxidation}} \underset{(B)}{2e{\bullet}\overset{O}{\overset{||}{C}}-O{\bullet}nn}$$

$$(B) \xrightarrow[\text{Release of Energy}]{\text{Deactivation}} 2CO_2 + Energy \quad (1.45a)$$

$$\underline{Overall\ Equation}: \underset{--}{O_2} + 2CO \longrightarrow 2CO_2 + Energy \quad (1.45b)$$

$$Overall\ equation: 2C_6H_5C_3H_7 + 8\underset{--}{O_2} \longrightarrow 2C_6H_5CHO + 4CO_2 + 6H_2O + Energy(8) \quad 1.130$$

Stage 11:

$$\underset{--}{O_2} \rightleftharpoons 2nn{\bullet}O{\bullet}en$$

$$2nn{\bullet}O{\bullet}en + (K) \rightleftharpoons 2\ C_6H_5\overset{O}{\overset{||}{C}}{\bullet}e + 2nn{\bullet}OH$$

Via Enolization $\longrightarrow$ $2\ C_6H_5-\overset{OH}{\overset{|}{C}}=O$ 1.121

$$\underline{Overall\ equation:}\ 2C_6H_5C_3H_7 + 9\underset{--}{O_2} \longrightarrow 2C_6H_5COOH + 4CO_2 + 6H_2O + Energy(8)$$

1.131

Via Non-Enolization

$$Overall\ equation: 2C_6H_5C_3H_7 + 9\underset{--}{O_2} \longrightarrow 2C_6H_5COOH + 6H_2O + 4CO_2 + Energy(8) \quad (1.110)$$

From Stage 2, the two systems in parallel commences in series to the end to give the final products- one for the fraction that did not enolize and the other for the fraction that enolized.

Via Enolization

$$\underline{Overall\ equation:}\ 2C_6H_5C_2H_5 + 6\underset{--}{O_2} \longrightarrow 2C_6H_5COOH + 2CO_2 + 4H_2O + Energy(5)$$

(1.122)

Via No Enolization

$$\underline{Overall\ equation}: 2C_6H_5C_2H_5 + 6\underset{--}{O_2} \longrightarrow 2C_6H_5COOH + 4H_2O + 2CO_2 + Energy(5) \quad (1.101)$$

Like the case above, two fractions emerged right from Stage 1- one that did not reject H and the other that rejected H. This is almost a complete parallel system from the beginning to the end. ***In this way, all the observations in Equation 1.91 can now be readily explained very differently <u>in the sense that you can never find a case where two members of a family are the same.</u>***

Just like the case of ethyl benzene, when enolization takes place, the number of stages is increased by one and that is the time it took it to re-arrange. Oxidation of ketone to enol does not take place according to Equation 1.91. ***Yet, the equation has been very useful in just providing a picture of what was observed in the laboratories that which has generally been the case since antiquity. In our world, they say SEEING IS BELIEVING; they say THE ONLY THING PERMANENT IS CHANGE; they also say WITH GOD ALL THINGS ARE POSSIBLE; they say ....too countless to list. They are all "absolutely meaningless". That is the world in which we live in. As we move deeper and deeper into CHEMISTRY, one will begin to know that some of these are true while some are not, because BELIEVING IS SEEING, and THE ONLY THING PERMANENT IS***

TIME BECAUSE IT IS ALWAYS THERE and WITH THE ALMIGHTY INFINITE GOD, ALL THINGS THAT WORK ACCORDING TO THE LAWS OF NATURE ARE POSSIBLE; IT IS WE HUMANS THAT MAKE THINGS THAT WORK AGAINST THE LAWS OF NATURE POSSIBLE…. too countless to list. Indeed, WITH THE ALMIGHTY INFINITE GOD, NOTHING IS IMPOSSIBLE.

We have seen the oxidation of alipharomatic ketones and its unstable enols. Let us complete the section with the oxidation of enols from aliphatic enols such as vinyl alcohol.

Oxidation of vinyl alcohol

Stage 1: $\underline{\underline{O_2}} \rightleftharpoons 2nn\bullet O\bullet en$

$$2nn\bullet O\bullet en + \begin{matrix} H & & H \\ | & & | \\ C & = & C \\ | & & | \\ H & & OH \end{matrix} \rightleftharpoons 2\ \begin{matrix} H & & H \\ | & & | \\ C & = & C \\ | & & | \\ H & & O\bullet en \end{matrix} + 2nn\bullet OH$$

(A)

$$(A) \rightleftharpoons 2\ \begin{matrix} H \\ | \\ C \\ | \\ H \end{matrix} = C = O + 2H\bullet e + Energy$$

(B)

$$2H\bullet e + 2HO\bullet nn \longrightarrow 2H_2O \qquad 1.132a$$

Overall equation: $2H_2C = CH(OH) + \underline{\underline{O_2}} \longrightarrow 2H_2C = C = O + 2H_2O + Energy$

1.132b

Stage 2: $\underline{\underline{O_2}} \rightleftharpoons 2nn\bullet O\bullet en$

$$2nn\bullet O\bullet en + (B) \rightleftharpoons 2\ \ e\bullet \begin{matrix} H \\ | \\ C \end{matrix} = C = O + 2nn\bullet OH$$

(C)

$$\longrightarrow 2\ \begin{matrix} H \\ | \\ C \\ | \\ OH \end{matrix} = C = O$$

(D) 1.133a

$$\underline{Overall\ equation}: 2H_2C = CH(OH) + 2\underset{--}{O_2} \longrightarrow H(OH)C = C = O + 2H_2O + Energy \quad 1.133b$$

Stage 3:

$$\underset{--}{O_2} \rightleftharpoons 2nn{\bullet}O{\bullet}en$$

$$2nn{\bullet}O{\bullet}en + (D) \rightleftharpoons 2\ \begin{matrix} H \\ | \\ C = C = O \\ | \\ O \bullet en \end{matrix} + 2nn{\bullet}\ OH$$

(E)

$$(E) \rightleftharpoons 2\ \ O = C = C = O + 2H{\bullet}\ e + Energy$$

(F)

$$2H{\bullet}\ e + 2nn{\bullet}\ OH \longrightarrow 2H_2O \quad 1.134a$$

$$\underline{Overall\ equation}: 2H_2C = CH(OH) + 3\underset{--}{O_2} \longrightarrow 2O = C = C = O + 4H_2O + Energy \quad 1.134b$$

Stage 4:

$$O = C = C = O \rightleftharpoons n{\bullet}\ \overset{\overset{O}{||}}{C} - \underset{\underset{O}{||}}{C}\ {\bullet}e$$

(G)

$$\rightleftharpoons e{\bullet}\ \overset{\overset{O}{||}}{C}\ {\bullet}n + \overset{\overset{O}{||}}{C}:$$

(H)

$$(H) \xrightarrow[Release\ of\ Energy]{Deactivation} :\overset{\overset{O}{||}}{C} + Energy \quad 1.135a$$

$$\underline{Overall\ equation}: 2H_2C = CH(OH) + 3\underset{--}{O_2} \longrightarrow 4CO + 4H_2O + Energy \quad 1.135b$$

Stage 5:

$$\underset{--}{O_2} \xrightleftharpoons[\text{Oxygen molecule}]{\text{Equilibruim State Existence of Oxidising}} 2en\cdot\ddot{\underset{..}{O}}\cdot nn$$

$$\underset{(Stabilized)}{2en\cdot O\cdot nn + 2C=O} \xrightleftharpoons{\text{Oxidation}} \underset{(B)}{2e\bullet \overset{O}{\overset{\|}{C}} - O\bullet nn}$$

$$(B) \xrightarrow[\text{Release of Energy}]{\text{Deactivation}} 2CO_2 + Energy \quad (1.45a)$$

$$\underline{Overall\ Equation}: \underset{--}{O_2} + 2CO \longrightarrow 2CO_2 + Energy \quad (1.45b)$$

Stage 6: Same as Stage 5 above for the second two moles of CO.

$$\underline{Overall\ equation}: 2H_2C=CH(OH) + 5\underset{--}{O_2} \longrightarrow 4CO_2 + 4H_2O + Energy(5) \quad 1.136$$

The (B) formed in Stage 1, a Ketene is an ELECTROPHILE. So also, is (D). The (F) in Stage 3, a diketone is also another type of ELECTROPHILE, which is very unstable and cannot be oxidized. Hence, based on the operating conditions, it decomposes instantaneously in Stage 4 to give two moles of carbon monoxide with release of energy.

What has been encountered thus far, look as an ART. It is not, but THE NEW REAL AND IMAGINARY SCIENCE. Our world is a COMPLEX world, where there exists the real and imaginary, both co-existing with each other for any to exist (The Law of Duality in Figure 1.3). The last two stages above are in parallel. On the whole, as simple as the vinyl alcohol may seem, that which is very unstable, six stages were involved, with energy generated from all the stages. For this reason, it was made unstable to rearrange to acetaldehyde at STP.

Oxidation of Acetaldehyde

Stage 1: $\underset{--}{O_2} \rightleftharpoons 2nn\bullet O\bullet en$

$$2nn\bullet O\bullet en + 2\ \underset{\text{Acetaldehyde}}{\overset{H}{\overset{|}{\underset{CH_3}{\underset{|}{C}}}}=O} \rightleftharpoons 2\ \underset{(A)}{\overset{H}{\overset{|}{\underset{e\bullet CH_2}{\underset{|}{C}}}}=O} + 2nn\bullet OH$$

$$\longrightarrow 2\ \underset{(B)}{\overset{H}{\overset{|}{\underset{HOCH_2}{\underset{|}{C}}}}=O} \quad 1.137a$$

$$\underline{Overall\ equation}: 2H_3C(CO)H + \underline{O_2} \longrightarrow 2HOCH_2(CO)H \qquad 1.137b$$

Stage 2:

$$\underline{O_2} \rightleftharpoons 2nn{\bullet}O{\bullet}en$$

$$2nn{\bullet}\,O\,{\bullet}en + 2\ \begin{matrix} H \\ | \\ C=O \\ | \\ HOCH_2 \end{matrix} \rightleftharpoons 2\ \begin{matrix} H \\ | \\ C=O \\ | \\ en{\bullet}\,OCH_2 \end{matrix} + 2nn{\bullet}\,OH$$

(B) (C)

$$(C) \rightleftharpoons 2\ O=\begin{matrix} H \\ | \\ C \end{matrix}-\begin{matrix} H \\ | \\ C \end{matrix}=O + 2H\,{\bullet}e + Energy$$

(D)

$$2H{\bullet}e + 2nn{\bullet}OH \longrightarrow 2H_2O \qquad 1.138a$$

$$\underline{Overall\ equation}: 2H_3C(CO)H + 2\underline{O_2} \longrightarrow 2HCO(HCO) + 2H_2O + Energy$$

(1.138*b*)

Stage 3:

$$\underline{O_2} \rightleftharpoons 2nn{\bullet}O{\bullet}en$$

$$2nn{\bullet}\,O\,{\bullet}en + (D) \rightleftharpoons 2\ O=\begin{matrix} {\bullet}e \\ C \\ | \\ H \end{matrix}-C=O + 2nn{\bullet}\,OH$$

(E)

$$(E) \rightleftharpoons 2\ O=\overset{{\bullet}e}{C}-\underset{{\bullet}n}{C}=O + 2H\,{\bullet}e$$

(F')

$$2H{\bullet}\,e + 2nn{\bullet}\,OH \rightleftharpoons 2H_2O$$

$$(F) \xrightarrow[Release\ of\ Energy]{Deactivation} 2\ O=C=C=O + Energy$$

(G)

1,139a

$$\underline{Overall\ equation}: 2H_3C(CO)H + 3\underline{O_2} \longrightarrow 2O=C=C=O + 2H_2O + Energy$$

1.139*b*

If in Stage 2, O=(H)C •e had been released in place of H•e, then the situation would have been different, and the overall equation after stage 3 would have been as shown below. Instead of O=C=C=O as the real product, CO_2 and H_2C=O would have been obtained.

<u>*Overall equation*</u>: $2H_3C(CO)H \ + \ 3\underline{O_2} \longrightarrow 2H_2C = O \ + \ 2H_2O \ + \ 2CO_2 \ + \ Energy$ 1.139*c*

One of the products in that Stage 2 would have been formic acid (HCOOH) in which the H removed like the cases seen so far, would have been the one from the OH more radical-pushing group, and not H behind the C center. *On the other hand, the H on the OH group is very unstable. For if held in equilibrium state of existence, it is will be more neither here nor there as shown below, unlike the second H atom.*

$$H - \overset{\overset{\displaystyle O}{||}}{C} - O \bullet nn \ + \ H\bullet e \longleftrightarrow H - \overset{\overset{\displaystyle O\bullet nn}{|}}{C} = O \ + \ H\bullet e$$

EQUILIBRATED STATE OF REARRANGEMENT 1.140

Hence, unlike with other members of carboxylic acids, the Equilibrium state of existence of formic acid is different from the Equilibrium state of existence of the other members of the family as shown below for acetic acid and formic acid.

$$H\overset{\overset{\displaystyle O}{||}}{C} - OH \rightleftharpoons n\bullet \overset{\overset{\displaystyle O}{||}}{C} - OH \ + H\bullet e \ ; \quad H_3C\overset{\overset{\displaystyle O}{||}}{C} - OH \rightleftharpoons H_3C\overset{\overset{\displaystyle O}{||}}{C} - O\bullet nn \ + \ H\bullet e$$

<u>Equilibrium States of Existence</u> 1.141

Though which ever H is removed by the oxidizing agent, the same products are obtained, it important to note the new concept shown in Equation 1.140, since we must understand how NATURE operates, as they will all be applied downstream for other things, without the need to ask questions.

It has been important to recall a situation where how Combustion and Oxidation, two phenomena which we use every second in the world, take place could ever be explained. It is *very funny* that we eat every day, yet we don't know how Combustion takes place!!! It may look *very funny*, but it is not. Now continuing from Stage 3 above, the followings are obtained.

<u>Stage 4:</u>

$$O = C = C = O \rightleftharpoons n\bullet \overset{\overset{\displaystyle O}{||}}{C} - \underset{\underset{\displaystyle O}{||}}{C}\bullet e$$

(G)

$$\rightleftharpoons \quad e\bullet\overset{\overset{\large O}{||}}{C}\bullet n \;(H) \quad + \quad \overset{\overset{\large O}{||}}{C}:$$

$$(H) \xrightarrow[\text{Release of Energy}]{\text{Deactivation}} :\overset{\overset{\large O}{||}}{C} \quad + \quad \text{Energy} \qquad (1.135a)$$

$$\underline{\text{Overall equation}}: 2H_3C(CHO) \; + \; 3\underset{--}{O_2} \longrightarrow 4CO \; + \; 4H_2O + Energy \qquad (1.135b)$$

Stage 5:

$$\underset{--}{O_2} \underset{\text{Oxygen molecule}}{\overset{\text{Equilibruim State Existence of Oxidisin g}}{\rightleftharpoons}} 2en\cdot\ddot{\underset{\cdot\cdot}{O}}\cdot nn$$

$$2en\cdot O\cdot nn + \underset{(Stabilized)}{2C=O} \overset{\text{Oxidation}}{\rightleftharpoons} \underset{(B)}{2\,e\bullet\overset{\overset{O}{||}}{C}-O\bullet nn}$$

$$(B) \xrightarrow[\text{Release of Energy}]{\text{Deactivation}} 2CO_2 \; + \; Energy \qquad (1.45a)$$

$$\underline{\text{Overall Equation}} : \underset{--}{O_2} + 2CO \longrightarrow 2CO_2 \; + \; Energy \qquad (1.45b)$$

$$\underline{\text{Overall equation}}: 2H_3C(CO)H \; + \; 5\underset{--}{O_2} \longrightarrow 4CO_2 \; + \; 4H_2O \; + \; Energy(5) \qquad 1.142$$

$$\underline{\text{Overall equation}}: 2H_2C=CH(OH) \; + \; 5\underset{--}{O_2} \longrightarrow 4CO_2 \; + \; 4H_2O \; + \; Energy(5) \; (1.136)$$

Almost unlike what was done with aromatic cases, one can clearly see the distinction between their oxidations, since no rearrangement took place here. Even then, they are essentially the same. It is easier to oxidize the ketones than their enols. They are all sources of energy.

From all the considerations, so far, one can see the world of differences between Combustion and Oxidation, wherein different kinds of flammable products can be obtained. One can imagine the numbers of new concepts forced to be introduced in order to provide the mechanisms for Combustion and Oxidation. Worthy of note, is that none of all the chemical reactions so far considered can take place CHARGEDLY, i.e., Ionically, Covalently, Electrostatically and Polarly. They can only take place radically, something which has been unknown since antiquity. Though few knew it, they just did not know what RADICALS are, because they have been so CONDITIONED to the world of ELECTRONS. International World Standard Organizational bodies will tell the world that the real name for Ethylene is ETHENE, but the world will not listen because of CONDITIONING. In the next chapter, we are going to be seeing for the first time what CHARGES are, not as it is known in present-day Science. Without Radicals, Charges cannot be formed. ***Present-day Science has paid little or no attention to presence of VACANT ORBITALS and PAIRED UNBONDED RADICALS carried by some Atoms, when we know that "NOTHING IS THERE FOR NOTHING".*** What each ATOM carries along THEIR BOUNDARIES, are where ALL ACTIONS AND REACTIONS in our world take place. It is only along the BOUNDARIES that all CHEMICAL AND POLYMERIC reactions take place using RADICALS. Nuclear reactions take place in the NUCLEUS of an atom, using NEGATRONS (β-RAYS). ANTI-NEGATRONS (Positrons), NEUTRONS, ANTI-NEUTRONS, PROTONS, ANTI-PROTONS, α-RAYS AND ANTI-α-RAYS, as partly shown in Appendix (I) and fully shown downstream in the New Frontiers with respect to so many things including Thunder and Lightning, Light, Mass and so on.

With the foundations laid so far, one can now handle the combustion and oxidation of higher hydrocarbons and other families. Though CO does not carry H, it can be oxidized, but not combusted, i.e. not all compounds that undergo combustion can be oxidized and verse visa. The physical state of a compound determines the level of combustion for hydrocarbons- the gaseous state being the strongest, a state where it is easier to exist in Equilibrium state of existence, where it exists. It is also important to note that some compounds such as sulfur exist which do not have to exist in Equilibrium state of existence for it to combust. For the first time, one of the causes of a particular type of cancer- Lung cancer for example has been shown. With this knowledge, one can now provide simple mechanism for breaking the so-called cell and cure the disease. With full understandings of mechanisms of systems- the essence of the new foundations, many new doors will be opened. Indeed, based on all the above and the ready availability of technology, all diseases have cures, since the solutions to all problems are here.

1.3.17 Ozone and its in-situ environmental synthesis

Since oxidizing oxygen molecule has been shown to be one of the products of combustion of some compounds depending on the **operating conditions**, there is need to know how ozone can be obtained from them. On the other hand, though the Atmospheric Structure of the gaseous area surrounding the planet is well known (Troposphere, Stratosphere [Main Ozone layer], Mesosphere, Thermosphere, and Exosphere), little or nothing is known about the molecular structures of ozone and the ozone layer.[47-49] Herein, only those with respect to ozone will first be provided. These include the ***Radical Structure*** which is the activated state and the ***Polar***

Structure which is the stable and natural State. The polar state is resonance stabilized and the radical state is obtained from the polar state via activation. Shown below is its synthesis from oxidizing oxygen molecule and molecular oxygen in attempt to show that ozone is a mixture of one mole of an element of oxidizing oxygen and one mole of molecular oxygen and not three moles of the element of oxidizing oxygen which carries one nucleo-non-free-radical and one electro-non-free-radical or three moles of the oxygen atom, which carries only two nucleo-non-free-radicals or one mole of oxidizing oxygen molecule and two moles of oxygen atom, etc.

Stage1:

$$\underline{O_2} \xrightleftharpoons[\text{Equilibrium State of existence}]{\text{Oxidising Oxygen in its}} 2nn\cdot O\cdot en$$
$$(\text{Oxidising Oxygen molecule})$$

$$2nn\cdot O\cdot en + \underset{(Air)}{2O=O} \xrightleftharpoons[\text{Molecule in Air}]{\text{Activation of Oxygen}} \underset{(A)}{2(nn\cdot \ddot{\underset{..}{O}}-\ddot{\underset{..}{O}}-\ddot{\underset{..}{O}}\cdot en)}$$

$$(A) \xrightarrow[\text{Release of Energy}]{\text{De-Activation}} 2(\ddot{\underset{..}{O}}=\underset{..}{O}^{+}-\ddot{\underset{..}{O}}:^{-}) + Energy$$
$$(2 \text{ molecules of Ozone}) \qquad 1.143a$$

$$\underline{\text{Overall Equation}}: \underline{O_2} + \underset{(Air)}{2(O=O)} \xrightarrow[\text{Stage}]{\text{Productive}} 2(Ozone) + Energy \qquad 1.143b$$

$$\ddot{\underset{..}{O}}=\underset{..}{O}^{+}-\ddot{\underset{..}{O}}:^{-} \xrightleftharpoons[\text{Heat}]{\text{Activation by}} nn\cdot \ddot{\underset{..}{O}}-\ddot{\underset{..}{O}}-\ddot{\underset{..}{O}}\cdot en$$
$$(\text{Real Structure of Ozone}) \qquad 1.143c$$

Worthy of note are the followings-

(i) The source of oxidizing oxygen molecule is very different from that of molecular oxygen. The oxidizing oxygen molecule is that obtained from for example two moles of HOOH and not one mole based on what a Stage is and how chemical reactions take place. It can be obtained from many other sources, such as fuming nitric acid, potassium permanganate, and much more.

(ii) For ozone to be obtained, two moles of molecular oxygen are required for one mole of oxidizing oxygen molecule, otherwise under the operating conditions, it cannot be obtained.

(iii) Note the real structure of ozone. It carries a ***polar bond*** in its stable state of existence. The charges shown are polar in character and not ionic. When activated by heat, they carry radicals without breaking any of the rules of nature.

(iv) Heat may be required for the activation of molecular oxygen in the second step above or this can be done by the oxidizing oxygen which carries some amounts of energy in their rings depending on the size of the two-membered imaginary ring.

(v) Ozone unlike oxidizing oxygen molecule does not have Equilibrium state of existence, but Activated state of existence. Because of the distance between the active centers when both are activated and its stable state of existence, it is therefore less harmless than oxidizing oxygen. The ozone layer is a polymeric giant molecule and how obtained can be envisaged based on the foundations laid so far.

$$:\ddot{O}=\underset{..}{O}^{\oplus} - \underset{..}{\ddot{O}} - (\underset{..}{\ddot{O}} - \underset{..}{\ddot{O}} - \underset{..}{\ddot{O}})_n - \underset{..}{\ddot{O}} - \underset{..}{\ddot{O}} - \underset{..}{\ddot{O}}{:}^{\odot}$$

<u>THE OZONE LAYER</u> 1.144

Unlike most polymers, there is no initiator carrying the chain and there is no terminating agent. It is a highly polar system. It is Nucleophilic in character. Through the paired unbonded radicals, it repels all the negatively nuclearly charged rays from the SUN, but attracts the positively nuclearly charged rays from the SUN and stable to the neutral ones.

Worthy of note as has already been said, is that in all stages of Equilibrium mechanisms just as in all the other two mechanisms, chemically it is the electro-radical free or non-free, that is, the male, that commences all the initial attacks on a stable compound in a stage and never the female radical. What this implies is very deep to comprehend.

1.4 **Conclusions**

This marks the beginning of something new, wherein the Natural Sciences (The trios) have been redefined. In view of these new definitions, very new concepts, which have general applications, have begun to be introduced herein. These new concepts include States of Existences, Stage-wise Operations, Mechanisms of Systems, Operating Conditions and more. All these resulted in establishing new classifications for Radicals.

Every system in our world has a domain and boundary, just like any atom in the Periodic Table. Nature as embraced by the three disciplines of Mathematics, Physics and Chemistry is nothing else other than first and foremost ***Orderliness.*** For the first time, combustion has been clearly distinguished from oxidation, by application of the new concepts to provide the mechanisms for combustion and oxidation of hydrogen, methane and ethane and for combustion of benzene and toluene and more. For the first time, the structures of oxidizing oxygen molecule have been provided and distinguished from that of molecular oxygen.

Combustion and oxidation can be used either alone or together depending on the operating conditions and the types of compounds involved. For the case of combustion of hydrogen and higher hydrocarbons, both are uniquely involved. Why the need for use of batteries does not arise when air, water, hydrogen and its catalyst or other chemical catalysts such as hydrides are used in place of other presently known sources of energy such as fuel, solar, wind, waterfalls and more, has been explained. It is a semi-perpetual process. For the first time, one of the real causes of lung cancer has begun to be shown. Also shown is the real structure of ozone and one of its methods of synthesis. So also, was the structure of the ozone layer provided. In order to provide support for the new concepts, the subjects contained herein are indeed numerous to list. In general, notice that new forms of chemical equations, which embrace Mathematics (Stage-wise operations as we shall see downstream) and Physics (Ring theory with respect to strain energy in springs, gravity, heat and more as we shall see downstream using most recent data), have begun to be introduced. It is too early now to use very current data in advanced areas of our development, because of the first need to first lay new unquestionable foundations. Obviously, if no questions can be raised, then the need to turn the wheels backwards in order to move forward, no longer arises. How can our world move forward if the mechanisms of the experimental data or observations obtained since antiquity cannot be explained without questions and exceptions? Down the road, these tools will find useful applications not only to Chemical engineering processes and systems in particular, such as in polymer, petroleum, petrochemical, environmental, biomedical, enzymatic engineering, but indeed to all other disciplines.

1.5 ***Foods for Thought***

Most of us, including the author and present-day professors in chemistry and related disciplines cannot answer the question below without reading the concepts of the new science begun to be shown above. It is very important to go through the questions, if knowledge is important to you, ***noting that knowledge is not exclusive to any human being, no matter who you are, since no human is an island to itself.***

PB1.1. What are the new classifications for the followings, comparing it with present-day knowledge?

a) Operating conditions.
b) Radicals.
c) States of Existence.
d) Mechanisms of systems
e) Stage-wise operations.

PB1.2. Provide the Equilibrium states of existence for the following compounds radically only:

a) Hydrogen molecule, (H_2).
b) Sodium chloride, (NaCl).
c) Bromine molecule, (Br_2).
d) Water, (H_2O).
e) Methane, (CH_4).
f) Ferric chloride, ($FeCl_3$).
g) Ethyl alcohol, (C_2H_5OH).
h) Ethene glycol (i.e. Ethylene glycol), ($HO(CH_2)_2OH$).
i) Glycerol, [$HOCH_2(CHOH)CH_2OH$].
j) Ammonia, (NH_3).

PB1.3. Provide also the Decomposition and Combination states of existence for the compounds in **PB1.2**

PB1.4. Of the following compounds, which of them can exist in Equilibrium state of existence? If some can, under what conditions?

a) Molecular Oxygen.
a) Hydrogen
b) Nitrogen.
c) Carbon dioxide.
d) Silicane.
f) Oxidizing oxygen molecule.
g) Hydrogen peroxide.
h) Nitric oxide.
i) Carbon monoxide.
j) Silicon dioxide.

PB1.5. Distinguish between the following signs using examples:

i) $A + B \rightarrow C$, where A and B are radical components.
ii) $A + B \rightleftarrows C$, where A, B, and C are compounds.
iii) $C \rightarrow A + B$, where A and B are radical components.
iv) $A + B \rightleftharpoons C + D$, where A, B, C, and D are compounds.
v) $A \longleftrightarrow B$, where A and B are compounds.
vi) $A \longleftrightarrow B + C$, where B and C are radical compounds.
vii) $A + B \rightarrow C + D$, where A, B, C and D are compounds.
viii) $A \rightarrow B + C$, where A and B are compounds and C a radical component.
ix) $A + B \rightarrow C$, where A and C are radical components and B is a Monomer.

PB1.6. Distinguish between the followings-

i) Molecular air oxygen and Molecular oxidizing oxygen.
ii) Active and passive catalysts.
iii) Combustion and Oxidation.
iv) Hydrogen molecule and other metallic hydrides.
v) Aliphatic carbon dioxide and Aromatic carbon dioxide.

PB1.7. Describe the mechanisms of combustion of the first four aliphatic hydrocarbons, showing the general equation for their combustion.

PB1.8. Explain the mechanisms for the combustion of the first three members of alipharomatic hydrocarbons and show one of the origins of cancer in their use as fuel.

PB1.9. Based on the New Science, provide very concisely the new classifications for metals in the Periodic Table. {Complete solution is in Chapter 2}

PB1.10. a) How is hydrogen oxidized and combusted?
b) How can methane be oxidized to produce methanol?
c) Can water be oxidized? If it cannot, under what conditions can it be oxidized to produce energy for many applications including driving a car?

PB1.11. a) Provide the mechanisms for the forced combustion of hydrogen, that which leads to the oxidation of water semi-perpetually.
b) Provide the mechanism for the non-forced combustion of hydrogen.
c) Of the two methods above, that is a) and b), which is the more environmentally friendly?

PB1.12. Imagine if our desire is to produce another source of fuel to produce energy for many applications, such as bio-diesel fuel for diesel built engines. Being bio-energy source in character, one can use one of the bio-sources, such as fats and oils from plants and animals. Therefore, answer the following questions:

a) Distinguish between fats and oils as an expert.
b) Explain how the bio-fuel can be produced from fat or oil.
c) Explain how the bio-fuel is combusted to produce energy.
d) Indeed, is the bio-fuel environmentally friendly when used in opened systems such as driving a car? Explain.
e) Then, how do we make driving a car a closed system, since it has been said that all systems must be closed if we have to sustain our environment.

PB1.13. a) What is ozone? How is it produced in situ in our environment?
b) What is the ozone layer? How is it produced in the upper layer of the ionosphere? Can you provide how it is destroyed?

PB1.14. The new concepts introduced herein, have been found to apply so far from the grain of sand of considerations presented herein to both the Real Sciences and Metaphysical Sciences. Explain what this means using States of Existence, Stage-wise operations of all things in humanity, Mechanisms of all Systems and Operating conditions.

PB1.15. Show the mechanisms for the full oxidation of Butyl benzene, identifying the types of products obtained in all the stages.

References

1. J. J. Thomson, *"On the structure of the Atom? An Investigation of the Stability and Periods of Oscillation of a number of Corpuscles arranged at equal intervals around the Circumference of a Circle; with Application of the Results to the Theory of Atomic Structure "*, Philosophical Magazine, Series 6, Volume 7, Number 39, March 1904, pp. 237-265.

2. J. J. Thomson, *"Cathode Rays",* Philosophical Magazine, 44, 293 (1897) [facsimile from Stephen Wright, Classical Scientific Papers, Physics (Mills and Boon, 1964).]

3. J. J. Thomson, *"On the Numbers of Corpuscles in an Atom",* Philosophical Magazine, Vol 11, June 1906, pp. 769-781.

4. J. J. Thomson, *"Rays of positive electricity"*. Proceedings of the Royal Society A 89, 1913, 1-20 [As excerpted in Henry A. Boorse & Lloyd Motz, Eds. The World of the Atom, Vol 1. (New York; Basic Books, 1966)].

5.Amedeo Avogadro, *"Essay in a Manner of Determining the Relative Masses of the Elementary Molecules of Bodies, and the Proportions in which They Enter into These Compounds",* Journal de Physique 73, 58-76 (1811) [Alembic Club Reprint No. 3].

6. Joseph Louis Gay-Lussac, *"Memoir on the Combination of Gaseous substances with Each Other"*, Memories de la Societe d'Arcueil 2, 1809, 207 [from Henry A. Boorse and Lloyd Motz, eds., The World of the Atoms, Vol.1 (New-York: Basic Books 1966) (translation: Alembic Club Reprint Nov.4)].

7. John Dalton, *"a New System of Chemical Philosophy"* [excerpts], Manchester, 1808.

8. Neils Bohr, *"On the Constitution of Atoms & Molecules"*, Philosophical Magazine, Series b, Volume 26, July 1913, pp. 1-25.

9. Neils Bohr, *"Atomic Structure"*, Nature, March 24, 1921.

10. J.R. Partington, *"The origins of the Atomic Theory"*, Annals of Science, Vol 4., No.39 July 1939.

11. G. B. Stones, *"The Atomic View of Matter in the XVth, XVIth, and XVIIth centuries"*, Isis, Vol. 10, part 2, No. 34, (January 1928).

12. Dmitri Mendeleev, *"The Periodic Law of the Chemical Elements",* Journal of Chemical Society (London) 55, 1889, 634-656.

13. D. Mendeleev, *"The Principles of Chemistry"*, 3rd English Ed. Longmans, Green, and Co., London, 1905.

14. Dmitri Mendeleev, *"On the Relationship of the Properties of Elements to their Atomic Weights"*, Zhurnal RussKee Fiziko-Khimicheskee Obshchestvo 10, 60-77 (1889); abstracted in Zeitschrift fur Chemie 12, 1869, 405-406.

15. J. A. R. Newlands, *"On relations among the equivalents"*, Chemical News 7, Feb 7, 1863, 70-72.
16. J. A. R. Newlands, *"On the Periodic law"*, Chemical News 38, 1878, 106-107.
17. Allan Frankln, *"Are There Really Electrons? Experiment and Reality"*, Physics Today, vol. 50, no.10 (October 1997), pp. 26-33.
18. Jed Z. Buchwald and Andrew Warwick, Eds, *Histories of the Electron: The Birth of Microphysics,* Cambridge, MA MIT Press, 2001.
19. Fitzgerald, George Francis, The Columbia Encyclopedia, 6th Ed. New York: Columbia University Press, 2001.
20. L. H. Haarsma, K. Abdullah, and G. Gabrielse, *"Extremely Cold Positrons Accumulated in Ultrahigh Vacuum"*, Physical Review Letters, Vol. 75, No. 5, July 1995, pp806-809.
21. W. Purcell et al, *"A Galactic Cloud of Antimatter"*, OSSE Compton Observatory, NASA, May 1st 1997.
22. Y. C. Jean, P. E. Mallon, R. Zheng et al, *"Positron Studies of Polymeric Coatings"*, Radiation Physics and Chemistry, Vol. 68, No. ¾, Oct/Nov. 2003, pp. 395-402.
23. C. R. Vane, S. Datz, E. F. Deveney, P. F. Dittnev, H. F. Krause, R. Schuch, H. Gao, and R. Hutton, *"Measurements of Positrons from Pair Production in Coulomb Collisions of 33-Tev Lead Ions with fixed Targets"*, Phys. Rev. A 56, 1997, 3682.
24. C. M. Surko, *"New Directions in Antimatter Chemistry and Physics"*, C. M. Surko and F. A. Gianturoo, eds. (Kluwer Scientific Publishers, the Netherlands).
25. G. Mazzitelli, F. Sannibale, P. Valente, M. Vescovi, P. Privitera, V. Verzi, *"Beam Instrumentation for the Single Electron DA-Ne Beam Test Facility"*, Proceedings DIPAC 2003-Mainz, Germany, pp. 59-61.
26. W. F. Henning, *"The Future of Nuclear Physics in Europe and the Demands on Accelerator Techniques"*, Proceedings DIPAC 2003-Mainz, Germany, pp. 3-6.
27. V. Simanzhenkov, R. Idem, *"Crude Oil Chemistry"*, Marcel Dekker, Inc., 2003, p45.
28. G. Von Eibe, and B. Lewis, *"Mechanism of the thermal Reaction between Hydrogen and Oxygen"*, Jour. Chem. Phys., vol.10, no. 6 June 1942, pp. 366-393.
29. A. C. Egerton, A. J. Everett, and N. P. W. Moore, *"Sintered Metals as Flame Traps"*, Fourth Symposium (International) on Combustion, The Williams & Wilkins Co. (Baltimore), 1952, pp. 689-695.

30. L. Isadore, Drell and Frank E. Belles, *"Survey of Hydrogen Combustion Properties"* Report 1383, National Advisory Committee for Aeronautics, Cleveland, Ohio, April 26, 1957, pp. 1161-1194.
31. J. O. M. Bockris and T. N. Veziroglu, *"A solar-hydrogen energy system for environmental compatibility"* Environmental Conservation, Vol. 12, No.2, 1985, pp.105-118.
32. T. N. Veziroglu and F. Barbir, *"Hydrogen the wonder Fuel"*, Int. J. Hydrogen Energy, Vol. 17, No. 6, 1992, pp. 391-404.
33. T. N. Veziroglu and Frano Barbir, *"Transportation Fuel- Hydrogen", E*nergy Technology and the Environment, Vol. 4, Wiley Interscience, 1995, 2712-2730.
34. J. M. Norbeck, J. W. Heffel, T. D. Durbin, B. Tabbars, J. M. Bowden, and M. C. Montano, *"Hydrogen Fuel for Surface Transportation"*, SAE, Warrendale, PA, 1996.
35. R. M. Santilli, *"Alarming Oxygen Depletion Caused by Hydrogen combustion and Fuel Cells and their Resolution by Magnesas^{TM}"*, International Hydrogen Energy Forum 2000, Munich, Germany, September 11-15, 2000.
36. F. Barbir, T. N. Veziroglu and H. J. Plass, Jr., *"Environmental Damage Due to Fossil Fuels Use"*, Int. J. Hydrogen Energy 10, 739 (1990).
37. Fontes, E. Nilsson, E. *"Modelling the fuel cell",* Industrial Physicist 2001, 7(4), 14-17.
38. C. R. Noller, "*Textbook of Organic Chemistry*", W. B. Saunders Company, (1966), p 66.
39. Steven S. Zumdahl, *"Introductory Chemistry"* 4th Ed. Houghton Mifflin Company, 2000, p 204.
40. E. de Barry Barnett and C.L. Wilson, *"Inorganic Chemistry- A textbook for Advanced Students",* The English Language Book Society and Longmans Green Co Ltd, 1959.
41. J.C. Geankoplis, "*Transport Processes and Unit Operations*" 3rd Ed. Prentice Hall P T R Englewood Cliffs, New Jersey 07632, (1993) p 588.
42. J. M. Smith, *"Chemical Engineering Kinetics"*, 2nd Ed. New York: McGraw -Hill Book Company, 1970.
43. Free Radical, The Columbia Encyclopedia, 6th Ed. New York: Columbia University Press, 2001.
44. Radical, in chemistry, The Columbia Encyclopedia, 6th Ed. New York: Columbia University Press, 2001.

45. C. R. Noller, "*Textbook of Organic Chemistry*", W. B. Saunders Company, (1966), pg. 17.
46. C.R. Noller, "*Textbook of Organic Chemistry*", W.B. Saunders Company, (1966), pg. 66.
47. James M. Rosen, Raymond M. Morales, Norman T. Kjome, V. W. J. H. Kirchhoff, and F. R. daSilver, "*Equatorial aerosol-ozone structure and variations as observed by balloon-borne backscattersondes since 1995 at Natal, Brazil (6°S)*", Jour. of Geophysical Research, Vol. 109, D03201, 2004.
48. D. Demirhan, C. Kahya, S. Topcu, S. Ineecik, "*Tropopause Height's Influence on Vertical Ozone Structure in Ankara, Turkey*", Jour. of Geophysical Research, Vol.5, 07063, 2003.
49. Karoline Wiesner, "*Electronic Structure and Core-Hole Dynamics of Ozone- Synchrotron-radiation based studies and ab-initio calculations*", Comprehensive Summaries of Uppsala Dissertations from the Faculty of Science and Technology 921, ACTA UNIVERSITATIS UPSALIENSIS UPPSALA 2003, 47pp.
50. C. R. Noller, "*Textbook of Organic Chemistry*", W. B. Saunders Company, (1966), pg. 388.

Chapter 2

New Classifications for Bonds and Charges and Their Impacts.

Like for radicals[1], herein contains new classifications for charges and bonds. There are real and imaginary bonds and charges. Ionic, and covalent bonds and charges are real, while electrostatic and some or all polar bonds and charges are imaginary. Only ionic charges can be isolated. Some of the electrostatic bonds that help to conduct electric currents in only fluids have begun to be shown. Just like X- and Y- chromosomes in animals and plants, so also there exist male and female macro-molecules, their established structures carrying what are called Activation centers (X- and Y-). Included herein for them, are new concepts on ringed compounds based on Functional centers (all X-) carried by some, Activation centers visibly (X- and Y-) and invisibly (all X-) present, Strain energies carried by all of them and their points of scission for some of them. These have been applied to start distinguishing between different polymerization systems. How alternating, isotactic and syndiotactic placements are obtained in Addition polymerization systems have begun to be explained. Poisonous and non-poisonous carbon monoxide have also been distinguished. Structures of important elements of zinc and some compounds have begun to be shown. The new concepts introduced herein are too countless to list. Each of all these new concepts is more than nobel, novel, scholarship, magnus opium in character according to many compliments from important universal journals.

2.0 Introduction

The existence of ionic charges and bonds has long been known. Based on universal data, there is much yet to know about them. The same too applies to covalent, electrostatic and polar charges and bonds. The lack of knowing exactly what these and more of them are, has for many years given rise to impossible concepts and confusions too numerous to list. For example, "hydrogen bonding" between two polymeric chains[2-4] by electrostatic attraction (called secondary forces) does not exist because hydrogen already bonded covalently can no longer carry any other additional bond in the absence of unbonded radical(s) or vacant orbital(s) in its last shell the closest to the nucleus. On the other hand, no polar bond can be formed between H and for example O. Polar bonds go with covalent sigma bond. Electrostatic bonds can only be formed if $H^{\oplus}$ can be made to sit on another center with paired unbonded radicals to form an imaginary positive charge, which still is not hydrogen bonding, but an electrostatic bond. However, with respect to "hydrogen bonding", there is indeed ***a force of attraction coming from the attraction of the electropositive end*** (metals that carry no paired unbonded radicals (e.g. Ionic metals) ***to the electronegative end of a non-metal*** (e.g. O, N, Cl which all carry paired unbonded radicals) of another molecule. This bond is not dotted, but invisible in character. It is a secondary and strong force compared to other intermolecular attractions, such as so-called van der Waals dispersion forces.[5-9] Without these invisible secondary forces of which that generated from hydrogen is a large member, beautiful crystals or blocks or strands of for example ice block, sodium chloride, polypeptides and DNAs cannot be obtained. These forces are better classified as ***Ionic Electropositive/Electronegative*** forces of attraction.

Secondly, the + and – signs we see on our batteries are not ionic charges, but *polar (if solids) or electrostatic (if fluids) charges* as will shortly be explained for fluids only.

Thirdly, though the Polymer Industry is one of the largest industries (largely based on the development of the Petroleum Industry) if not indeed the largest, that has greatly helped to advance technology today, the journey is still a long one, because even the advancements of these technologies are still ***95% ART.*** Unknown in the Science of today is the existence of male and female macromolecules as exist in living systems as one has already started to identify in Chapter 1. These like the chromosomes of males and females carry Activation/Functional centers. These centers are carried by some elements, Addition monomers (ringed and non-ringed), and some ringed compounds. There are different types of Addition polymerization systems (Bulk, Solutions, Suspensions and Emulsions) depending on the types of Initiators used for polymerization, the types of environment and the types of monomers involved. **Free Cations** cannot be used as initiators for several reasons, the most important of which is charge balancing. There is a difference between Cations and other Positive charges. Non-free Anions can be use freely where possible. Some of the initiators, which today for example are so-called *Anionic ion-paired initiators*[10] (specific for male monomers), *Cationic ion-paired coordination initiators*[10] (specific for female monomers), *Ziegler-Natta (Z/N) initiators*[10] (specific for male and female monomers) and more, do not carry ions. Secondly, the names given to them are all wrong. What they do and how they do them are unknown. However, most of them carry charges, which are different from ions. Unlike ions and radicals, they cannot be isolated. Knowing their real names and what they are, will open new doors. The mechanisms by which these initiators are obtained are still unknown.[11-13] Also, how for example *isotactic and syndiotactic* polypropylenes are obtained from propylene monomer using Z/N types of combinations are still unknown.[14-17] The same applies to all monomers. *Why propylene cannot be polymerized radically* are still unknown. Yet unknown, is that it can be polymerized free-radically. *Why vinyl chloride and vinyl acetate cannot be polymerized chargedly, but only radically* are still unknown. Unknown also is that there are some female monomers such as ethylene and butadiene that have both male and female characters (Bisexual), since they can be polymerized using either male or female free radical initiators under different operating conditions. The unknowns are too numerous to list. If all these unknowns were known, the world we are living in today would not have been what it is. The consumers are the losers, paying hundreds of times what they should have been, for all products. What manufacturers should have spent 1dollar to produce, is produced with 100 dollars due lack of understanding how NATURE operates. All what they do is just an ART and not a SCIENCE. All industries are OPENED, instead of being CLOSED creating great problems to the environment. How can CO_2 we breathe out all the time be said to be the cause of GLOBAL WARMING, when indeed the real WASTES in our environment are the human beings themselves, the Caretakers that generate the so-called wastes which if CLOSED can be recycled? How can the positive charge on for example $H^{\oplus}$ (called a PROTON) be removed leaving what behind, when only a radical can be removed from the H atom (H•e), leaving a cation – the PROTON (the positive charge on the carrier) behind? How can the PROTON here outside the NUCLEUS be the same as the SO-CALLED PROTON inside the NUCLEUS of H or any other Atom? These are parts of what are done in present-day Science, such as in Chemistry,

Nuclear Physics, Nuclear Engineering, Electrical Engineering, Chemical Engineering and so on. Though Electrical Engineering has been one of the greatest Engineering disciplines in its ability to make our world a global village, little or nothing is known about how current or sound is transmitted. They think that current is a function of charge, that which is absolute nonsense, because charges cannot be removed from their carriers to move from hole to hole. All that we have been doing are to discover something new, with little or no understanding of what has been discovered, after which the discoverer is awarded a prize!!! The Rich becomes richer and the Poor poorer, leaving no Future behind for younger generations.

2.1 New Classifications for Charges and Bonds

Without the new classifications for Radicals[1], the new classifications for Bonds and Charges and other phenomena that follow cannot be explained.

Notice that so far[1], no strong reference has been specifically made of charges. They all have their origins from radicals. All the foundations laid so far in present-day Science are in order, except that ***the eye of the needle has not yet been fully applied***. With the use of the eye of the needle, no chemical reactions take place **ionically,** because ionically, species where possible *can only exist in equilibrium state and no other state because of the driving forces that favour their existences.*

Water-Ionic

$$H_2O \rightleftharpoons H^{\oplus} + HO^{\ominus}$$

(Empty last shell) (Full last shell)

2.1a

Sodium Chloride-Ionic

$$NaCl \rightleftharpoons Na^{\oplus} + Cl^{\ominus}$$

(Empty last shell) (Full last shell)

2.1b

Magnesium hydroxide-Ionic

$$Mg(OH)_2 \rightleftharpoons Mg^{2\oplus} + 2\,{}^{\ominus}OH$$

(Empty last shell) (Full last shell)

2.1c

$$Mg(OH)_2 \not\rightleftharpoons HO - Mg^{\oplus} + {}^{\ominus}OH$$

(Non-empty last shell) (Full last shell)

[Impossible existence] **2.1d**

$$Mg(OH)_2 \rightleftharpoons HO - Mg\bullet e + nn\bullet OH$$

[Only Favoured existence] 2.1e

Ferric Chloride-Non-Ionic

$$FeCl_3 \xrightleftharpoons[\text{Possible}]{\text{No}} Fe^{3\oplus} + 3Cl^{\ominus}$$

(Non-empty last shell) (Full last shell)

[Impossible existence] 2.2a

$\rightleftharpoons Cl_2Fe•en + Cl•nn$ (Real state) 2.2b

Chlorine- Non-Ionic

$$Cl_2 \not\rightleftharpoons Cl^{\oplus} + Cl^{\ominus}$$

(Non-empty last shell) (Full last shell)

[Impossible existence] 2.2c

Ammonium hydroxide-Non-ionic

$$NH_4OH \not\rightleftharpoons H_4N^{\oplus} + {}^{\ominus}OH$$

(***Overfilled*** last shell) (Full last shell)

[Impossible existence] 2.2d

For chemical reactions to take place productively under Equilibrium mechanism, Combination state of existence[1] must be involved in the last step and this is not possible ionically wherein ***no Combination and Decomposition States of existences exist for all compounds***. Covalently, Electrostatically and Polarly, ***Equilibrium, Combination and Decomposition States of existences do not exist, because these charges cannot be isolatedly placed.***

For ionic bonds to be formed, there are *three driving forces required-*

(i) The last shell or boundary of the cation (the male) must be empty. For example, HO-$Mg^{\oplus}$ is not a cation because Mg is divalent and its last shell is not empty (See Equation 2.1d above). $Fe^{3\oplus}$ above is not a cation because the last shell is not empty. The charges carried on the center cannot be isolated from that on chlorine in $FeCl_3$, the bonds being σ-covalent bonds. π-Covalent charges cannot also be isolatedly placed. One has not shown how the structure of ammonium hydroxide is represented when bonds are carried in present – day Science. But as shown in Equation 2.2d, the last shell of N is carrying ten radicals, when the maximum allowed by NATURE (One of Boundary Laws) is eight radicals. For N to carry ionic positive charges, the last of N must be empty, that which is impossible.

(ii) The last shell or boundary of the anion (the female) must be full and most importantly be Polar (i.e., must carry paired unbonded radicals). Examples of such centers include Cl, O, N, in $:\ddot{\underset{..}{Cl}}:^{\odot}$, $H-\ddot{\underset{..}{O}}:^{\odot}$ *and* $^{\odot}:\underset{..}{N}H_2$ respectively. The last shell (The boundary) of Cl, O and N can never carry more than eight radicals, based on the Periods in the Periodic Table ***(The greatest invention since antiquity)*** to which they belong. This maximum is indicated by Group VIIIB elements – the inert gases in the Periodic Table (Limitations) for every Period in the Table. These are laws based on Domains and Boundaries on everything in our world since every atom on which all things depend on has domain and boundary.[1] Hence, a structure such as shown below in today's Science for ammonium hydroxide does not exist.

(Ammonium hydroxide)

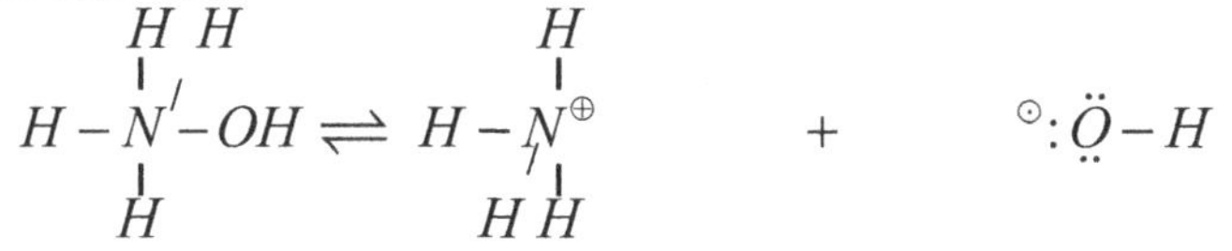

$$\begin{array}{c} H\ H \\ |\ / \\ H-N-OH \\ | \\ H \end{array} \rightleftharpoons \begin{array}{c} H \\ | \\ H-N^{\oplus} \\ /\ | \\ H\ H \end{array} \quad + \quad ^{\odot}:\ddot{\underset{..}{O}}-H$$

(Wrong Structure) (A) (B)

(Overfilled Last Shell) 2.3

The last shell of Nitrogen though an electronegative element, is more than full with two more radicals than the required maximum of 8 radicals against one of the laws of Nature. One will shortly show the real structure of ammonium hydroxide.

(iii) No hybridization of the radicals in sub-atomic orbitals can take place in the central atoms, as takes place in for example C centers (sp^3, sp^2, sp hybridizations) or vanadium or aluminium etc. centers. They do not take place with H, Na, Mg, Ca, or indeed all Group IA, IIA and IIIA metals, noting that ***H is a metal of the gaseous type***. Based on the provisions for ionic bond formation, one can observe that not all metals are ionic in character. Only Group IA and IIA elements are Ionic Non-Transition metals and Group IIIA elements are Ionic-Transition metals, cases where the last shell or boundary can be emptied without hybridization. Electronegative atoms such as N, O, Cl and so on cannot also be hybridized as is wrongly believed to be the case for N. Many non-ionic metallic atoms undergo hybridization.

Ionic charges are the only ones amongst charges, which can be isolated. For example, consider the case of Sodium chloride solution (Brine).
<u>Ionic Sodium Chloride Solution</u>

$$NaCl \underset{\text{of Existence of NaCl}}{\overset{\text{Ionic Equilibrium State}}{\rightleftharpoons}} Na^{+} + :\ddot{\underset{..}{Cl}}:^{-}$$

(*Empty*) (*Full*)

$$H_2O \underset{\text{State of Existence of Water}}{\overset{\text{Ionic Equilibrium}}{\rightleftharpoons}} H^{+} + {}^{-}:\ddot{\underset{..}{O}}H$$

(*Empty*) (*Full*)
(Both charges can be isolated) 2.4

$$H_3CCl \underset{\text{Existence}}{\overset{\text{Equilibrium State of}}{\rightleftharpoons}} H_3C\bullet e + Cl\bullet nn$$

REAL STATE OF EXISTENCE

$$H_3CCl \underset{\text{State of Existence}}{\overset{\text{Impossible Equilibrium}}{\rightleftharpoons}} H_3C^{\oplus} + Cl^{\ominus}$$

IMPOSSIBLE STATE OF EXISTENCE 2.5

The C center cannot carry ionic charges, but can carry covalent and electrostatic charges which cannot be isolated. While Radicals can also be isolated such as in electrolysis (on the anodes and cathodes) and in all chemical reactions, other types of charges cannot be isolated and these include the followings-

Covalent charges,
Electrostatic charges and } **cannot be isolated**
Polar charges

While Ionic and Covalent charges are **<u>REAL</u>**, the Electrostatic and Polar charges are **<u>IMAGINARY</u>**, because of the natural limitations placed on the boundary where only specified maximum numbers of radicals can reside as shown by their radical (not electronic) configurations[1], which are in place (2 for Period 1, 8 for Periods 2 and 3, 18 for Period 4, 36 for Period 5 and so on). ***While Real charge can repel and attract, Imaginary charges cannot repel or attract. Radicals cannot also repel and attract,*** since for example two H atoms (H •e) can add together in Equilibrium mechanism in the last step to form H_2 (Combination state of Existence for H_2) as already seen in Chapter1 in some stages where H_2 is produced.

Figure 2.1 shows the new classification for Bonds. Those for Charges are shown in Figure 2.2.

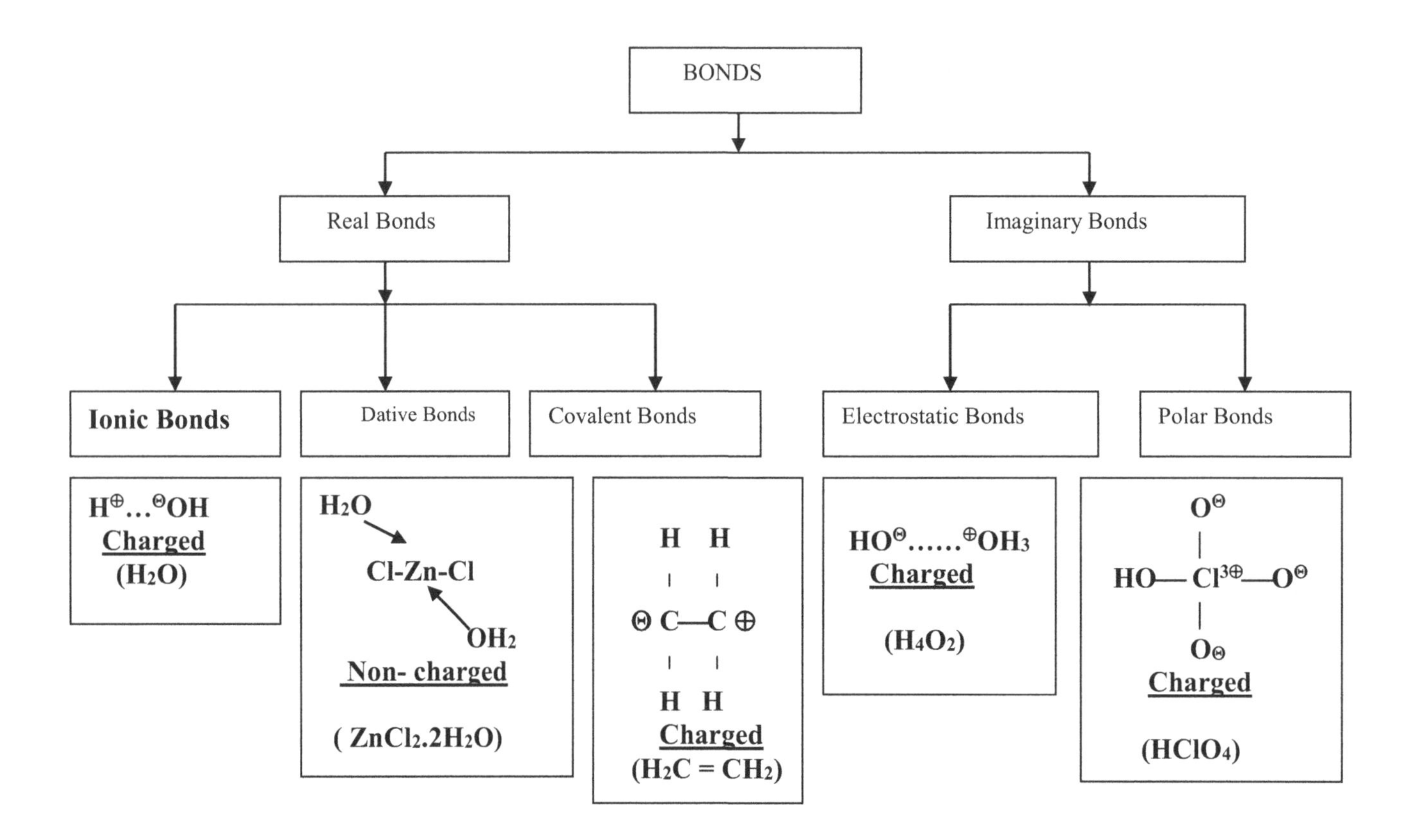

Figure 2.1 New Classifications for Bonds

How can one for example isolate a negative or positive covalent, or electrostatic or polar charge in their carriers? How can one remove the charges from their carriers? There are different types of the four kinds of charges-Ionic, Covalent, Electrostatic and Polar. In lithium alkyls (e.g. $LiC(CH_3)_3$), covalent bonds between the C and H centers, C and C centers are different from those between Li and C centers. In lithium hydride (LiH) just like in H_2, no charges can be carried by the centers, because lithium and hydrogen belong to the same family noting that hydrogen is an ionic metal of the gaseous type, electropositive in character. Like other ionic metals, $H^{\ominus}$ does not exist and is not a hydride. This looks like what present-day Science also calls "Electron" and they use the same name for radicals!!! To keep for example Na as $Na^{\ominus}$ is impossible. The same applies to Li, K, Mg and so on for all metals. To keep H as $H^{\ominus}$ will require very harsh operating conditions far above what is in the Nucleus. In fact, it does not exist as will be shown in Appendix (I). ***The hydride is H·n.*** In LiH, each atom can only be isolated as radicals and not charges. Because of the electropositivity of hydrogen compared to the other metals in Group IA and IIA of the Periodic Table (Electronegativity of H = 2.1, Lithium = 1.0 and sodium = 0.9), and its multiple characters (such as used in Atomic Energy and in the Nucleus), and its size, hence its behaviour is very unique amongst all atoms. All anions are ***non-free in character and polar***, because of presence of paired unbonded radicals in the last shell. All negative charges are ***free in character and non-polar***, because of absence of paired unbonded radicals in the last shell. All cations are ***free in character and non-polar,*** because the last shell is empty.

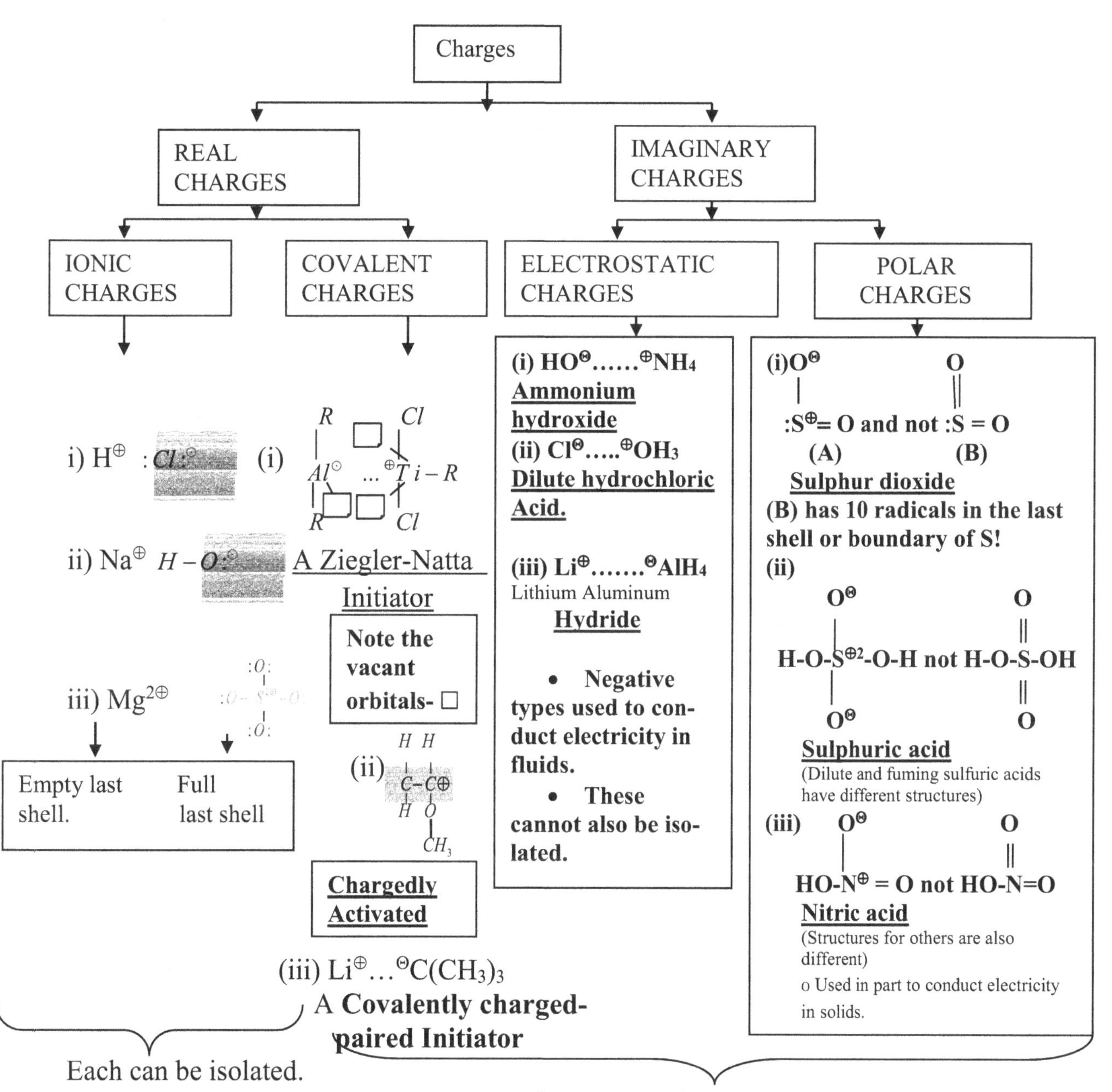

Figure 2.2 New classifications for Charges

All positive charges are ***free or non-free in character,*** since their last shell is not empty and some carry paired unbonded radicals, making them ***polar in character such as the one carried on the center of Fe in $FeCl_2$.***

It is important to note the examples used in the figures, most of which are new in character. One of the real types of initiators obtained based on the ratio of catalyst/co-catalyst (AlR_3:$TiCl_4$ at ratio of 2:1) involved, has been shown in the figure. The Al and Ti centers have vacant orbitals still left on them which are used as reservoirs for the monomers in their activated states. These initiators carry ***elastic covalent bonds***. *These are mistakenly called Zeigler/Natta Initiators*. These are indeed ***Covalently charged-paired initiators,*** in which Z/N initiators is one of them, almost like LiC_4H_9 above (which in present-day Science is also *mistakenly called Anionic-ion-paired initiator*). One did not use anionic or cationic or positively or negatively charged paired initiator in the new underlined name above, because their two centers are active. They (unlike cases where ionic metals are involved) can be used to carry male or female monomers, the positive center for females (Nucleophiles) and the negative center for males (Electrophiles). Though the two initiators above belong to the same family, they are still uniquely different in many respects. This is NATURE. These and what are newly called ***Ionically charged-paired initiators*** such as $H_3CO^{\ominus}$.....$^{\oplus}$Na have two centers that are also active only radically, each for males and females of different types. These and ***Radical-paired initiators*** are the only **Paired–media groups of charged- and radical-paired** with two active centers. All the other Paired-media initiators have only one active center as shown for ammonium hydroxide above- the anionic center. Note that the real structure of the ammonium hydroxide has been shown above and how it was obtained will shortly be shown. The paired bonds are called **Electrostatic bonds** and not ionic bonds as is currently believed to be the case. This belongs to the family of Initiators called ***Electrostatically anionically charged-paired initiators*** in the New Frontiers. (In the New Frontiers, there are still three other electrostatically charged-paired initiators.) The negative center is anionic. It is said to be anionic, because there **are those** called ***electrostatically negatively charged-paired initiators*** in the New Frontier. They are so called, because the two of them cannot be used for the polymerization of the same monomer due to charge balancing. One can begin to start seeing some of the Initiators used in Nature for polymerization systems. There are also what are called **Free-media groups of ionic and radical initiators.** There are only two groups. These can only be used when *they are the only ones present in system without the opposite pair*. For example, when **Nucleo-free-radicals** are used, the electro-free or non-free-radicals must not be present in the system, otherwise, **Propagation step** (Growth) during polymerization cannot take place. We have **Nucleo-free-radicals** generating initiators such as Benzoyl peroxide (C_6H_5COOOC-C_6H_5), Azobisisobutyro-nitrile [$(CH_3)_2$ CNC-N = N-CCN$(CH_3)_2$]; **Electro-free-radical** generating initiators such as Sodium cyanide (NaCN), Titanium trichloride ($TiCl_3$); **Nucleo-non-free-radical** generating initiators such as Benzoyl peroxide, Azobisisobutyronitrile; and **Electro-non-free-radical** generating initiators such as Iron (Fe) itself. It is during the **Initiation step** (Birth) that one can know whether these initiators will polymerize a monomer or not. If **one monomer unit** cannot be carried by the initiator, then there is no initiation. When carried signifies BIRTH. For example, a Nucleo-free-radical (female) when present alone in the system cannot polymerize a monomer such as propylene a female (Real name is Propene), because it has transfer species of the first kind of the first type which will prevent it being initiated during Initiation step as will shortly be shown. Ethylene (Real name ethene) does not carry this transfer species. When an Electro-free-radical is used on the propene, Initiation step will be favoured and propagation step (Growth) follows. When an Electro-non-free-radical is used, Initiation step will not be favoured, because when used via Combination mechanism (Addition polymerization), ***the equation will not be radically balanced,*** since the active centers of activated propene are carrying free-radicals and not non-free-radicals as will be shown downstream. Now we can see that we are beginning to use the EYE OF THE NEEDLE. What in general we neglect or attach no importance to are where the so-called mysteries of life are embedded in. One used the word so-called, because they are not, since the only mystery in humanity is THE ALMIGHTY INFINITE GOD. In our world, we neglect the POOR, anything that is very small, and so many too countless to list! We even neglect ZERO, when without ZERO no number exists.

2.2 Activation Centers and Male/Female Macromolecules

In the last Chapter, the concept of the existence of what are called Activation centers was introduced. Therein, it was said that there are three kinds of Activation centers with different types, in which only two of them were shown- the visible and invisible π-bond and Vacant orbitals/Paired unbonded radicals in the last shell (the Boundary) present on some atoms whether bonded or not. The third one will shortly be seen.

Compounds or systems that have little or nothing to use to exist in Equilibrium state of existence at all operating conditions, can be activated if they have activation centers. Only compounds that have activation centers can be made to undergo Addition polymerization via Combination mechanisms[1] to give polymers depending on the operating conditions. *In polymer science and engineering for example, unknown to all the related Scientists and Engineers including the author is that where radicals are involved for initiation, over ninety-five percent of them obtained from specific catalysts such as benzoyl peroxide are female in character, all of which can polymerize specific female monomers such as ethylene, vinyl chloride, vinyl acetate, butadiene and more under **mild to very harsh** operating conditions and male monomers such as acrylamide, acrylonitrile, acrolein and more under **normal to very mild** operating conditions.* Some activation centers are shown below for some elements and compounds.

(A) Zinc dust element and Zinc metallic atom-Female

$$\boxed{:} - Zn - \square \xrightleftharpoons[(heat)]{Activation} nn\bullet\ Zn\ \bullet en$$

$$\downarrow \qquad\qquad Zinc\ Dust$$

$$Vacant\ orbitals\ in\ the\ last\ shell \qquad (An\ element\ of\ Zn) \qquad 2.6a$$

$$: Zn - \square \xrightleftharpoons[(By\ a\ neighbour)]{Excitation} en\bullet Zn \bullet en$$

$$Zn\ metal\ (Being\ electropositive) \qquad 2.6b$$

Presence of a particular neighbour as will be shown in the New Frontiers is required for excitation while heat is required for activation. ***(Remember that one is always excited differently when a neighbour is around).*** These atoms seem to know their neighbours via for example their electronegativity/electropositivity. Chargedly, it cannot be activated due to *electrostatic forces of repulsion* on the same center carrying paired unbonded radicals. If any charge can be carried on the same center, they can only be either covalent or electrostatic when paired or polar and never ionic. When excited, the paired unbonded radicals separate, carrying with them the electropositive or electronegative character of the atom, ***i.e., one radical which is always the male in the orbital where they are oppositely placed as male and female, pushes the female away and places her in a reversed position and moves to another vacant orbital such as the case above, or the male leaves the orbital to another vacant orbital in a reversed position such as in Carbon. In fact, amongst all atoms, only CARBON has the***

ability to do this. All these are based on the radical configuration (not electronic configuration) of the atom as well as the electronegativity or electropositivity of the atom as shown above and shown below for oxygen and carbon atoms respectively.

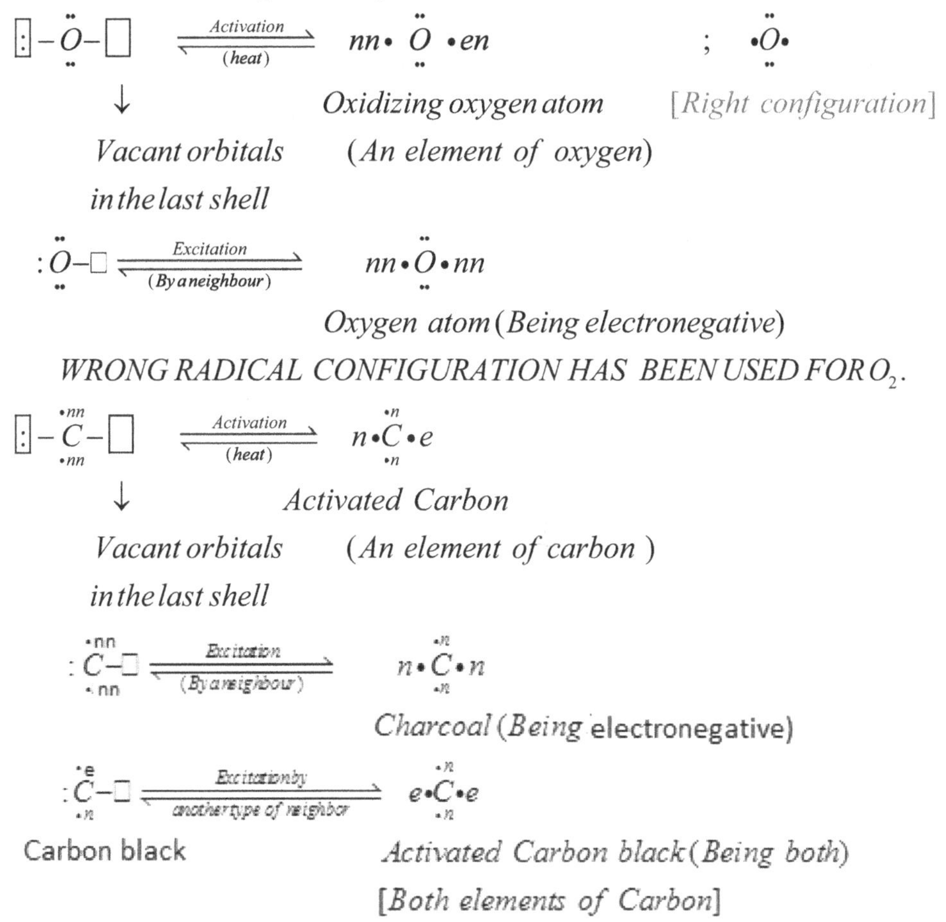

$$\boxed{:}-\ddot{\underset{\bullet\bullet}{O}}-\Box \xrightleftharpoons[(heat)]{Activation} nn\bullet \ddot{\underset{\bullet\bullet}{O}} \bullet en \quad ; \quad \bullet\ddot{\underset{\bullet\bullet}{O}}\bullet$$

↓ *Oxidizing oxygen atom* [*Right configuration*]

Vacant orbitals in the last shell (*An element of oxygen*) 2.7*a*

$$:\ddot{\underset{\bullet\bullet}{O}}-\Box \xrightleftharpoons[(By\ a\ neighbour)]{Excitation} nn\bullet\ddot{\underset{\bullet\bullet}{O}}\bullet nn$$

Oxygen atom (*Being electronegative*)

WRONG RADICAL CONFIGURATION HAS BEEN USED FOR O_2. 2.7*b*

$$\boxed{:}-\overset{\bullet nn}{\underset{\bullet nn}{C}}-\Box \xrightleftharpoons[(heat)]{Activation} n\bullet\overset{\bullet n}{\underset{\bullet n}{C}}\bullet e$$

↓ *Activated Carbon*

Vacant orbitals in the last shell (*An element of carbon*) 2.7*c*

$$:\overset{\bullet nn}{\underset{\bullet nn}{C}}-\Box \xrightleftharpoons[(By\ a\ neighbour)]{Excitation} n\bullet\overset{\bullet n}{\underset{\bullet n}{C}}\bullet n$$

Charcoal (*Being* electronegative)

$$:\overset{\bullet e}{\underset{\bullet n}{C}}-\Box \xrightleftharpoons[another\ type\ of\ neighbor]{Excitation\ by} e\bullet\overset{\bullet n}{\underset{\bullet n}{C}}\bullet e$$

Carbon black *Activated Carbon black* (*Being both*)

[*Both elements of Carbon*] 2.7*d*

For Zn atom (Solid) **when excited,** the nucleo-non-free-radical left behind or pushed is reversed, while during **activation,** the electro-non-free-radical is not reversed when it moves to a vacant orbital. When the right radical configuration is used for Oxygen atom, it cannot be excited or activated. Hence, what are shown above are impossible, despite the fact that it is showing an oxidizing oxygen element. For C, both roles are played when excited and activated, because of its electronegativity and its metallic and non-metallic character. C can also be used in its ground state. Notice that, only electro-radicals (free- or non-free-) as has already been said in Chapter 1 move in all reactions. They are the "workers". ***Nucleo-radicals do not move, because our world is owned by MOTHER NATURE as "GIFT". In our world, have we heard of FATHER NATURE?*** All these we seem to neglect. An atom which has no vacant orbital cannot undergo these phenomena. One can observe how wonderful NATUTE operates. All these since antiquity, we the elites have neglected. We including the author think we see or know them, when indeed we cannot see and know nothing. All the examples above are Females, because they all have one activation center. In fact, all activation centers whether in males or females are all nucleophilic, that is, female in character.

Oxygen molecule itself cannot be activated chargedly, due to electrostatic forces of repulsion between the negative charge and paired unbonded radicals on the adjacently placed oxygen element. Acetylene for example, cannot be activated chargedly as is done in present-day Science, due to electrostatic forces of repulsion between the negative charge and the remaining π-bond. The same applies to N_2. ***In NATURE, being ADJACENTLY PLACED or CONJUGATEDLY PLACED*** is so important in all operations carried out in our world. It is therefore no surprise that "we are known by the friends we keep". Without ADJACENT PLACEMENT, how can current flow; how can messages be transmitted?

(B) Poisonous and Non-poisonous Carbon Monoxide-Female
(Using the Vacant orbital/Paired unbonded radicals)

$$\boxed{:}-\overset{\overset{O}{\|}}{C}-\square \underset{Heat}{\overset{Activation}{\rightleftharpoons}} e\bullet\overset{\overset{O}{\|}}{C}\bullet n \qquad or\ Covalently(None)$$

$$\downarrow \qquad\qquad \downarrow$$

Vacant orbital — *The origin of poison(Both radicals on same center)*

Female and Stable Non – poisnous — *Activated State Poison*

2.8a

(Using the Visible π-bond)

Stage 1:

$$Mg \overset{Excited}{\rightleftharpoons} e\bullet Mg\bullet e$$

$$e\bullet Mg\bullet e \ + \ O=C: \underset{Electric\ furnace}{\overset{Activated\ in}{\rightleftharpoons}} e\bullet Mg-O-\overset{\bullet n}{\underset{\bullet e}{C}}\bullet e$$

$$(A)$$

$$(A) \rightleftharpoons e\bullet Mg-O\bullet nn \ + \ :\underset{\bullet e}{C}\bullet e$$

$$(B) \qquad \underline{Coke}$$

$$(B) \xrightarrow[Release\ of\ Heat]{Deactivation} Mg^{\oplus} - {}^{\odot}O \ \ [And\ not\ Mg=O]$$

$$(C)$$

$$\underline{Overall\ equation}: \ Mg \ + \ CO \xrightarrow{Electric\ furnace} MgO \ + \ Coke \qquad 2.8b$$

(C) Carbon Dioxide-Female

$$\underset{Female}{O=C=O} \underset{Heat}{\overset{Activation}{\rightleftharpoons}} \underset{(Radically)}{e\bullet \overset{O}{\overset{\|}{C}}-O\bullet nn} \text{ (Both radicals on different centers)} \text{ and } \underset{(Covalently)}{\oplus \overset{O}{\overset{\|}{C}}-\ddot{\underset{..}{O}}\ominus} \quad 2.8c$$

More or less heat will be required for each of them, based on their known properties and operating conditions, but for the reaction in the middle. For this reaction, the radicals are on two different centers just like the last case. In CO_2, the activation center is the π-covalent bond, anyone of the two, one at a time. In CO the first case, the activation center is provided by the presence of a vacant orbital in the last shell of carbon (as long as radical donating materials do not occupy it) and paired unbonded radicals are present in another orbital. In the second reaction, the only π-bond present can only be activated in the presence of some ionic metals at very high temperatures. In that Stage 1 of Equation 2.8b, notice for the first time the structure of Magnesium oxide. ***Metals in general cannot carry double bonds, because of their strong electropositive character. Polar bonds are carried in its place.*** Also, note for the first time ***the structure of Coke a metallic carbon.*** All the activation centers so far identified are female or X in character.

(D) Real Structure of SO_2-Female

$$:\overset{O^-}{\overset{|+}{S}}=O \underset{(Heat)}{\overset{Activation}{\rightleftharpoons}} en\bullet \overset{O^{\odot}}{\overset{|}{S^{\oplus}}}-O\bullet nn \; ; \quad \oplus \overset{O^{\odot}}{\overset{|}{\underset{..}{S}^{\oplus}}}-\ddot{\underset{..}{O}}\ominus \text{ (Covalently – cannot exist)}$$

$$\underset{\text{(Radically)}}{\underset{\text{Electro – non free radical}}{\downarrow} \quad \underset{\text{Nucleo – non free radical}}{\downarrow}} \quad 2.9$$

In the SO_2 above, a female, it is the π-bond, rather than the polar bond that is activated, because of the limitation imposed by the boundary of the S atom, the Central atom. Its last shell cannot carry more than eight radicals. ***Hence for example some of the structures of groups used in some present-day Templates (Software) are not correct.*** Chargedly, it cannot be activated due to electrostatic forces of repulsion between the negative charge and paired unbonded radicals on the S center. The Polar negative charge, that is imaginary, unlike the Covalent negative charge cannot repel or attract. Note the type of radicals generated by the activation centers so far. The case above is full non-free (that is non-free male and female radicals). The case of CO_2 is half free or non-free (free on C center and non-free on the oxygen center). The case of CO is full free, however on same carbon center. Their use as monomers will depend on the type of initiator, operating conditions, noting that one cannot prevent electrostatic forces of repulsion from taking place when real charges are carried and adjacently placed.

Indeed they (the COs) may add to themselves to form living deadly cells, such as shown below. Ringed ones such as shown below cannot readily be formed, unless the ring is so large. Four or six membered rings for this multi-ketonic ring cannot exist because of the very large stain energy in such rings. Even three or four membered rings with only one CO group find it difficult to exist. The very large living cells are the deadly ones, like the first case shown below. However, the chain will keep growing until it becomes very large to block flow of fluids in some systems or do something else. As will be seen in the New Frontiers, a three-membered ring with only one ketone group in the ring cannot exist, because of what is called **Maximum Required Strain Energy (Max.RSE).** Any ring that is about to have it, cannot exist as ring. There is also the **Minimum Required Strain Energy (MRSE).** This is the energy that must be added to a ring to unzip the ring if the ring has a point of scission, like cyclopropane.

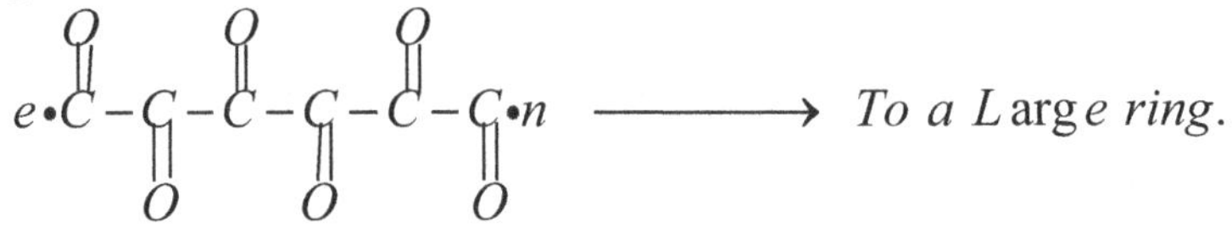

$$e\bullet\overset{\overset{O}{\|}}{C}-\underset{\underset{O}{\|}}{C}-\overset{\overset{O}{\|}}{C}-\underset{\underset{O}{\|}}{C}-\overset{\overset{O}{\|}}{C}-\underset{\underset{O}{\|}}{C}\bullet n \longrightarrow \textit{To a Large ring.}$$

Living cells

2.10a

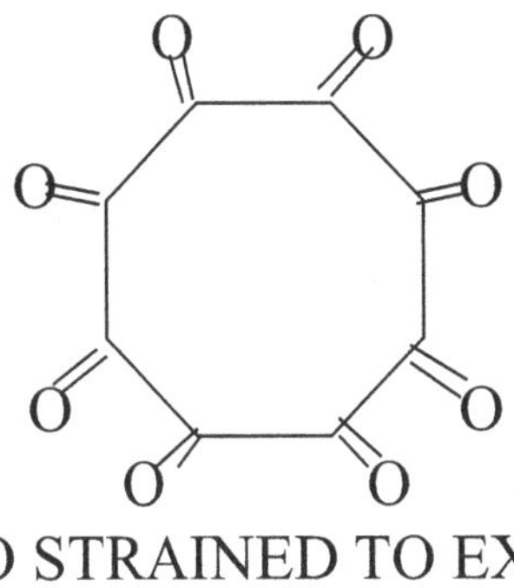

TOO STRAINED TO EXIST [Eight membered ring]

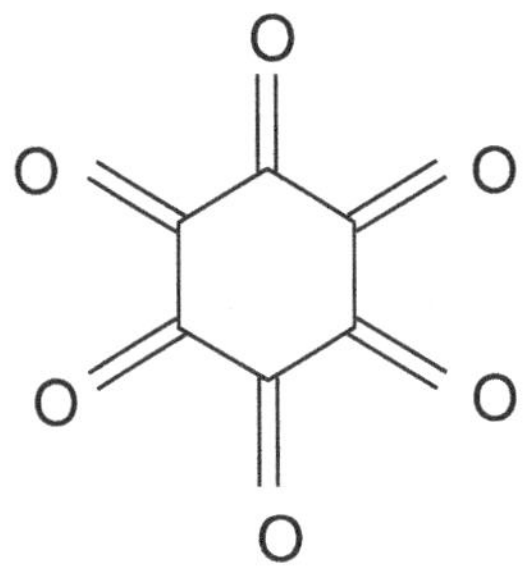

TOO STRAINED TO EXIST [Six membered ring] 2,10b

(E) Olefin monomers

Now, consider the activation of **Olefins**, starting with ethylene. When activated chargedly or radically, the followings are obtained.

$$\underset{(A)}{\begin{array}{c} H\ \ \ H \\ |\,X\,| \\ C=C \\ |\ \ \ | \\ H\ \ \ H \end{array}} \xrightleftharpoons[\text{Radically}]{\text{Free-}} \underset{(Any\ Center)}{e\bullet \overset{H}{\underset{H}{C}} - \overset{H}{\underset{H}{C}} \bullet n\,;} \xrightleftharpoons{\text{Chargedly}} \underset{(Any\ Center)}{\oplus \overset{H}{\underset{H}{C}} - \overset{H}{\underset{H}{C}} \ominus} \qquad 2.11a$$

(Less female and full free)-***Ethylene [ethene].***

$$\underset{(B)}{\begin{array}{c} H\ \ \ CH_3 \\ |\,X\,| \\ C=C \\ |\ \ \ | \\ H\ \ \ H \end{array}} \xrightleftharpoons[\text{Radically}]{\text{Free-}} \underset{(Fixed\ Centers)}{n\bullet \overset{H}{\underset{H}{C}} - \overset{CH_3}{\underset{H}{C}} \bullet n\,;} \xrightleftharpoons{\text{Chargedly}} \underset{(Fixed\ Centers)}{\ominus \overset{H}{\underset{H}{C}} - \overset{CH_3}{\underset{H}{C}} \oplus}$$

(Female and full free)- ***Propylene [propene].*** 2.11b

$$\underset{(C)}{\begin{array}{c} H\ \ \ C_2H_5 \\ |\,X\,| \\ C=C \\ |\ \ \ | \\ H\ \ \ H \end{array}} \xrightleftharpoons[\text{Radically}]{\text{Free-}} \underset{(Fixed\ Centers)}{n\bullet \overset{H}{\underset{H}{C}} - \overset{C_2H_5}{\underset{H}{C}} \bullet e\ ;} \xrightleftharpoons{\text{Chargedly}} \underset{(Fixed\ Centers)}{\ominus \overset{H}{\underset{H}{C}} - \overset{C_2H_5}{\underset{H}{C}} \oplus}$$

(More female and full free)-***1-Butene.*** 2.11c

$$\underset{(D)}{\begin{array}{c} H\ \ H\ \ \ \ \ \ \ \ H \\ |X|\ \ \ \ \ \ X| \\ C=C-C=C \\ |\ \ \ \ \ \ \ \ |\ \ \ | \\ H\ \ \ \ \ \ \ H\ \ H \end{array}} \xrightleftharpoons[\text{Radically}]{\text{Free-}} \underset{\begin{array}{c}(Any\ Center)\\ \text{Resonance stabilized}\end{array}}{n\bullet \overset{H}{\underset{H}{C}} - \overset{H}{C} = \underset{H}{C} - \overset{H}{\underset{H}{C}} \bullet e\,;} \xrightleftharpoons{\text{Chargedly}} \underset{\begin{array}{c}(Any\ Center)\\ \text{Cannot be resonance stabilized}\end{array}}{\ominus \overset{H}{\underset{H}{C}} - \overset{H}{\underset{\oplus}{C}} - \underset{H}{C} = \overset{H}{\underset{H}{C}}}$$

(Less female and full free)- ***1, 3-butadiene.*** 2.12a

$$\begin{array}{c} H \quad H \qquad H \\ |\,X\,| \qquad X\,| \\ C=C-C=C \\ | \qquad | \quad | \\ H \qquad H \quad CH_3 \end{array} \underset{Radically}{\overset{Free-}{\rightleftharpoons}} \begin{array}{c} H \quad H \qquad H \\ | \quad | \qquad | \\ n\bullet C-C=C-C\bullet e \\ | \qquad | \quad | \\ H \qquad H \quad CH_3 \end{array} \overset{Chargedly}{\rightleftharpoons} \begin{array}{c} H \quad H \qquad H \\ | \quad | \qquad | \\ \ominus C-\underset{\oplus}{C}-C=C \\ | \qquad | \quad | \\ H \qquad H \quad CH_3 \end{array}$$

(E) — (Fixed Centers) Resonance stabilized — (Fixed Centers) Cannot be resonance stabilized

(More female and full free)- ***1, 3-methyl butadiene.*** 2.12b

$$\begin{array}{c} H \quad H \\ |\,X\,| \\ C=C \\ | \quad | \\ H \;\; :\ddot{C}l: \end{array} \underset{Radically}{\overset{Free-}{\rightleftharpoons}} \begin{array}{c} H \quad H \\ | \quad | \\ e\bullet C-C\bullet n\,; \\ | \quad | \\ H \;:\ddot{C}l: \end{array} \underset{CHARGEDLY}{\overset{IMPOSSIBLE}{\rightleftharpoons}} \begin{array}{c} H \quad H \\ | \quad | \\ \oplus C-C\ominus \\ | \quad | \\ H:\ddot{C}l: \end{array}$$

Electrostatic forces of repulsion

(F) — (Fixed Centers)

(Female and full free)- ***Vinyl chloride.*** 2.13

$$\begin{array}{c} H \quad H \\ |\,X\,| \\ C=C \\ | \quad | \\ H \;\; :\ddot{O}: \\ \quad |\,X \\ \quad C=O \\ \quad | \\ \quad CH_3 \end{array} \underset{Radically}{\overset{Free-}{\rightleftharpoons}} \begin{array}{c} H \quad H \\ |\,X\,| \\ e\bullet C-C\bullet n \\ | \quad | \\ H \;:\ddot{O}: \\ \quad |\,X \\ \quad C=O \\ \quad | \\ \quad CH_3 \end{array} \underset{CHARGEDLY}{\overset{IMPOSSIBLE}{\rightleftharpoons}} \begin{array}{c} H \quad H \\ |\,X\,| \\ \oplus C-C\ominus \\ | \quad | \\ H \;:\ddot{O}: \\ \quad |\,X \\ \quad C=O \\ \quad | \\ \quad CH_3 \end{array}$$

Electrostatic forces of Repulsion

(G) — (Fixed Centers)

(Female and full free)- ***Vinyl acetate.*** 2.14a

$$\begin{array}{c} H \quad H \\ |\,Y\,| \\ C=C \\ | \quad |\,X \\ H \quad C=O \\ \quad | \\ \quad O \\ \quad | \\ \quad CH_3 \end{array} \underset{Radically}{\overset{Free-}{\rightleftharpoons}} \begin{array}{c} H \quad H \\ |\,Y\,| \\ e\bullet C-C\bullet n\,; \\ | \quad |\,X \\ H \quad C=O \\ \quad | \\ \quad O \\ \quad | \\ \quad CH_3 \end{array} \overset{Chargedly}{\rightleftharpoons} \begin{array}{c} H \quad H \\ |\,Y\,| \\ \oplus C-C\ominus \\ | \quad |\,X \\ H \quad C=O \\ \quad | \\ \quad O \\ \quad | \\ \quad CH_3 \end{array}$$

(H) — (Fixed Centers) — (Fixed Centers)

(Male and full free)- ***Methyl acrylate.*** 2.14b

$$\underset{(I)}{\overset{CH_3}{\underset{H}{C}}=C=O} \;(Y,\ X) \xrightleftharpoons{Radically} n\bullet\overset{CH_3}{\underset{H}{C}}-\overset{O}{\overset{\|}{C}}\bullet e\;,\; e\bullet \overset{}{\underset{\underset{CH_3}{H-C}}{C}}-O\bullet nn \;;\; \xrightleftharpoons{Chargedly} \ominus\overset{CH_3}{\underset{H}{C}}-\overset{O}{\overset{\|}{C}}\oplus\;,\;\oplus\underset{\underset{CH_3}{H-C}}{C}-O\ominus$$

Male Center Female Center (Fixed Centers)   *Male Center Female Center (Fixed Centers)*

(Male, full free and half non-free)-***Methyl ketene.*** 2.15

(A), (B), and (C) are mono-olefins, female in character based on the routes which they undergo during polymerizations, a fact which is unknown in present-day Science. (C) is more female than (B) and (B) is in turn more female than (A), because C_2H_5 group is more radical-pushing than CH_3 group which in turn is more radical-pushing than H. Indeed, while (B) and (C) cannot be productive when attacked by female radical initiators represented as **N.n**, because of the presence of transfer species of the first kind of the first type (in this case H) on CH_3 and C_2H_5 groups, (A) can be productive when attacked by either male or female initiators (i.e. (A) is bisexual in character). Shown below is the case for propylene.

Why Nucleo-Free-radical Initiation of Propylene Is Impossible

$$\underset{\substack{Initiator\\(Female)}}{N\cdot n} + \underset{(B)}{\overset{H\ \ CH_3}{\underset{H\ \ H}{C=C}}} \xrightarrow{Activation} N\cdot n + \underset{Activated\ monomer}{e\cdot\overset{CH_3\ H}{\underset{H\ \ H}{C-C}}\cdot n} \longrightarrow NH + \underset{(B)^N}{n\cdot\overset{H\qquad H}{\underset{H\ \ H\ \ H}{C-C=C}}}$$

$$(B)^N + (B) \xrightarrow{activation} (B) + (B)^N \quad NO\ \ INITIATION$$

2.16

When $(B)^N$ is used in place of N•n, the same monomer is continuously obtained making its initiation impossible. $(B)^N$ will continue to exist in the system. This transfer species does not exist in ethylene.

When (A)-ethylene a female is polymerized with female initiators, it is well known to require the use of very harsh operating conditions (high temperatures and pressures), ***and only in the absence of free male initiators in the reactor or system or the small world during polymer growth***. Why this is so has long been unknown. Its symmetric structure is a very important factor during activation requiring more energy than when non-symmetric as will be vividly seen in the New Frontiers. When male initiators are used, these require very normal operating conditions just as almost applies to propene in its Vapor-Phase-Polymerization using Coordinated Initiators in

Fluidized-bed reactors[18], because ***the route is natural to the monomers.***

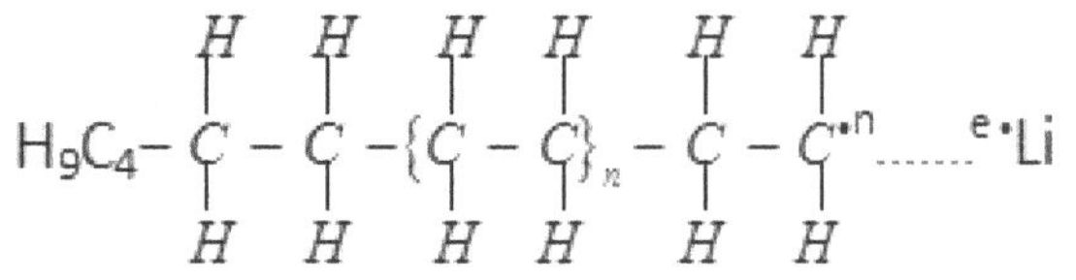

$$H_9C_4 - \underset{H}{\overset{H}{C}} - \underset{H}{\overset{H}{C}} - \{\underset{H}{\overset{H}{C}} - \underset{H}{\overset{H}{C}}\}_n - \underset{H}{\overset{H}{C}} - \underset{H}{\overset{H}{C}}^{\bullet n} \ldots\ldots {}^{e}\bullet Li$$

$$\text{Chain can be killed by using a foreign terminating agent.} \qquad 2.17a$$

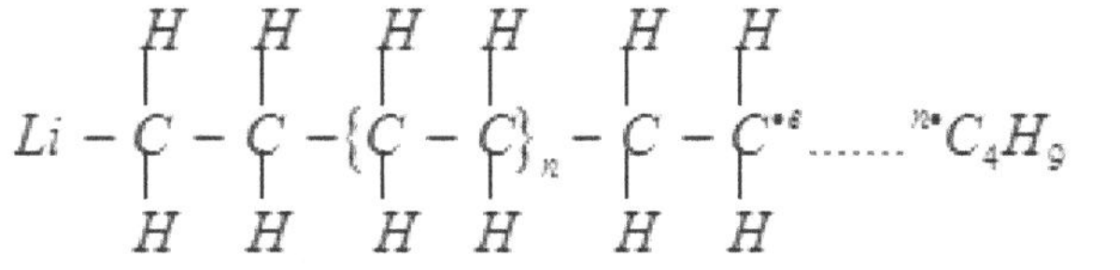

$$Li - \underset{H}{\overset{H}{C}} - \underset{H}{\overset{H}{C}} - \{\underset{H}{\overset{H}{C}} - \underset{H}{\overset{H}{C}}\}_n - \underset{H}{\overset{H}{C}} - \underset{H}{\overset{H}{C}}^{\bullet e} \ldots\ldots {}^{n}\bullet C_4H_9 \qquad \text{Impossible existence}$$

Paired centers cannot exist between two C centers. 2.17b

<u>Addition Polymerization of Ethylene using LiC_4H_9</u>

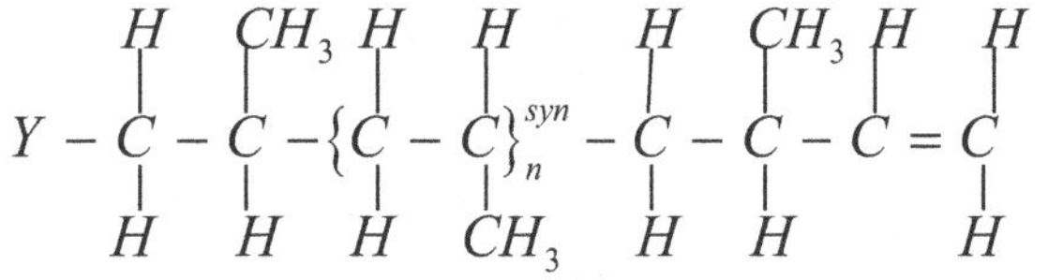

$$Y - \underset{H}{\overset{H}{C}} - \underset{H}{\overset{CH_3}{C}} - \{\underset{H}{\overset{H}{C}} - \underset{CH_3}{\overset{H}{C}}\}_n^{syn} - \underset{H}{\overset{H}{C}} - \underset{H}{\overset{CH_3}{C}} - \overset{H}{C} = \underset{H}{\overset{H}{C}}$$

$$Y \equiv \text{Positively charged part or Electro} - \text{free} - \text{radical part of}$$
$$\text{a cordination Initiator. Chain killed by what is called "Termination by Starvation"}$$

<u>Addition Polymerization of Propylene (Propene)</u> 2.17b

For the ethylene, shockingly enough, one has used the so-called Anionic ion-paired initiator, to show that the carrier of the chain is indeed C_4H_9 when it should be Li a male center, because the monomer is female in character. But. based on the second equation above, a coordination center cannot be formed between two Carbon centers. Since ethylene has no transfer species and since Nature abhors a vacuum, hence $H_9C_4^{\theta}$ is made to carry the chain for which higher operating conditions than normal will be required, noting that the monomer is symmetric and bisexual in character. To kill the chain whether the optimum chain length has been reached or not requires the use of a foreign terminating agent. If the chain is starved however, Li may close the chain. The coordination used above for propene is that with two vacant orbitals on the counter center where they are placed one at time as will shortly be shown using Paired-media types of initiators. When only one vacant is present, only isotactic placements will be obtained. ***Termination by starvation*** as will fully be seen in the New Frontiers for various applications was used to kill the chain. In ***Free-media systems***, only atactic placement can be obtained for propene since no *real or imaginary coordination* is present. Because of the irregular placement of CH_3 group, the polymer obtained is not crystalline, but amorphous in character. (C) is almost exactly like (B). Lesser operating conditions will be required for (C) than for (B), since (C) is more Female than (B).

The route where initiation was not favoured, is said to be ***unnatural route.*** Ethylene favours ***both natural and unnatural routes***, because of absence of transfer species. So also, is the case with vinyl chloride (F) where syndiotactic placements are obtained under mild operating conditions in the route not natural to the monomer, because vinyl chloride is far less Nucleophilic than ethylene. That is, vinyl chloride is closer to a male in character, in view of the presence of Chlorine in place of H on the active C center as will clearly be seen in the New Frontier with Halogenated hydrocarbons. How it favours syndiotactic placement (i.e., very regular placement) in the absence of vacant orbitals or coordination centers which is not there to be used in FREE-MEDIA systems of polymerization, will shortly be explained (Electrodynamic forces of repulsions). The imaginary coordination is electrodynamic in character coming from the Cl center. It is the regular placement that makes the polymeric product crystalline.

(D) and (E) are di-olefins with (E) being more female than (D). Like (A), (D) is symmetric and bisexual, since both routes are favoured by them. Unlike the mono-alkenes already considered (A), (B) and (C)- the Alkenes (D) and (E)- Di-alkenes are resonance stabilized only radically, because of the presence of two π-bonds conjugatedly placed. When activated, only one activation center can be activated one at a time as already said in the last Chapter either radically or chargedly, whether the two centers are of the same Nucleophilic capacity or not. Unlike what is presently known in present-day Science, Resonance stabilization cannot take place chargedly as shown in Equations 1.12a and 2.12b. They only take place radically, because charges cannot be removed from their carriers. Therefore, when activated chargedly, only the center activated can be used as shown in the equations. That is, only 1,2- or 3,4-mono-forms can be obtained during polymerization along the chain chargedly. In Equation 2.12a, the center activated is 3,4-mono-form. It is the same in Equation 2.12b. In Chloroprene, it is the 3,4-mono-form, while in isoprene, it is also the 3,4-mono-form as shown below. The two monomers are very important as we know them today. Isoprene in particular is the monomer in natural rubber. Polyisoprenes made synthetically are said to differ from natural rubber in that they contain a small amount of 3,4-mono-form along the chain, instead of all 1,4- mono-forms in natural rubber. Based on the new foundations, the presence of 3,4-mono-form can readily be prevented. The 3,4 mono-form can only appear at the beginning of the growing chain when the initiator is weak.

$$\begin{matrix} H_X Cl & & {}_X H \\ C^1 = C^2 - C^3 = C^4 \\ H & H & H \end{matrix} \xrightleftharpoons[Radically]{Free-} \begin{matrix} H \quad Cl \quad H \\ e\bullet C - C = C - C\bullet n \\ H \qquad H \quad H \end{matrix} ; \; \xrightleftharpoons{Chargedly} \begin{matrix} H \quad Cl \quad H \\ {}^{\oplus}C - \overset{\ominus}{C} = C - C \\ H \qquad H \quad H \end{matrix}$$

(A) Chloroprene — (Fixed Center) Resonance stabilized — (Fixed Center −1,2-) Electrostatic forces of repulsion

(More female and full free)- ***1, 3-Chloroprene.*** 2.18a

$$\begin{matrix} Cl_X H & & {}_X H \\ C^1 = C^2 - C^3 = C^4 \\ H & H & H \end{matrix} \xrightleftharpoons[Radically]{Free-} \begin{matrix} Cl \quad H \quad H \\ n\bullet C - C = C - C\bullet e \\ H \qquad H \quad H \end{matrix} ; \; \xrightleftharpoons{Chargedly} \begin{matrix} H \quad H \quad :\ddot{Cl}: \\ C = C - \overset{\oplus}{C} - C \ominus \\ H \qquad H \quad H \end{matrix}$$

Electrostatic forces of repulsion

(A) 1−Chloro−1,3−Butadiene — (Fixed Center) Resonance stabilized — (Fixed Center −1,2−) Cannot be activated chargedly

(Less female and full free) 1-Chloro-1,3-Butadiene 2.18b

$$\begin{matrix} H_X CH_3 & & {}_X H \\ C^1 = C^2 - C^3 = C^4 \\ H & H & H \end{matrix} \xrightleftharpoons[Radically]{Free-} \begin{matrix} H \quad CH_3 \quad H \\ n\bullet C - C = C - C\bullet e \\ H \qquad H \quad H \end{matrix} \xrightleftharpoons{Chargedly} \begin{matrix} H \quad CH_3 \quad H \\ C = C - \overset{\ominus}{C} - C\oplus \\ H \qquad H \quad H \end{matrix}$$

(B) Isoprene — (Fixed Centers) Resonance stabilized — (Fixed Centers −3,4−) Cannot be resonance stabilized

(Less female and full free)- ***1, 3-Isoprene*** 2.19

Vinyl chloride cannot be activated chargedly. So also, chloroprene cannot be activated chargedly, because of electrostatic forces of repulsion. The less nucleophilic center is the center carrying Cl a radical-pulling internally located, unlike the case of 1-chloro-1,3-butadiene also shown above in Equation 2.18b. Also for this case, the less nucleophilic center is the center carrying Cl externally located. This monomer is less nucleophilic than chloroprene, and this in turn is less female than isoprene which has transfer species that is shielded from any attack because the CH_3 is internally located unlike the case of pentadiene (E) of Equation 2.12b where the CH_3 group is externally located. p-chloro-styrene unlike all the cases above, can be activated chargedly, because the Cl atom is distantly placed to the active center carrying the negative charge. One can clearly see the heavy use of the EYE OF THE NEEDLE here. The current nomenclature for numbering, that which is in order, has been used all above, but slightly differently. Though very little has been revealed about Resonance stabilization phenomena based on what is contained in the New Frontier, one can observe that universally, little or nothing is known about these phenomena. Yet, they (The Elites in this discipline) will claim that they know!!! This is the world in which we live in. These phenomena take place all the time in our environment- during Thunder and Lightning (Heavy current), Electrical Engineering systems, Communication between human beings, Communication between systems and so on. Yet, we including the author think that we know.

The 1,4-mono-form can never be obtained chargedly as wrongly represented in present-day Science, unless from a ringed compound. The I,4-mono-form can only be obtained free-radically. On the other hand, in addition to the syndiotactic and isotactic placements obtained for mono-Alkenes, are Cis- and Trans- placements for Dienes. This is not new, but different. Sulphur dioxide a female is non-symmetric but bisexual. All these contain X centers, the female center. Free-radical initiators cannot be used for it, because the equation will *not be radically balanced in Combination mechanism.* The initiators that can be used when polymerized alone are non-free-radical generating initiators, since when sulfur dioxide is activated only radically, the active centers carry non-free-radicals as already shown in Equation 2.9. Since it is a Nucleophile (Female), the natural route is electro-non-free-radical route as shown below.

$$Fe\bullet en + n(\overset{\overset{O\odot}{|}}{\underset{\cdot\cdot}{S^{\oplus}}} = O)_m \xrightarrow[\text{Mechanism}]{\text{Combination}} Fe-(O-\overset{\overset{O\odot}{|}}{\underset{\cdot\cdot}{S^{\oplus}}})^{syn}{}_{m-1} - O - \overset{\overset{O\odot}{|}}{\underset{\cdot\cdot}{S^{\oplus}}}\bullet en$$

Addition Polymerization of Sulfur dioxide 2.20a

$$Na\bullet e + n(\overset{\overset{O}{\|}}{C} = O)_m \xrightarrow[\text{Mechanism}]{\text{Combination}} Na-(O-\overset{\overset{O}{\|}}{C})^{syn}{}_{m-1} - O - \overset{\overset{O}{\|}}{C}\bullet e$$

Addition Polymerization of Carbon dioxide 2.20b

As will shortly be explained, the oxygen externally located can only be placed syndiotactically (syn) as shown above, due to what is called ***electro-dynamic forces of repulsion.*** Nucleo-non-free-radically, it can be poly-merized, but under higher operating conditions, because the route is not natural to it. To kill the chain, unlike the case of propylene where termination by starvation was used as shown in Equation 2.17b, here this cannot be done, since there is no transfer species. External agents must be used to kill the chain. Included above, for the purpose of showing that CO_2 is a potential monomer and the fact that the choice of Initiator is the most important step in Polymerization systems, is the polymerization of CO_2 using NaCN as initiator generating source. The choice of Sodium (which is natural to the monomer a Female) as a carrier for the chain is not good, because the chain is still living from that end. External terminating agents must be used here too. It should not be that carrying OH group, since the chain will also be living from that end. The H should be removed and replaced with an R group, e.g., CH_3. So also, is the Na. If we use Fe for CO_2 and Na for SO_2, initiation will never be favoured, via Combination mechanism, because of radical balancing.

Why (F) vinyl chloride and (G) vinyl acetate cannot be polymerized chargedly, has been shown above in Equations 2.13 and 2.14a. This was also shown for chloroprene and 1-Chloro-1,3-butadiene in Equations 2.18a and 2.18b. Cl and O in (F) and (G) are next-door neighbours to the C center. Because of the presence of paired unbonded radicals on the Cl and O centers, the negative charge cannot stay, due to *electrostatic forces of repulsion*. According to universal data, Cl, $COOCH_3$, and many others are "radical-pulling" groups, while CH_3, C_2H_5, and many others are "radical-pushing" groups. All these are different from so-called **electron withdrawing and donating groups**[19] which proper names are radical withdrawing and radical donating compounds because they exist and are molecules, but not groups. They are different from radical-pulling and radical-pushing groups. As it seems however, they are used synonymously, thereby creating a state of confusion.

(G) and (H) of Equation 2.14a and 2.14b are isomers. Yet, while (G) is female carrying two X activation centers isolatedly placed of different capacities, (H) is male carrying male (Y) and female (X) activation centers adjacently located to themselves also of different capacities. The C = O center is far more nucleophilic than C = C center. Hence the C = C center is always the first to be activated all the time. If the activator is very strong, activation of C = O may in addition be possible, since (H) cannot be resonance stabilized as will be shown in the New Frontiers. The same new concepts above apply to the chromosomes of living systems as established by their true structures today. These are clearly shown in Figure 2.3 for an example of male hormones with X and Y centers on the male part and only X centers on its female part; and for the female hormones with only X centers on all of them.

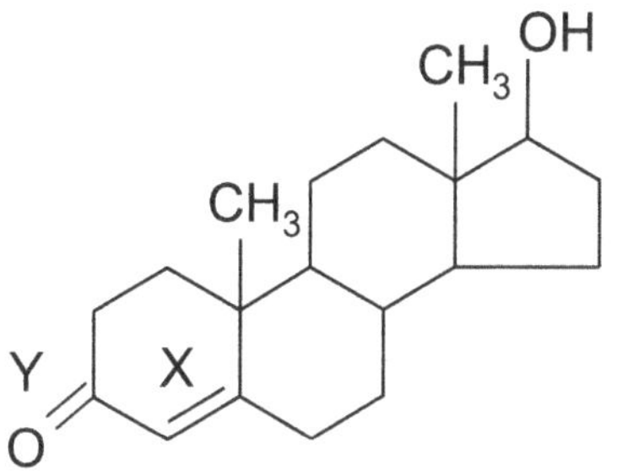

Testosterone (Y) 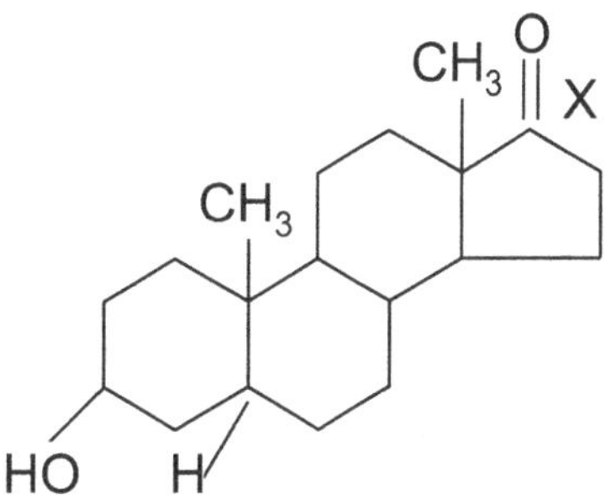 Androsterone (X)

Some Androgenes
(Male hormones)

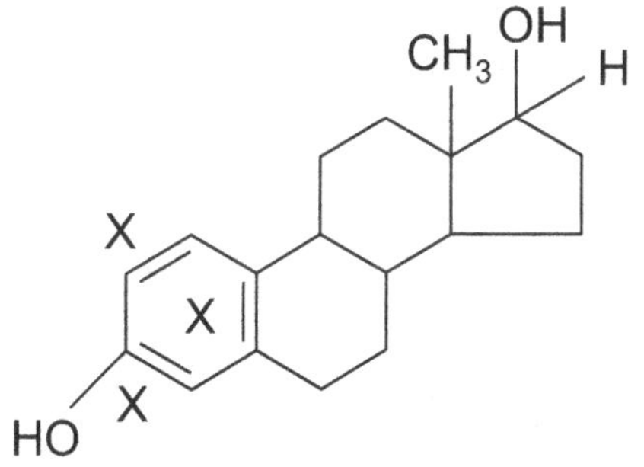

Estradiol (X) 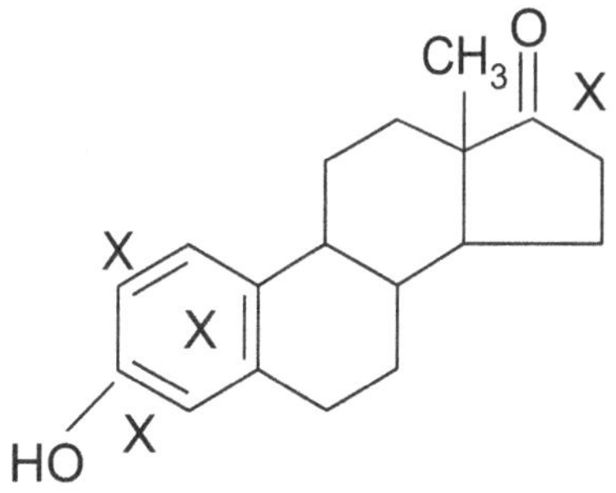 Estrone (X)

Some Estrogenes
(Female hormones)

Figure 2.3 Activation centers in male and female hormones.

All the Xs in the female hormones have different capacities. When a female radical is around, it activates the Y center, while when a male radical is around, it activates the X center. More will be said about it downstream, since it is very important in our lives. While present-day Science did a tremendous job to show that Males carry both male and female sexes and Females carry only female sexes, the Science was an ART, because what the male and female sexes are, were not known. How they knew it must have been via well-organized laboratory data. Without their knowing it, what is contained in the New Frontiers would not have been possible.

(F), (G), and (H) of Equations 2.13, 2.14a and 2.14b respectively are members of family of monomer called Olefins. (G) is vinyl acetate, while (H) is methyl acrylate. (I) a ketene is male with X and Y centers cumulatively placed. There are indeed many male compounds such as acrylonitrile, acrylamide, and acrolein.

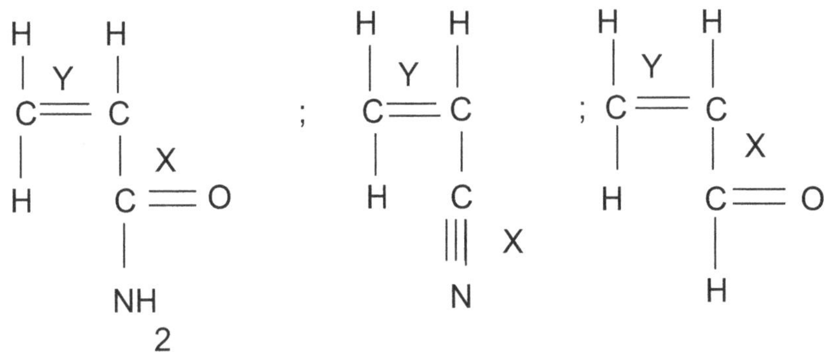

```
H    H             H    H            H    H
| Y  |             | Y  |            | Y  |
C == C          ;  C == C         ;  C == C
|    | X           |    |            |    | X
H    C == O        H    C            H    C == O
     |                  |||  X            |
     NH2                N                 H
```

Acrylamide ; Acrylonitrile ; Acrolein 2.21

Activation centers shown so far are some of the ***visible ones***. In all the Electrophiles or Males seen so far excluding Benzoquinone shown in Chapter 1 and Male chromosomes, their X centers which is C = O. is far more nucleophilic than their Y centers. So also, are C≡N, C=NR, etc., centers. This is not the general rule, since as has been said, there are different kinds of Males of different types. So also, are the Females.

Functional center the third kind of activation centers with different types, is unique in character and is carried by some *Ringed Addition monomers* as shown below. These are no Functional groups as is believed in present-day Science where they think in general that these ringed monomers undergo Step polymerization. Functional centers are different from Functional groups carried by what are called Step monomers (e.g. COOH groups in adipic acid- $HOOC(CH_2)_4COOH$, OH groups in glycerol-$HOCH_2CH(OH)CH_2OH$). Functional center are not present in the human chromosomes. In its place are C=C activation centers, just like benzoquinone. Shown below are some ringed monomers or compounds. Some are carrying function centers such as (A), (B), (C) and (F), and some are not such as (D) and (E). While (A) has two functional centers, (B), (C) and (F) have only one functional center. All the functional centers are of different types, but female or nucleophilic in character.

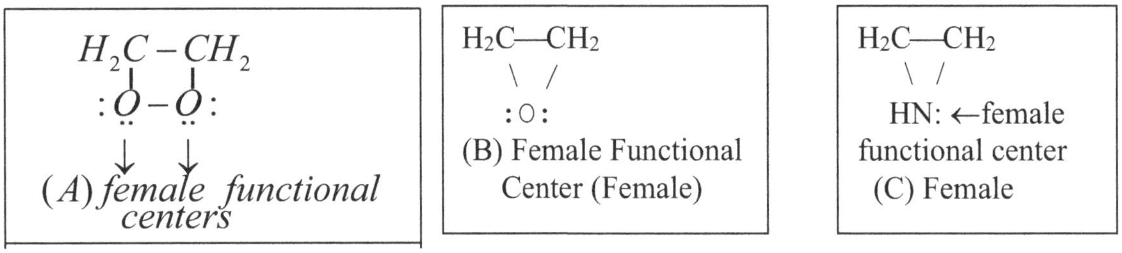

H_2C-CH_2 $:\ddot{O}-\ddot{O}:$ ↓ ↓ (A) female functional centers	$H_2C—CH_2$ \ / :O: (B) Female Functional Center (Female)	$H_2C—CH_2$ \ / HN: ←female functional center (C) Female

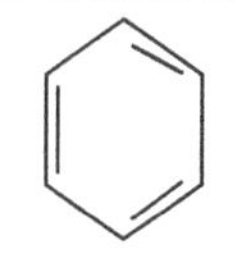

(D) Female with 3 Activation centers (No Functional Center)

$H_2C—CH_2$
\ /
CH_2
No Functional center
(E) Female

$H_2C—CH_2$
| |
H_2C C
\ /\\Y Male Activation center
X →: O: :O:
Female Functional center
(F) Male

2.22

The functional centers are marked by ***the presence of paired unbonded radicals*** on centers such as O, N, S, P and more. Rings are like Springs in Physics.[1, 20, 21] Some rings can very readily be opened if they have points of scission or have conjugatedly placed double bonds inside the ring such as in Cyclooctatetraene and benzene as will be partly shown herein and fully in the New Frontiers. In the former where points of scission exist, the ***Minimum Required Strain Energy (MRSE)*** must be provided for them, to use to unzip the ring. In benzene and octatetraene, where no point of scission exists, the MRSE already exists in such well strained rings. Their method of opening is completely different from all the others. Special initiators are required for them. For the others such as (A), (B), (C), (E), and (F) above, MRSE can be provided for some of them depending on the size of the ring. For example, while it can be provided for three-membered (E), four-membered, five-membered cycloalkanes which have no functional center, it cannot be provided for cyclohexane, cycloheptane and so on, because of the size of the rings. Unlike what is believed today in present-day Science, it is impossible to have a ring that is strainless (i.e. SE equals zero), otherwise the ring will not exist as a ring. For example, shown below are the linear isomers of some ringed compounds.

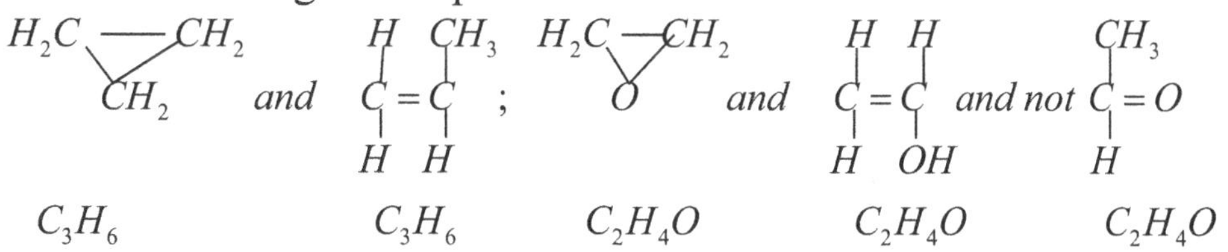

$H_2C—CH_2$ \/ CH_2 *and* H CH_3 | | C=C | | H H ; $H_2C—CH_2$ \/ O *and* H H | | C=C | | H OH *and not* CH_3 | C=O | H

C_3H_6 C_3H_6 C_2H_4O C_2H_4O C_2H_4O

2.23a

(benzene ring) and H H H | | | C = C – C ≡ C – C = C | | | H H H

C_6H_6 C_6H_6

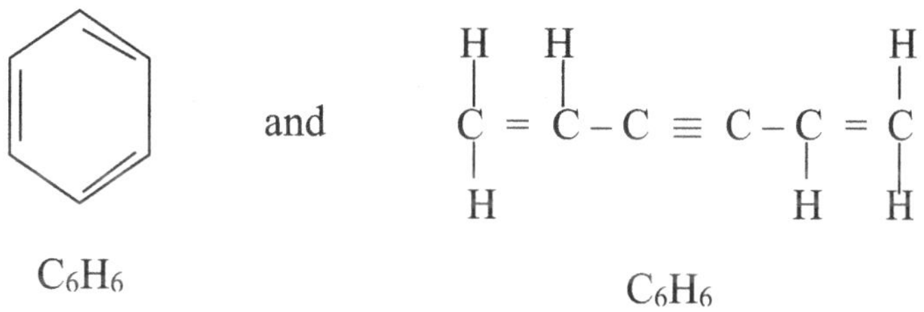

2.23b

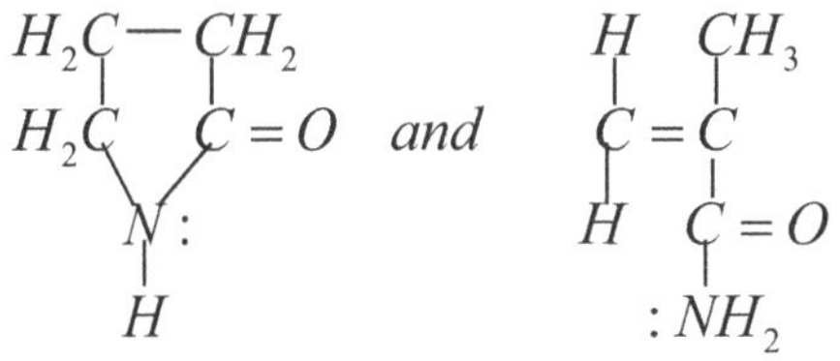

$$\begin{array}{ccc} \begin{matrix} H_2C - CH_2 \\ | \quad\quad | \\ H_2C \quad C=O \\ \diagdown \diagup \\ N: \\ | \\ H \end{matrix} & \text{and} & \begin{matrix} H \quad CH_3 \\ | \quad\quad | \\ C = C \\ | \quad\quad | \\ H \quad C=O \\ | \\ :NH_2 \end{matrix} \\ C_4H_7NO & & C_4H_7NO \end{array} \qquad 2.23c$$

For cyclopropane, the linear isomer is propene. Propene has a π-bond which cannot be seen in cyclopropane. But, it is there. It is the SE which when compared with the propene, is an invisible π-bond and these bonds carry stored energy. In the epoxide (ethylene oxide), the linear isomer is the vinyl alcohol which contains a visible π-bond not seen in the ring- the invisible π-bond. The acetaldehyde is not the linear isomer, because the invisible π-bond is not a C = O bond. In Benzene, the linear isomer is a triene with a triple bond at the center. One of the π-bonds in the triple bond is the invisible π-bond in the ring. Hence, one can see that the strain energy in any ring can never be zero. In the five-membered Cyclic amide above the linear isomer is methyl acrylamide. For the four-membered ring, the linear isomer is acrylamide. One can observe vividly here that the SE in the ring is a C =C type of center which is Y in the linear isomer but X in the ring, in addition to the visible X center on N- or on O- the paired unbonded radicals. One can also see that the SE for all cyclo-alkane rings regardless the size of the ring, has a fixed value. Each ring independent of size shares this fixed energy to exist. If the SE for a family such as cycloalkanes is 25kcals/mole, that is the energy carried by any of the rings, independent of size. The same applies to all families, noting that there are some families where only one size exists. For example, eight-membered cyclic sulfur containing ring (Rhombic sulfur) is the only one in its family, different from seven- or six- or five- membered cyclic sulfur containing ring. Each one of them, is the only member of its family, because as we move from one size to another size, the number of paired unbonded radicals on the S center has increased. This is unlike the case of moving from cyclopropane to cyclobutane or moving from four-membered cyclic amide to five-membered cyclic amide. Presence of π-bonds, paired unbonded radicals, invisible π-bonds are very important in ringed compounds, since these carry large amount of energy. Then how can a ring with the size of fifty be held to exist as a ring? In our world, we keep wasting TIME and MONEY, without using THE EYE OF THE NEEDLE, making stupid conclusions such as the fact that a ring like cyclohexane has zero strain energy. Using the mathematical language, one can observe that the SE above can easily be calculated, using the Bond energies of all the bonds, taking into cognizance the types of groups carried by the bonds. The SE is just the difference between the total energy of all the bonds in the ring and the total energy of all the bonds in the linear isomer. We have just only begun a journey which is endless. We thought we had already gone far. Now is time to turn the wheel backwards, dismantle all what have been done and start rebuilding; this time with ease and no pains and most importantly in Stages using all the three mechanisms since we are now beginning to see our States of existences. Obviously, Equilibrium is not one of them.

These strain energies which are ***Invisible π-bonds*** inside the ring are all Nucleophilic (FEMALE) in character. Never is there a time anyone of them is Electrophilic in character. (B) and (C) of Equation 2.22 (Ethylene oxide and Ethyleneimine) have two points of scissions those between C and O or N centers respectively and can readily be opened instantaneously or via the functional center positively or electro-free-radically, since that center (X) is nucleophilic. All these will depend on the operating conditions. If (B) for example was propylene oxide, there will only be one point of scission and only one route will be favoured by it, unlike the case of ethylene oxide which has no transfer species.

$$\ddot{N}\bullet nn + \underset{(I)}{\overset{H_2C—CH_2}{O}} \xrightarrow[\text{Opening of ring(Heat}]{\text{Ins tan tan eous}} \ddot{N}\bullet nn + nn\bullet O-\underset{H}{\overset{H}{C}}-\underset{H}{\overset{H}{C}}\bullet e \xrightarrow[\text{Mechanism}]{\text{Combination}} \ddot{N}-\underset{H}{\overset{H}{C}}-\underset{H}{\overset{H}{C}}-O\bullet nn$$

Unnatural route (*Favored*) (*II*)

$$E\bullet e + (I) \xrightarrow[\text{Center}]{\text{Via the Functional}} E\bullet e + (II) \xrightarrow[\text{Mechanism}]{\text{Combination}} E-O-\underset{H}{\overset{H}{C}}-\underset{H}{\overset{H}{C}}\bullet e$$

Natural route 2.24

$$\ddot{N}\bullet nn + \underset{(I)}{\overset{H_2C—CH(CH_3)}{O}} \xrightarrow[\text{Opening of ring(Heat}]{\text{Ins tan tan eous}} \ddot{N}\bullet nn + nn\bullet O-\underset{H}{\overset{H}{C}}-\underset{H}{\overset{CH_3}{C}}\bullet e \xrightarrow[\text{Mechanism}]{\text{Combination}} \ddot{N}H + \underset{H}{\overset{H}{C}}=\underset{H}{C}-\underset{H}{\overset{H}{C}}-O\bullet nn$$

Unnatural route (*II*) *Initiation not favored*

$$E\bullet e + (I) \xrightarrow[\text{Center}]{\text{Via the Functional}} E\bullet e + (II) \xrightarrow[\text{Mechanism}]{\text{Combination}} E-O-\underset{H}{\overset{H}{C}}-\underset{H}{\overset{CH_3}{C}}\bullet e$$

Natural route

$$\xrightarrow{+n(I)} E-(O-\underset{H}{\overset{H}{C}}-\underset{H}{\overset{CH_3}{C}})_n-O-\underset{H}{\overset{H}{C}}-\underset{H}{C}=\underset{H}{\overset{H}{C}} + H\bullet e \qquad 2.25$$

(*Termination by Starvation*)

While ethylene oxide favours both nucleo-non-free-radical and electro-free-radical routes, propylene oxide favours only the electro-free-radical route, propylene oxide being more nucleophilic than ethylene oxide. Propylene oxide has transfer species of the first kind of the first type, while ethylene oxide has none. In the route natural to them, both being Nucleophiles (Females), the ring was opened via the functional center carried by O-the paired unbonded radicals (two of them on oxygen and one on nitrogen). The growing polymer chain of propylene oxide, was killed by termination by Starvation to give dead terminal double bond polymers, just like the case of propylene. The transfer species released was the same transfer species that prevented the monomer during initiation in the route not natural to it. If a nucleo-free-radical had been used for ethylene oxide, the initiation step will not be favoured, because the equation will not be radically balanced, ***unless under Equilibrium conditions.*** If the same initiator was used on propylene oxide, initiation step will not be favoured for almost the same reason as shown below.

$$N\bullet n + \underset{(I)}{H_2C\overset{O}{-}CH(CH_3)} \xrightarrow[\text{Opening of ring (Heat}]{\text{Instantaneous}} N\bullet n + \underset{(II)}{nn\bullet O-\underset{H}{\overset{H}{C}}-\underset{H}{\overset{CH_3}{C}}\bullet e} \underset{\text{Mechanism}}{\overset{\text{Equilibrium}}{\rightleftharpoons}} NH + \overset{H}{C}=\underset{H}{C}-\underset{H}{\overset{H}{C}}-O\bullet nn$$

Unnatural route *Initiation not favored*

2.26

$$\ddot{N}\bullet nn + \underset{H}{\overset{CH_3}{C}}=\underset{H}{\overset{H}{C}} \underset{\text{Mechanism}}{\overset{\text{Equilibrium}}{\rightleftharpoons}} \ddot{N}\bullet nn + e\bullet\underset{H}{\overset{CH_3}{C}}-\underset{H}{\overset{H}{C}}\bullet n \underset{\text{Mechanism}}{\overset{\text{Equilibrium}}{\rightleftharpoons}} \ddot{N}H + n\bullet\underset{H}{\overset{H}{C}}-\underset{H}{C}=\underset{H}{\overset{H}{C}}$$

2.27

Under Equilibrium mechanism, the reaction is favoured as a step, without a productive stage being obtained. Hence, Propagation cannot take place via

Equilibrium mechanism, since no productive stage can be obtained. Invariably, there will be no productive stage equilibrium-wise and combination-wise. This is unlike the case shown below.

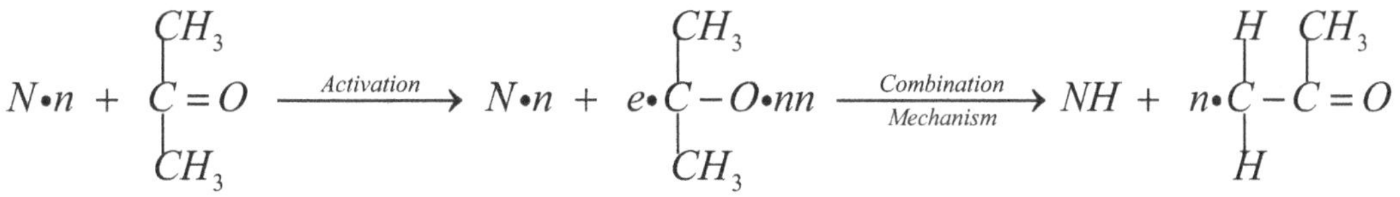

$$N{\bullet}n + \underset{CH_3}{\overset{CH_3}{C}}=O \xrightarrow{Activation} N{\bullet}n + e{\bullet}\underset{CH_3}{\overset{CH_3}{C}}-O{\bullet}nn \xrightarrow[Mechanism]{Combination} NH + n{\bullet}\underset{H}{\overset{H}{C}}-\overset{CH_3}{C}=O$$

$$\text{Re}\,active\ and\ Favoured - No\ Initiation$$

2.28

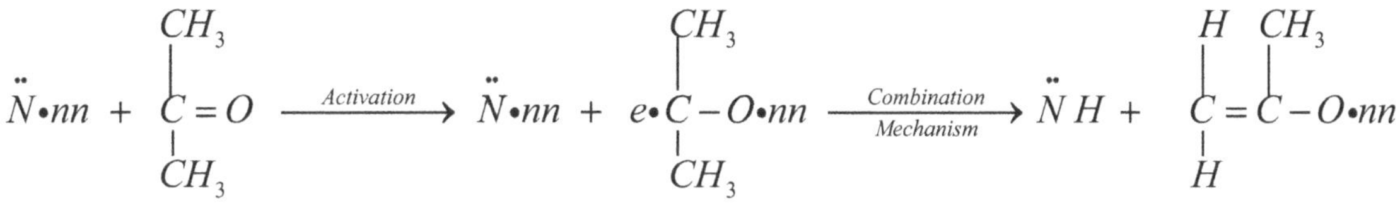

$$\ddot{N}{\bullet}nn + \underset{CH_3}{\overset{CH_3}{C}}=O \xrightarrow{Activation} \ddot{N}{\bullet}nn + e{\bullet}\underset{CH_3}{\overset{CH_3}{C}}-O{\bullet}nn \xrightarrow[Mechanism]{Combination} \ddot{N}H + \underset{H}{\overset{H}{C}}=\overset{CH_3}{C}-O{\bullet}nn$$

$$\text{Re}\,active\ and\ Favoured - No\ Initiation$$

2.29

Unlike the case above where no productive reaction took place, here, productive reaction took place to give another N•n or :N•nn and the same monomer will remain produced, clear indication that Initiation can never be favoured for all of them. It is remarkable to note how NATURE operates. One cannot force a monomer to favour the route which is not natural to it, when it has unshielded transfer species. The unshielded transfer species is like a sacrifice used to prevent what is not natural to be a part of the monomer. (A) of Equation 2.22 which may not exist because of the presence of Max.RSE in the ring, has only one point of scission, that between the two oxygen centers and can more readily be opened instantaneously because the O-O bond has the weakest energy and the size of the ring is small. It does not exist because the following is obtained in its place.

Stage 1:

$$\underset{(A)}{\begin{matrix} H_2C-CH_2 \\ | \quad\quad | \\ O+O \end{matrix}} \xrightleftharpoons[\text{Opening of Ring}]{\text{Instantaneous}} \underset{(B)}{en\bullet O-\overset{H}{\underset{H}{C}}-\overset{H}{\underset{H}{C}}-O\bullet nn}$$

$$(B) \rightleftharpoons H_2C=O + \underset{(C)}{e\bullet \overset{H}{\underset{H}{C}}-O\bullet nn} + Energy$$

$$(C) \xrightarrow[\text{Release of Energy}]{\text{Deactivation}} H_2C=O + Energy$$

$$\underline{Overall\ equation}: (C) \longrightarrow 2H_2C=O + Energy(2) \qquad 2.30$$

It will seem that formaldehydes are its linear isomer. It cannot be, because the invisible π-bond is not C=O, but C=C. Hence, the linear isomer is (HO)HC=CH(OH). When O is replaced with S, its existence is readily favoured, because, apart from the fact that the O center in the ring is more nucleophilic than the S center in the ring, S and C are equi-electropositive (Electronegativity of S = C = I =2.5). C, S and iodine (I) are the three atoms that form transition between metals and non-metals in the Periodic Table.

(E) of Equation 2.22 has three points of scission and can only be opened instantaneously because it has no functional or π-bond activation centers. It can easily be provided with the MRSE, being a three-membered ring. When opened, it will favour both nucleo-free-radical and electro-free-radical routes. As a nucleophile (Female), its natural route is either positively charged or electro-free-radical routes. When one H atom is changed to CH_3, the cyclopropane becomes more nucleophilic, now to favour only one route, like the others considered so far. In general, a point of scission in a ring as can be observed so far, in the absence of two hetero atoms placed side by side such as O-O, S-S, etc., is the point with ***the largest radical-pushing potential difference*** such as shown in Equations 2.25 and 2.26 for propylene oxide.

(F) of Equation 2.22 unlike others, is an Electrophile (Male) with X and Y centers and has only one point of scission, that between O and C=O centers. This is another kind of Male, unlike the ones encountered so far. Here the C=O which has always been X, is now the Y for this ring, still adjacently located to the X center which is now a functional center instead of a π-bond type of Activation center (C=C) seen so far. This is almost like the Male chromosomes in humans, but with a difference. The Male chromosomes and benzoquinones do not have functional centers. It may be possible that some other living species may have this kind. With this kind of Male, the C=O center still remains more nucleophilic than the functional center. The C=O center favours only the use of anionic (Not negatively charged center) and nucleo-non-free-radical generating initiators, being the Y center. The X center favours only the use of positively charged or electro-free-radical generating initiators. No matter what types of groups are carried by these monomers, no transfer species exist for this kind of monomers, unlike the case of Male chromosomes and benzoquinones. Benzoquinones have no transfer species, unless when the H atom is changed to alkylane groups, and the same too applies to the Male chromosomes. Hence the Y center in Benzoquinones is not different from those of Male chromosomes. When an initiator, such as LiC_4H_9 which is a Covalently charged-paired initiator with two active centers (in which only one can be used), is used on (F), no initiation is favoured anionically, because, the negative charge carried by one of the centers is not anionic as shown below.

$$H_9C_4^{\ominus}\ldots\ldots\ldots{}^{\oplus}Li \qquad \text{versus} \qquad RH_2CO^{\ominus}\ldots\ldots\ldots{}^{\oplus}Na$$

(A) Covalently Charged-paired Initiator (B) Ionically Charged-paired Initiator 2.31

In present-day Science, (A) is universally very wrongly called "ANIONIC-ION-PAIRED INITIATOR" as already said. Under such state of confusion which is in all we do, how can we move forward? This is why one said that the wheel of Knowledge must be turned backwards to correct all **the ills** we have done in the past up to the present moment, if we are to move forward. In (A) and (B), the two centers are active, but only one can be used, because the cationic centers cannot be used due to charge balancing. Now the monomer in question is a Male, wherein the negative center is ANIONIC, i.e. polar in character and the route natural to it is anionic. This is carried by (B) and not by (A). In (A), the negative center is non-polar. Therefore, the negative charge is not anionic. If (A) is used on the monomer anionically, Initiation will not take place, because the equation will not be CHARGEDLY BALANCED via Combination mechanism. It will only be chargedly balanced when (B) is used. Notice that for both (A) and (B), the positive centers are CATIONIC, since Li and Na are IONIC METALS and not Non-ionic metals such as Cu, Zn, Ti, Fe and so on. Cu, Zn, Fe are polar metals while Ti is a non-polar metal. All ionic metals are non-polar metals. Also, notice that the two centers on Li and Na have at least two vacant orbitals, reservoirs for only the Male monomers to reside in in their activated and oriented states of existence. The females do not have reservoir(s) on C and O centers to reside in. The active centers seem to communicate between themselves when a visitor is around. One can see how the so-called simple things we neglect (vacant orbitals, paired unbonded radicals (not paired unbonded electrons), types of charges, electrostatic and electrodynamic forces of repulsion and attraction, ... too countless to list, play major roles in all things in humanity. The bond carried by (A) is covalent and elastic in character, since it is between a metal and a non-metal which are both non-polar. The bond carried by (B) is ionic and also elastic in character. ***In present-day Science, nothing is known about RADICAL AND CHARGE BALANCING.*** Only STOICHIOMETRIC balancing is known for chemical balancing. For the first time, one is beginning to see very important concepts we neglect. It is no surprise why they are neglected, because when reaction take place via EQUILIBRIUM mechanism, only Stoichiometric balancing applies. But unknown is that when reactions take place via COMBINATION mechanisms, Stoichiometric, Radical, and Charge balancing must be applied. For Decomposition mechanism, Charge balancing does not arise, since charges cannot be isolatedly placed. Radical balancing does apply in the overall equation when radicals are to be formed as one the products. ***For example, based on the Decomposition states of existences of some unique symmetric compounds (e.g., Peroxides), radicals must be formed.*** Also for non-symmetric compounds and metallic reactions such as Alkylation, Radical balancing via Decomposition mechanism applies. Stoichiometric balancing applies to all of them.

For the (F) as with others, either the active center for the Male monomer go to the right center to activate it and add to it or the monomer is opened instantaneously due to presence of strong electrostatic forces provided by the charges and paired unbonded radicals. With the use of (B) which has additional paired unbonded radicals on the O center, the ring is opened instantaneously as shown below.

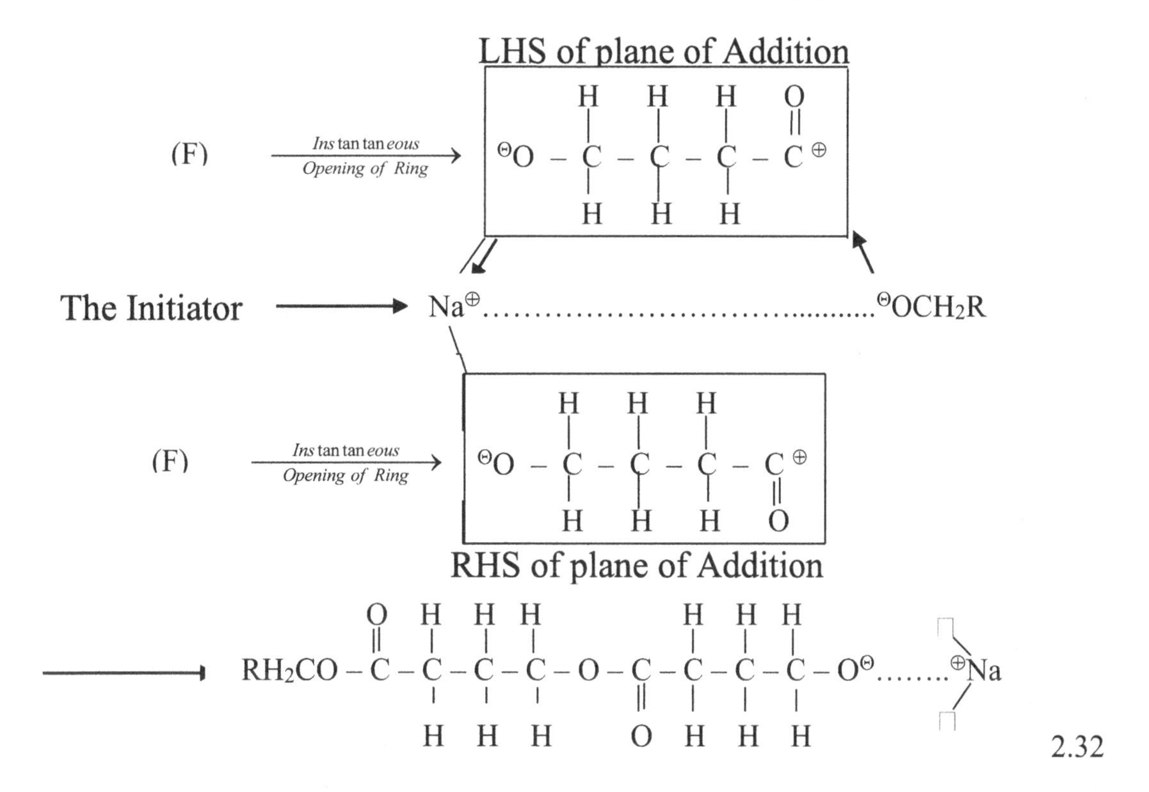

Notice the great significance of DIFFUSION CONTROLLED mechanism. They are done by Electrostatic and Electrodynamic forces of repulsion and attraction. Both Initiation step and addition of one monomer have been shown above for the purpose of showing the syndiotactic placement of the C=O group along the chain in view of the presence of two reservoirs for the monomers on the Na center. Worthy of note is that the monomer cannot be polymerized cationically, but only electro-free-radically. The centers can be paired radically. Thus, only the anionic route is found to be involved above, because the monomer is a MALE and NATURE does not operate indiscriminately or in an anyhow manner. Invariably, these chemical species carrying charges or radicals communicate. They are carrying things which they use to SEE, FEEL, HEAR, SMELL AND TASTE imaginarily. Cationically if it had been possible, (A) above cannot be used for this monomer, being an Electrophile (a Male). For a Female monomer such as ethylene oxide, the cationic center cannot be used, because of charge balancing. It can only be used electro-free-radically if the centers are paired radically. For this female monomer, the lithium center will only be the carrier of the chain radically. It took a long time during the development of the New Frontiers to find that no monomer can be polymerized CATIONICALLY, but only with positive charges and electro-radically. It took time to note that the (A) and (B) above can also be paired radically. It took time, because of what has been known to be the "established Science" universally.

The initiator above based on the charged and non-charged environment in the system, can be made to carry radicals and remain paired as shown below. for reasons shortly to be given.

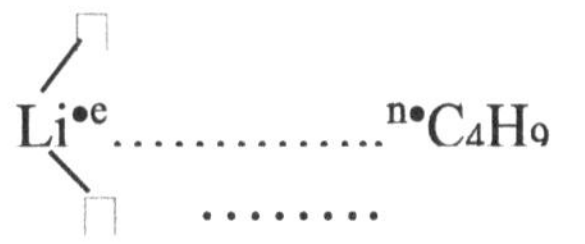

$$Li^{\bullet e}..............{}^{n\bullet}C_4H_9 \quad ; \quad Na^{\bullet e}..............{}^{nn\bullet}OCH_2R$$

(C) Covalently Radical-paired Initiator 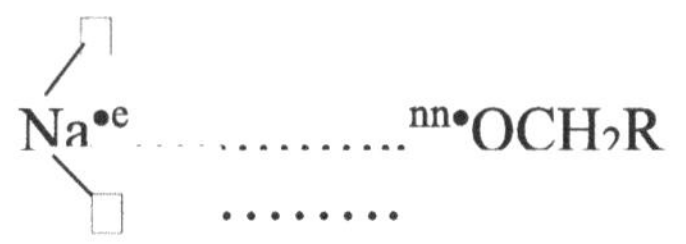 (D) Covalently Radical-paired Initiator

2.33

In a charged environment, the initiators shown in Equation 2.31 are the one, while in non-charged environment, what is shown above are formed. It should be noted however, that though before pairing can take place between two centers, charges must first be formed, because that is when forces of repulsion and attraction come into play; pairing can still place radically between two "specific" centers. The word "specific" is with respect to the electropositivity and electronegativity of the centers, and the polarity of the centers. ***When the two centers carry paired unbonded radicals i.e., both centers are polar, pairing radically cannot take place. When the carriers of both centers are of almost the same electronegativity or electropositivity and are non-metallic, pairing radically cannot also take place.*** There are still more additional driving forces required for pairing between two radical centers as will be shown downstream in the New Frontiers. (A) of Equation 2.31 cannot be used positively and anionically. (C) cannot be used nucleo-non-free-radically, because the equation will not be radically balanced. Electro-free and nucleo-free-radically (C) can be used. (D) will polymerize the Lactone nucleo-non-free-radically, because that is the route natural to it, but not (C). Electro-free-radically, (D) cannot be used, since the monomer is a Male and since the nucleo-non-free-radical center can be used, the route being natural to the monomer. Electro-free-radically, (C) can be used, because Nature abhors vacuum. For these families of monomers (Lactones, Lactams, Cyclic anhydrides, etc.) the same monomer unit must be obtained for them anionically or nucleo-non-free-radically, and positively or electro-free-radically. This is unlike the Male chromosomes where the anionic, or negatively charged or nucleo-free-radical or nucleo-non-free-radical routes are not favoured by the X center. In fact, its ring which is ketonic in character is well strained, and cannot be opened. It can however be expanded. It cannot be used as a monomer due to steric limitations, otherwise that center will favour electro-free-radical attack, in the absence of a nucleo-free-radical in the system, i.e., its environment.

As is wrongly believed in present-day Science, when an initiator such as dilute hydrochloric acid is used, the route is thought to be cationic, that which is not true. Shown below is 50% dilute hydrochloric acid solution.

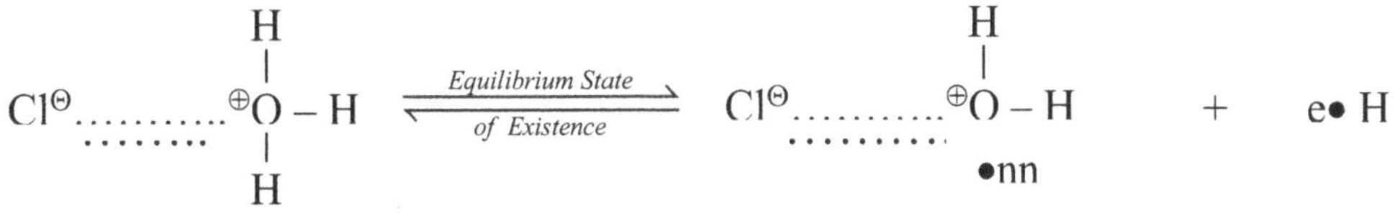

$$Cl^{\ominus}\ldots\ldots\ldots\ {}^{\oplus}O(H)(H) - H \xrightleftharpoons[\text{of Existence}]{\text{Equilibrium State}} Cl^{\ominus}\ldots\ldots\ldots\ {}^{\oplus}O(H)(\bullet nn) - H \quad + \quad e\bullet H$$

(A) Dilute HCl (1:1) (B)

Electrostatically Anionically-paired Initiator 2.34

This is an initiator in which the bond between the centers is electrostatic in character, because the positive charge on the oxygen center is **Imaginary.** The negative charge on the chlorine center is anionic and **Real** and that is the only active center on the initiator when in STABLE state of existence. When used under this operating condition, anionic route, the route natural to the monomer, is favoured by the monomer where the carrier of the chain is Cl.

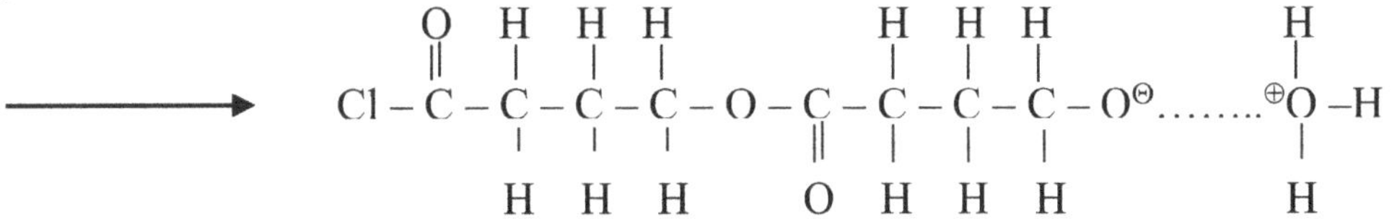

$$\longrightarrow \quad Cl - C(=O) - CH_2 - CH_2 - CH_2 - O - C(=O) - CH_2 - CH_2 - CH_2 - O^{\ominus}\ldots\ldots\ {}^{\oplus}O(H)(H) - H \qquad 2.35$$

In the absence of vacant orbitals and ***electrodynamic forces of repulsion***, no regular placement can be obtained. The C=O group will be randomly placed along the chain- i.e., atactic placement. When however, the initiator is in EQUILIBRIUM state of existence, it is (B) of Equation 2.34 that is used. When used, the carrier of the chain is H and ***the mechanism of addition is Equilibrium mechanism as opposed to Combination mechanism as shown below.***

Stage 2:

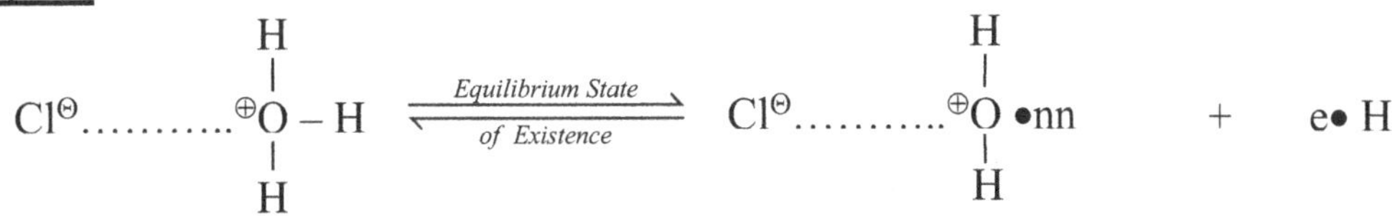

$$Cl^{\ominus}\ldots\ldots\ldots\ {}^{\oplus}O(H)(H) - H \xrightleftharpoons[\text{of Existence}]{\text{Equilibrium State}} Cl^{\ominus}\ldots\ldots\ldots\ {}^{\oplus}O(H)(H)\ \bullet nn \quad + \quad e\bullet H$$

(A) Dilute HCl (1:1) (B)

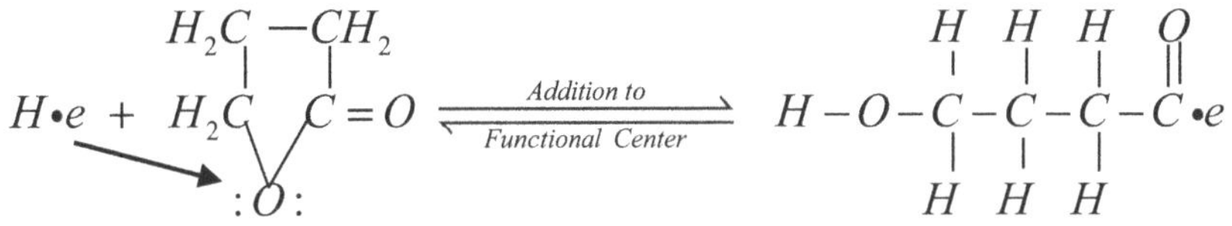

$$H\bullet e + \underset{(F)}{\begin{array}{c} H_2C - CH_2 \\ | \quad\quad | \\ H_2C \quad C=O \\ \diagdown \diagup \\ :\ddot{O}: \end{array}} \underset{\text{Functional Center}}{\overset{\text{Addition to}}{\rightleftharpoons}} \underset{(C)}{H-O-\overset{H}{\underset{H}{C}}-\overset{H}{\underset{H}{C}}-\overset{H}{\underset{H}{C}}-\overset{O}{\overset{||}{C}}\bullet e}$$

$$(B) + (C) \longrightarrow Cl^{\ominus}..........{}^{\oplus}\underset{H}{O}-\overset{O}{\overset{||}{C}}-\overset{H}{\underset{H}{C}}-\overset{H}{\underset{H}{C}}-\overset{H}{\underset{H}{C}}-O-H$$

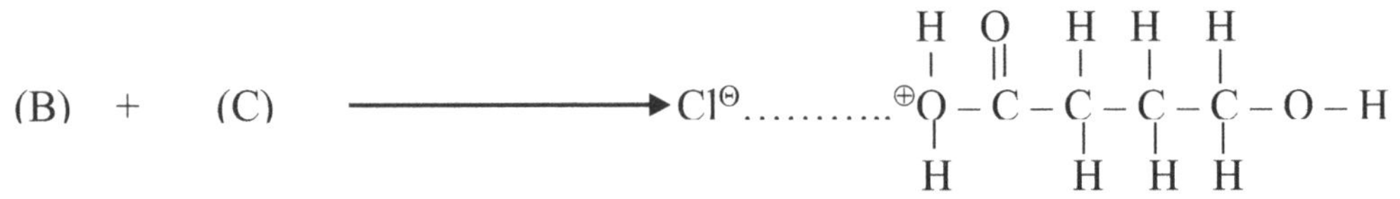

(D) INITIATION STEP

2.36

After the first stage, which is the preparation of the Initiator, the dilute HCl, this is then followed by the second stage which is the Initiation step, all via Equilibrium mechanism. In the third stage, (D) exists in Equilibrium state of existence and adds another. This like Step polymerization which takes place via Equilibrium mechanism as will partly be shown herein and fully in the New Frontiers, will take far longer time than polymerization via Combination mechanism. Hence, Step polymerization systems are of the orders of hours, while Addition polymerization systems via Combination mechanism are of the orders of seconds and minutes. ***For the first time, one can observe that Addition ringed monomers with functional centers undergo polymerization via Equilibrium mechanism.*** With respect to how polymers are made, present-day Science in their very low level of development know only Combination mechanism. Even what is only known by them about it cannot be explained. Even, what they call dilute acids, concentrated acids and fuming acids are unknown in present-day Science. The pain is that they including the author think they know. How can one continue to live in this kind of world every day? This has always been the pains in one's life. ***But the joy is that the more you think that you know nothing, the more you will continue to know. If you think you know too much, then you know nothing.*** The route which present-day Science thought was Cationic because H is the carrier of the chain or because acid was used, is Electro-free-radical route. Because the initiator is in Equilibrium state of existence, the Cl center can never be used and that is the route which is natural to the monomer. One can wonderfully observe how NATURE operates, because what we have just seen takes place in our everyday life unknown to us in line with the laws of Nature. In the absence of coordination in a Free-media system or electrodynamic forces of repulsion, regular placements can never be obtained.

One has seen the role of Electrostatic forces of repulsion. What then is Electrodynamic forces of repulsion? These forces come into manifestation visibly and more in Free-media systems. We have already seen a part of it in Equations 2.20a and 2.20b. When vinyl chloride or vinyl acetate is polymerized radically using benzoyl peroxide (Free-media systems), the polymers obtained are very crystalline, clear indication of regular placement of the groups. How the crystalline polymer is obtained, are unknown to present-day Science, and nobody cares to find out, because of fear of the unknown. When polymerized nucleo-free-radically, one of the initiators from benzoyl peroxide, a route which is not natural to the monomer, the followings are obtained.

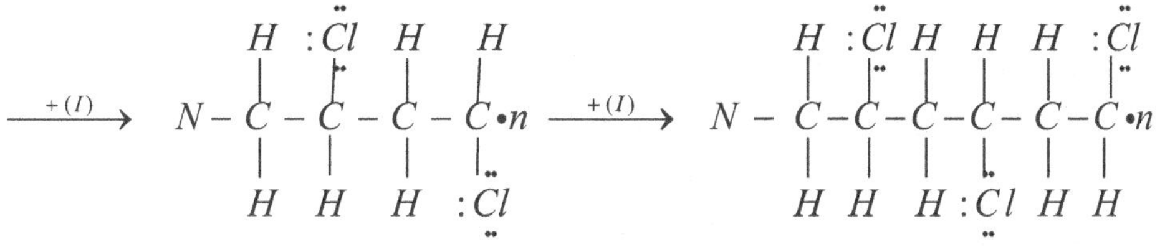

$$\xrightarrow{+(I)} \begin{array}{ccccccccc} & & H & & :\ddot{Cl} & & H & & H \\ & & | & & | & & | & & | \\ N & - & C & - & C & - & C & - & C \bullet n \\ & & | & & | & & | & & | \\ & & H & & H & & H & & :\ddot{Cl} \end{array} \xrightarrow{+(I)} \begin{array}{ccccccccccccc} & & H & & :\ddot{Cl} & & H & & H & & H & & :\ddot{Cl} \\ & & | & & | & & | & & | & & | & & | \\ N & - & C & - & C & - & C & - & C & - & C & - & C \bullet n \\ & & | & & | & & | & & | & & | & & | \\ & & H & & H & & H & & :\ddot{Cl} & & H & & H \end{array}$$

Syndiotactic Placement

ELECTRODYNAMIC FORCES OF REPULSION 2.37

Due to electrodynamic forces of repulsion, the vinyl chloride monomer units are sydiotactically placed along the chain. It is dynamic, because a whole body of monomer was dynamically placed as a result of forces of repulsion resulting from three paired unbonded radicals on the Cl centers. Those radicals are not there for nothing.

(D) of Equation 2.22, benzene as has been said, has far more than the MRSE, but less than the Max.RSE, otherwise it will not exist as a ring. It has no point of scission and therefore cannot be opened without first adding for example hydrogen via Activation (E.g. Hydrogenation). However, as will be shown downstream, based on universal data of its polymerization, it can readily be used as a monomer using special initiators, to give a polymer as strong as steel. With the new concepts on Rings, how rings can be opened, expanded, reduced and much more, have begun to be differently explained, noting that these are just introductions to the new theories on rings.

2.3 Important examples of some applications of the new concepts.

Shown finally below, are some applications of the new concepts considered so far. The applications are indeed provisions of Mechanisms of some important chemical and polymeric reactions something which has been lacking since the development of Science, i.e., since antiquity.

2.3.1 Nucleo–free radical polymerization of SO_2/propylene pair

How alternating placement of these two monomers along the chain are obtained, have never been explained. Yet we do not care to know, except to go to the laboratory collecting data all of which cannot be unquestionably explained! We use all what is gathered from the data to apply to other things by use of Semblance, Copying, Assumptions and so on!

Initiation Step: In the Initiation step, benzoyl peroxide has been chosen as the Initiator. Its decomposition is usually carried out at temperatures of the order of 50° and above.

Stage 1: [Decomposition mechanism]

C – O – O – C (benzoyl peroxide) ⟶ 2 C – O •nn
(A)

(A) ⟶ 2 •n (phenyl) + 2 e• C – O •nn
(C)

(C) ⟶ $2CO_2$ + Energy 2.38a

Overall equation: $(C_6H_5)_2(CO_2)_2$ ⟶ 2 n• C_6H_5 + $2CO_2$ + Energy
(2N.n) 2.38b

Worthy of note are the followings-

i. Benzoyl peroxide has been decomposed via **Decomposition mechanism** because of the *symmetric* structure to give only two products at a particular temperature. The two products are the nucleo-free radicals (The Initiator) and carbon dioxide.
ii. In the system, there is no male radical, that is, an electro-free radical to disturb the growth of a polymer chain during propagation step. Its presence may lead to existence of just molecules or telomers (Low molecular weight compounds).
iii. The Equilibrium state of Existence of the peroxide is similar to that of hydrogen peroxide. That shown in the first step of the stage above is its Decomposition state of existence.

$$H_5C_6-\overset{\overset{O}{\parallel}}{C}-O-O-\overset{\overset{O}{\parallel}}{C}-C_6H_5 \underset{\text{Existence of the peroxide}}{\overset{\text{Equilibrium State of}}{\rightleftharpoons}} H_5C_6-\overset{\overset{O}{\parallel}}{C}-O\cdot en + nn\cdot O-\overset{\overset{O}{\parallel}}{C}-C_6H_5 \quad 2.39$$

iv. Notice that while the first step in Stage 1 of Equation 2.38a is not radically balanced, the same applies to the overall equation, because the peroxide decomposing is symmetric. But the steps are balanced and inherently, since two moles (even) of radicals are obtained, all the equations are radically balanced. If it was non-symmetric, this initiator cannot be obtained as we saw in Chapter 1 when the mechanisms for Combustion were provided. Therein, their decompositions were done via Equilibrium mechanism. This reason why this is so, is because just as when one mole of the symmetric peroxides when decomposed via Equilibrium mechanism, no products can be obtained, so also no products can be obtained when non-symmetric peroxides are decomposed via Decomposition mechanism. For example, via Equilibrium mechanism, radicals can never be produced and the stage will not be productive as shown below.

Stage 1:

$$H_5C_6-\overset{\overset{O}{\parallel}}{C}-O-O-\overset{\overset{O}{\parallel}}{C}-C_6H_5 \underset{\text{Existence of the peroxide}}{\overset{\text{Equilibrium State of}}{\rightleftharpoons}} H_5C_6-\overset{\overset{O}{\parallel}}{C}-O\cdot en + nn\cdot O-\overset{\overset{O}{\parallel}}{C}-C_6H_5$$

$$H_5C_6-\overset{\overset{O}{\parallel}}{C}-O\bullet en \rightleftharpoons H_5C_6\bullet e + CO_2 + Energy$$

$$H_5C_6-\overset{\overset{O}{\parallel}}{C}-O\bullet nn \rightleftharpoons H_5C_6\bullet n + \underset{(A)}{e\bullet \overset{\overset{O}{\parallel}}{C}-O\bullet nn}$$

$$H_5C_6\bullet e + n\bullet C_6H_5 \longrightarrow H_5C_6-C_6H_5 \quad 2.40$$

$$(A) \underset{\text{Release of Energy}}{\overset{\text{Deactivation}}{\longrightarrow}} CO_2$$

UNREACTIVE STAGE 2.40

For the first time, we are encountering a stage via Equilibrium mechanism where no product can be obtained, because the last two steps in the stage have double headed single arrow. Such a stage cannot be interpreted mathematically, without making the assumption that one of the steps is the rate determining step as is done in present-day Science.

The Initiation of propylene has been shown not to be possible nucleo-free radically (See Equation 2.16). The same applies to sulfur dioxide as shown below.

$$\underset{(Nucleo-free\ radical)}{N\cdot n + nn\cdot O-\underset{..}{\overset{\overset{O^-}{|}}{S^+}}\cdot en} \xrightarrow[(Not\ radically\ balanced)]{Not\ Possible} \underset{(Nucleo-non-free\ radical)}{N-\underset{..}{\overset{\overset{O^-}{|}}{S^+}}-O\cdot nn} \quad 2.41$$

During Combination mechanism, unlike what goes on in Equilibrium mechanism where the arrow is neither here nor there, ***if the initiator is a nucleo-free-radical, the growing chain remains nucleo-free-radical, otherwise it will never grow***. In the equation above, we started with a nucleo-free radical on the left-hand side ending up with a nucleo-non-free radical on the right-hand side. This is not radically balanced. Hence the initiation above is not possible via Combination mechanism. For different reasons, both propylene and sulfur dioxide cannot be polymerized separately nucleo-free-radically. Electro-free-radically propylene can be polymerized. With sulfur dioxide, electro-free radically it cannot be polymerized, because the equation will remain radically unbalanced. Electro-non-free radically, SO_2 can be polymerized as already shown in Equation 2.20a. For propylene, it cannot be polymerized, because the equation will not be radically balanced. ***Propylene is more nucleophilic or female than sulfur dioxide.*** In view of the ways Nature operates wherein it abhors a vacuum where possible as the case here, sulfur dioxide and propylene combine to form what is called ***"a couple"*** as shown below.

$$n\cdot\underset{H}{\overset{H}{C}}-\underset{H}{\overset{CH_3}{C}}\cdot e \; + \; nn\cdot\ddot{\underset{..}{O}}-\underset{..}{\overset{O^{\odot}}{S^{\oplus}}}\cdot en \; \underset{diffuses\ to\ Less\ female}{\overset{The\ More\ female}{\rightleftharpoons}} \; n\cdot\underset{H}{\overset{H}{C}}-\underset{H}{\overset{CH_3}{C}}-\ddot{\underset{..}{O}}-\underset{..}{\overset{O^{\odot}}{S^{\oplus}}}\cdot en$$

$$(More\ Female)\xrightarrow{Diffuses}(Less\ female) \qquad (A\ Couple)-(A) \quad 2.42$$

Indeed, the driving forces favouring alternating placements are too early and numerous to list herein at this point in time. However, it was shocking to note that, ***it is the female that diffuses to add to the male using its electro-free-radical end after both have been activated,*** otherwise if the male diffuses, the couple formed where possible cannot be initiated for all cases where alternating placement takes place by Addition polymerization. One thought before this revelation from Chemistry, that it was the opposite. Herein again, are embedded some laws of Nature. Polymerization systems are complete embodiments of NATURE. This is to be expected, because all around us are POLYMERS, both in living and so-called non-living systems.

In Nature such as in frogs, it is the female that croaks and diffuses to the male to form the couple; otherwise at the end, a productive stage will not be formed. ***The initiator "sitting down watching",*** now comes to complete the Initiation step to balance the equation as follows.

$$\underset{\text{(Female Initiator)}}{\overset{\downarrow}{N\cdot n}} + (A) \xrightarrow{\text{Equation is balanced}} \underset{\text{(Living couple unit)}}{N-\overset{O^{\ominus}}{\underset{..}{\overset{|}{S}}}{}^{\oplus}-\ddot{\underset{..}{O}}-\overset{CH_3}{\underset{H}{C}}-\overset{H}{\underset{H}{C}}\cdot n} \qquad 2.43$$

INITIATION STEP

If the Initiator was an electro-free-radical (male), no couple will be formed and only propylene will be polymerized. ***For a couple to be formed, in order to have full alternating placement, both monomers must be unreactive to the initiator and the initiator must be in control of all the affairs.*** What is striking to note, is that if an initiator generating compound has the ability of producing both nucleo-free and nucleo-non-free-radical is placed in a system at a range of mild operating conditions containing a monomer which favours only the nucleo-non-free-radical route, the initiator knows what the monomer is and begins to work accordingly without breaking any law. All these take place, either via inherent use of some senses based on the operating conditions or other means. Countless observations which have been made as will be seen in the New Frontiers, clearly indicates, that all these species "communicate" differently all mathematically.

After the Initiation step where one couple unit is consumed, the propagation step commences as follows.

Propagation Step:

m Stages:

$$N-\overset{O^-}{\underset{..}{S^+}}-\ddot{\underset{..}{O}}-\overset{CH_3}{\underset{H}{C}}-\overset{H}{\underset{H}{C}}\cdot n + m(A) \xrightarrow[\text{Mechanism}]{\text{Combination}} N-(\overset{O^-}{\underset{..}{S^+}}-\ddot{\underset{..}{O}}-\overset{CH_3}{\underset{H}{C}}-\overset{H}{\underset{H}{C}}-)_m-\overset{O^-}{\underset{..}{S^+}}-\ddot{\underset{..}{O}}-\overset{CH_3}{\underset{H}{C}}-\overset{H}{\underset{H}{C}}\cdot n \qquad 2.44$$

(*Alternating polymer chain*)

The chain continues to grow ***until an optimum chain length based on the Glass-Transition temperature of the polymer (A theory with mathematical representations as obtained from polymer reaction engineering data universally)*** is reached at the operating temperature and pressure. The question now is- are the couples all formed in situ before propagation commences or are the couples formed in every stage of propagation before adding to the chain? Note that the couples are living from the electro-radical end only, i.e., the male end, just like a male being the head of a family. The nucleo-radical end of the couple is at the other end behind. Based on how Nature operates, a couple is first formed in every stage of propagation after which the chain adds to it.

Introducing a terminating agent, which can provide an electro-radical, kills the living growing polymer chain. Worthy of note is that in the example above two female monomers of different capacities have reacted together to form a couple under very unique operating conditions ***(The absence of the males).*** Indeed, from Co-polymerization reactions, there are too much to comprehend in terms of how Nature operates. Things we thought were abnormal can be found to be normal in the Natural world, but under very different operating conditions. All these observations from Chemistry can find many extensive applications in many disciplines including Social Sciences and Medical fields as we will see in the New Frontiers.

2.3.2. **Zeigler/Natta (Z/N) polymerization of propylene**:

Here one is going to use AlR_3 (Aluminium tri-alkyl) as catalyst and $TiCl_4$ (Titanium tetrachloride) as co-catalyst at a molar ratio of two to one, that is, excess AlR_3. AlR_3 commences the alkylation step strongly because it is in excess both metals being equi-electropositive, and for no other reason as will shortly be explained. Al and Ti are two very different metals. Al is a non-ionic, non-Transition, polar metal. When used, it becomes non-polar. Ti is a non-ionic, Transition, polar metal which when used becomes non-polar.

Limited Alkylation's Step: [Decomposition mechanism]

$$\underline{Stage 1}: AlR_3 \xrightarrow[State]{Decomposition} R\cdot n + e\cdot AlR_2$$

$$R_2Al\cdot e + TiCl_4 \xrightarrow{Abstraction} R_2Al-Cl + e\cdot TiCl_3$$

$$R\cdot n + e\cdot TiCl_3 \underset{Existence\ of\ RTiCl_3}{\overset{Equilibrium\ State\ of}{\rightleftharpoons}} R-TiCl_3 \qquad 2.45a$$

$$\underline{Overall\ Equation}: AlR_3 + TiCl_4 \longrightarrow R_2AlCl + RTiCl_3 \qquad 2.45b$$

$$\underline{Stage2}: AlR_2Cl \xrightarrow[State]{Decomposition} R\cdot n + e\cdot AlRCl$$

$$RClAl\cdot e + RTiCl_3 \xrightarrow{Abstraction} RAlCl_2 + e\cdot TiRCl_2$$

$$R\cdot n + e\cdot TiRCl_2 \underset{Existence\ of\ TiR_2Cl_2}{\overset{Equilibrium\ State\ of}{\rightleftharpoons}} R_2TiCl_2 \qquad 2.46a$$

$$\underline{Overall\ Equation}: AlR_3 + TiCl_4 \longrightarrow AlRCl_2 + R_2TiCl_2 \qquad 2.46b$$

The alkylation is <u>limited</u> (as opposed to <u>full</u> alkylation), because Al and Ti are *equi-electropositive* [two equal giants independent of their size]. As such, AlR_3 being in excess commenced the alkylation step, since in general, regardless the concentrations or ratios of the components, the more electro-positive one, commences the alkylation step. Being equi-electropositive, alkylation cannot go beyond the two stages above, because none wants to be cheated. Notice how NATURE has operated here. The mechanism above is another type of ***Decomposition mechanism, which can be interpreted into mathematical equations without the need for making necessary and unnecessary assumptions.*** *T*he initiator is then prepared as follows using **<u>Equilibrium mechanism</u>**.

<u>Stage 1</u>:

$$AlR_3 \underset{Existence\ of\ AlR_3}{\overset{Equilibrium\ State\ of}{\rightleftharpoons}} R\cdot e + n\cdot AlR_2$$

$$R_2TiCl_2 \underset{Existence\ of\ R_2TiCl_2}{\overset{Equilibrium\ State\ of}{\rightleftharpoons}} R\cdot n + e\cdot TiRCl_2$$

$$R\cdot e + n\cdot R \underset{\substack{Existence\ at\ betwen\ melting\ and\\ boiling\ point\ of\ hydrocarbon}}{\overset{Equilibrium\ State\ of}{\rightleftharpoons}} R-R$$

$$e\cdot TiRCl_2 + n\cdot AlR_2 \longrightarrow Cl-\overset{\overset{Cl}{|}}{\underset{\underset{R}{|}}{Ti^{\oplus}}}\square\ \square \cdots\cdots\cdots \square\ \square\ {}^{\ominus}Al\begin{matrix}R\\R\end{matrix}$$

$$(A)-(Zeigler\ /\ Natta\ Initiator) \qquad 2.47a$$

$$\underline{Overall\ overall\ equation}: 2AlR_3 + TiCl_4 \longrightarrow R-R + RAlCl_2 + Cl_2RTi^{\oplus}\cdots\cdots{}^{\ominus}AlR_2 \qquad 2.47b$$

Firstly, notice that both components are in equilibrium state of existence. Secondly, notice the equilibrium state of existence of AlR_3; it is that where-in the aluminium center is carrying a female radical, a nucleo-free radical while the carbon center is carrying the male radical, an electro-free radical. Why did NATURE make it that way when Al is more electropositive than C? That is a research work. With R_2TiCl_2, it is R that is held as a female radical as opposed to Cl. These are clear indications that these ***non-ionic metallic centers have different personalities or properties.*** Though all these cannot be expressed chargedly, in the last step of the stage, charges are obtained in the reactions between two metallic centers, not isolatedly placed, but paired. Thirdly, notice the equilibrium state of existence of the alkane formed. This somehow partly dictates the operating conditions of the system because alkanes have different equilibrium states of existence at and below melting points, and at between melting points and boiling points, and at and above boiling point as indicated by the plots of boiling points and melting points versus number of carbon atoms for normal alkanes.[22]

$$H_4C \underset{\text{of Existence}}{\overset{\text{Equilibrium State}}{\rightleftharpoons}} H\bullet e + n\bullet CH_3 \quad (At\ all\ temperatures)$$

$$H_6C_2 \underset{\text{of Existence}}{\overset{\text{Equilibrium State}}{\rightleftharpoons}} H\bullet e + n\bullet C_2H_5 \quad (At\ Bpt.\ and\ above)$$

$$\left.\begin{array}{l} H_6C_2 \underset{\text{of Existence}}{\overset{\text{Equilibrium State}}{\rightleftharpoons}} H_3C\bullet e + n\bullet CH_3 \\ H_6C_2 \underset{\text{of Existence}}{\overset{\text{Equilibrium State}}{\rightleftharpoons}} H\bullet e + n\bullet C_2H_5 \end{array}\right\} (Between\ Mpt.\ and\ Bpt.)$$

$$H_6C_2 \underset{\text{of Existence}}{\overset{\text{Equilibrium State}}{\rightleftharpoons}} H_3C\bullet e + n\bullet CH_3 \quad (At\ Mpt.\ and\ below)$$

2.48a

$$H_8C_3 \underset{\text{of Existence}}{\overset{\text{Equilibrium State}}{\rightleftharpoons}} H\bullet e + n\bullet C_3H_7 \quad (At\ Bpt.\ and\ above)$$

$$\left.\begin{array}{l} H_8C_3 \underset{\text{of Existence}}{\overset{\text{Equilibrium State}}{\rightleftharpoons}} H_3C\bullet e + n\bullet C_2H_5 \\ H_8C_3 \underset{\text{of Existence}}{\overset{\text{Equilibrium State}}{\rightleftharpoons}} H\bullet e + n\bullet C_3H_7 \end{array}\right\} (Between\ Mpt.\ and\ Bpt.)$$

$$H_8C_3 \underset{\text{of Existence}}{\overset{\text{Equilibrium State}}{\rightleftharpoons}} H_3C\bullet e + n\bullet C_2H_5 \quad (At\ Mpt.\ and\ below)$$

2.48b

Fourthly, using the eye of the needle notice the vacant orbitals carried by the centers. These are coming from the established radical (not electronic) configurations of Al and Ti along their boundaries. Ti has five vacant orbitals, two from 3d and three from 4p sub-atomic orbitals. Only two of them have been shown above, since we do not need them for the polymerization of propylene a female monomer. Nevertheless, only two closest ones can be used. They are reservoirs for male monomers. The aluminium center has only one vacant orbital left. This is the reservoir for the female monomer kept in an activated state all the time in a fixed position as shown below for the down location of the vacant orbital. Its activation is provided by the covalent charges carried by the metallic centers as above.

$$\begin{matrix} & H & CH_3 \\ & | & | \\ \odot & C - & C \oplus \\ & | & | \\ & H & H \end{matrix}$$

Activated Fixed Position of propylene for Isotactic placement when located below the horizontal axis, the axis of linear addition. When located above it is upsidedly placed. 2.49

With the monomer oriented in this manner inside the vacant orbital placed at the bottom, the positive titanium end diffuses to the negative end of the activated monomer while the positive end of the monomer at the same time diffuses to the aluminium center. Hence the covalent bond between the two metallic centers is of the elastic type. With only one vacant orbital as reservoir, only isotactic placement a regular structure is obtained.

Initiation Step:

Stage 1:

$$\underset{(Z/N\ Initiator)}{(A)} + \underset{(monomer)}{H_2C = CH(CH_3)} \xrightarrow[Mechanism]{Combination} \underset{(B)}{Cl - \overset{Cl}{\underset{R}{Ti}} - \overset{H}{\underset{H}{C}} - \overset{CH_3}{\underset{H}{C^{\oplus}}} \ldots\ldots\ldots {}^{\odot}_{\square}Al \begin{matrix} R \\ R \end{matrix}} \qquad 2.50$$

With (B) formed, Propagation step commences via same Combination mechanism. The (B) formed is like the birth of a child into the world. In the propagation step, the growth of the child begins.

Propagation Step:

M-Stages:

$$(B) + M[H_2C = CH(CH_3)] \xrightarrow[Mechanism]{Combination} Cl - \overset{Cl}{\underset{R}{Ti}} - (\overset{H}{\underset{H}{C}} - \overset{CH_3}{\underset{H}{C}})^{iso}_{M} - \overset{H}{\underset{H}{C}} - \overset{CH_3}{\underset{H}{C^{+}}} \ldots\ldots {}^{-}_{\square}Al \begin{matrix} R \\ R \end{matrix}$$

$(C) -$ Living Polymer Chain 2.51

The chain either continues to grow until the optimum chain length dictated by the *Glass-Transition temperature* of the polymer is reached as already indicated or it is terminated prematurely for specific applications.

Universally, most or all-polypropylene manufacturing industries spend unnecessary millions of moneys just for premature Termination alone, because of lack of knowledge of the mechanisms of the polymeric reactions.

Because of Termination, some companies in developing countries are forced to shut down for months when the availability of H_2 one of the commonly used Terminating agents becomes a problem.[23] Unknown, is that the H_2 does not add to close the living chain, in order to kill it. Because of its smaller size compared to that of the monomer and its stable character in the absence of H_2 catalyst, it disallows the monomer from occupying the vacant orbital and the chain kills itself as has already been shown for syndiotactic placement and shortly be recalled for isotactic placement. The same hydrogen molecules intermittently put into the system to terminate chain growth, though used without being consumed, is wasted for nothing. If the mechanism was well understood, termination can be carried out by Starvation, that is, cutting off the supply of the monomer. This will of course require a redesign of the reactors either better placed in trains or otherwise. Even a Nation's Petrochemical industry was awarded a patent for the recommendation of using a molecule such as methane in place of H_2 from the author which was experimented on without recognizing the origin of the ideas! Polyvinyl chloride manufacturing industries and academia have spent millions of moneys trying to find out if vinyl chloride can be polymerized chargedly, without the realization that the solution was right in front of their nose. One can imagine the billions and trillions of moneys that have been wasted since the beginning of acquisition of Knowledge, just because what RADICALS are, has never been known. This has always been the MISSING LINK, the gap between the ART and SCIENCE as shed in Chapter 1. That link has affected all things in humanity including Religion. It has affected the use of Nuclear energy in a very different way. It is preferable to "die" for the **Ignorance (Stupidity)** of humanity than to die for their **Sins** and that has already been accomplished many centuries ago. Why should anyone die for the sins of Humanity when the Laws of Nature including the First and most important Law (All inside CHEMISTRY) have already been set in motion long before humans were created? Indeed, all disciplines including Religion were there long before humans were created.

Termination by Starvation:

Stage 1:

$$(C) \xrightarrow[Monomer]{No} Cl-\underset{R}{\overset{Cl}{Ti}}-(\underset{H}{\overset{H}{C}}-\underset{H}{\overset{CH_3}{C}})_M^{iso}-\underset{H}{\overset{H}{C}}-\underset{H}{C}=\underset{H}{\overset{H}{C}} + HAlR_2 \qquad 2.52$$

The Dead Polymer

$$HAlR_2 \underset{\text{State of Existence}}{\overset{\text{Its Equilibrium}}{\rightleftharpoons}} H\cdot n + e\cdot AlR_2 \qquad 2.53$$

The mechanism herein like in propagation is still Combination mechanism, wherein in the last step of the stage, the living polymer releases H as a cation and this adds to the AlR_2 by combination followed by deactivation with release of energy. The Equilibrium state of existence of the AlR_2H is shown above. Its use is not required above since the reaction is taking place chargedly. A <u>dead terminal double bonded polymer</u> (A gigantic monomer) is produced. Depending on the size of the chain and the operating conditions, this can be re-activated to give branched polymers using a living chain and monomers.

If the ratio of the Ti/Al components is equal to or greater than 2, another type of Initiator which is not of the Z/N type is obtained as shown below.

<u>Limited Alkylation's Step</u>: [Decomposition mechanism]

$$\underline{Stage 1}: TiCl_4 \xrightarrow[\text{State}]{\text{Decomposition}} Cl\cdot nn + e\cdot TiCl_3$$

$$Cl_3Ti\cdot e + AlR_3 \xrightarrow{\text{Abstraction}} Cl_3Ti-R + e\cdot AlR_2$$

$$Cl\cdot nn + e\cdot AlR_2 \underset{\text{Existence of } RTiCl_3}{\overset{\text{Equilibrium State of}}{\rightleftharpoons}} Cl-AlR_2 \qquad 2.54a$$

$$\underline{Overall\ Equation}: AlR_3 + TiCl_4 \longrightarrow R_2AlCl + RTiCl_3 \qquad 2.54b$$

$$\underline{Stage 2}: TiCl_3R \xrightarrow[\text{State}]{\text{Decomposition}} Cl\cdot nn + e\cdot TiRCl_2$$

$$RCl_2Ti\cdot e + ClAlR_2 \xrightarrow{\text{Abstraction}} R_2TiCl_2 + e\cdot AlRCl$$

$$Cl\cdot nn + e\cdot AlRCl \underset{\text{Existence of } TiR_2Cl_2}{\overset{\text{Equilibrium State of}}{\rightleftharpoons}} RAlCl_2 \qquad 2.55a$$

$$\underline{Overall\ Equation}: AlR_3 + TiCl_4 \longrightarrow AlRCl_2 + R_2TiCl_2 \qquad 2.55b$$

Since, $TiCl_4$, is in excess, like the case before this, alkylation started with it. The alkylation is <u>limited</u> (as opposed to <u>full</u> alkylation), because Al and Ti are *equi-electropositive* [two equal giants independent of their size]. The initiator is next prepared as follows using Equilibrium mechanism.

Equilibrium mechanism.

Stage 1:

$$RAlCl_2 \underset{\text{Existence of } AlR_3}{\overset{\text{Equilibrium State of}}{\rightleftharpoons}} Cl\cdot nn + e\cdot AlRCl$$

$$Cl\bullet nn + TiCl_4 \underset{\text{to vacant orbital}}{\overset{\text{Electrostatic Addition}}{\rightleftharpoons}} n\bullet TiCl_5$$

$$RClAl\bullet e + nn\bullet TiCl_5 \longrightarrow Cl(R)Al^{\oplus}\text{..........}{}^{\ominus}TiCl_5$$

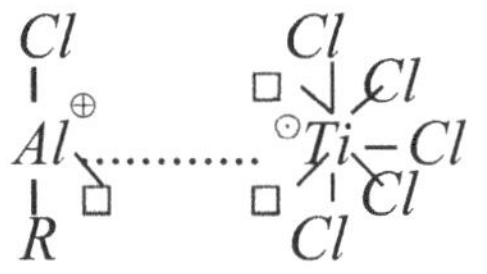

$$(A) - (\text{Electrostatically Positively Charged – paired Initiator}) \qquad 2.56a$$

Overall overall equation: $AlR_3 + 2TiCl_4 \longrightarrow R_2TiCl_2 + RClAl^{\oplus}.........{}^{\ominus}TiCl_5$ 2.56b

Worthy of note are the followings-

i) The products obtained during the limited alkylation, are essentially the same whether AlR_3 started it or not.

ii) The Equilibrium, Combination and Decomposition states of existence of the compounds and their intermediates involved.

iii) The order of States of existence of these compounds. Where the Equilibrium state of existence of $TiCl_4$ was suppressed by the presence of $AlRCl_2$ in Stage 1 above, clearly indicates that $TiCl_4$ is a stable salt not in general, but in the presence of the Al salts. Also, the order of Equilibrium state of existence of the Aluminium salts is as follows- $AlR_3 > AlClR_2 > AlCl_2R > AlCl_3$; while that of Titanium is as follows- $TiR_4 > TiR_3Cl > TiR_2Cl_2 > TiRCl_3 > TiCl_4$. Yet their individual states of existence are all different, based on the components held.

$$AlR_2Cl \rightleftharpoons R_2Al\bullet e + nn\bullet Cl \quad ; \quad TiR_3Cl \rightleftharpoons R_2ClTi\bullet e + n\bullet R$$

$$AlRCl_2 \rightleftharpoons RClAl\bullet e + nn\bullet Cl \quad ; \quad TiR_2Cl_2 \rightleftharpoons RCl_2Ti\bullet e + n\bullet R$$

$$AlR_3 \rightleftharpoons R_2Al\bullet n + e\bullet R \quad ; \quad TiR_4 \rightleftharpoons R_3Ti\bullet e + n\bullet R$$

$$AlCl_3 \rightleftharpoons Cl_2Al\bullet e + nn\bullet Cl \quad ; \quad TiCl_4 \rightleftharpoons Cl_3Ti\bullet e + nn\bullet Cl$$

$$TiR_4 > AlR_3 \; ; \; AlCl_3 > TiCl_4 \qquad 2.57$$

iv) The fact that the bond between the two metallic centers is not covalent like the Z/N case, but electrostatic, i.e., imaginary. Therefore, only one center is active and that center is the real center which is carrying a positive charge and not a cation. This is very much unlike the case of **Covalently charged –paired Initiators** which the Z/N type belongs to. For the positive side, unlike the Z/N case, there are two reservoirs on the Titanium side for the Female monomer to reside in during polymerization. Hence when used, syndiotactic placements are obtained as already shown. The electrostatic bond is far more elastic in character, since it is imaginary. Since, the positive center is not cationic, hence the initiator has been called **Electrostatically positively charged-paired initiator.**

v) From which one can see one of the origins of Complex numbers in Mathematic for this initiator inside Chemistry.

z	=	x	+	iy
Initiator		Real		Imaginary
		Active center		Inactive center

Since the Z/N case (A of Equation 2.47) has two active centers (one for male and the other for female), and the case above has only one active center specific for females- the Aluminium center, the carriers of the polymeric chains are important to note, since the life of a chain is dependent on the carriers. Isotactically, for the Z/N case, the carrier is Ti, while syndiotactically for the same combination, the carrier is Al. The two initiators from $AlR_3/TiCl_4$ combinations which could not be identified in present-day Science were all said to be Z/N types of Initiators- that which leads to a state of confusion.

2.3.3 Z/N type of Initiator from VCl_3/AlR_3 Combinations.

How Vanadium trichloride and Aluminium trialkyl have been used over the years to give an initiator with coordination ability provided by vacant orbitals has never been known. Even with the forehand knowledge of the Z/N case above, it will still remain unknown. The case considered herein is slightly different from the $TiCl_4/AlR_3$ combination, because of the presence of the paired unbonded radicals carried by the vanadium center in VCl_3. The presence obviously will affect the Alkylation step as follows.

Decomposition mechanism

Alkylation of VCl_3

Stage 1

$$AlR_3 \xrightarrow[\text{of Existence}]{\text{Decomposition State}} R\bullet n \; + \; e\bullet AlR_2$$

$$R_2Al\bullet e \; + \; :VCl_3 \xrightarrow[\text{Transfer species}]{\text{Abstraction of}} R_2AlCl \; + \; en\bullet \ddot{V}Cl_2$$

NOT RADICALLY BALANCED

2.58

The stage above could not take place, because the equation is not radically balanced. Then, how can it be made to be radically balanced without making any assumption? With us, YES, but with NATURE, NO, because NATURE does not work with assumptions. Where it is not possible, NATURE will indicate it, like the case above. But unlike the case of Initiation where a route cannot be favoured, here, there is a solution, since NATURE has prevented the existence of a vacuum here by already providing vacant orbitals on the V center for many applications.

Transformation of the VCl_3

Stage 1:

$$:VCl_3 \xrightleftharpoons[\text{of Existence}]{\text{Equilibrium State}} en\bullet \underset{\bullet\bullet}{V}Cl_2 + nn\bullet Cl$$

$$Cl_2\underset{\bullet\bullet}{V}\bullet en + \diamond - \underset{\bullet\bullet}{V}Cl_3 \xrightleftharpoons[\text{Vanadium Center}]{\text{Activation of}} Cl_2\underset{\bullet\bullet}{V}\bullet en + n\bullet VCl_3\bullet e$$

$$\rightleftharpoons Cl_2\underset{\bullet\bullet}{V} - VCl_3\bullet e \quad (A)$$

$$(A) + Cl\bullet nn \longrightarrow Cl_2\underset{\bullet\bullet}{V} - VCl_4 \quad (B)$$

$$\underline{\text{Overall equation:}} \quad 2\underset{\bullet\bullet}{V}Cl_3 \longrightarrow Cl_2\underset{\bullet\bullet}{V} - VCl_4 \qquad 2.59$$

Two moles of VCl_3 under mild heat has been transformed into one mole of Di-vanadium hexachloride (B) with only one center carrying paired unbonded radicals. The (B) is now adequately ready to be used for alkylation without breaking one of the laws of NATURE-Radical balancing. ***Since Al is more electropositive than V, regardless the concentration of the vanadium salt, alkylation will always begin with Al. In addition, alkylation must be full and not limited.*** In present-day Science, one can see so far that little or nothing is known about paired unbonded radicals, and vacant orbitals. They think they are there for nothing, without asking the simple question - Why do some atoms have them and some don't? Why do some have paired unbonded radicals with no vacant orbitals? As already said NOTHING IS THERE FOR NOTHING. Some will suggest that the paired unbonded radicals in one VCl_3 could be used to fill the vacant orbital of another VCl_3 to form a dimer. However, this is not the case as we shall see.

Alkylation of (B)

Stage 1:

$$AlR_3 \xrightarrow[\text{of Existence}]{\text{Decomposition State}} R_2Al{\bullet}e + n{\bullet}R$$

$$R_2Al{\bullet}e + Cl_2\underset{\bullet\bullet}{V}-VCl_4 \xrightarrow[\text{Transfer Species}]{\text{Abstraction of}} R_2AlCl + Cl_2\underset{\bullet\bullet}{V}-\underset{\bullet e}{V}Cl_3$$

$$Cl_2\underset{\bullet\bullet}{V}-\underset{\bullet e}{V}Cl_3 + R{\bullet}n \rightleftharpoons Cl_2\underset{\bullet\bullet}{V}-VCl_3R$$

2.60a

$$\underline{Overall\ equation}: AlR_3 + Cl_2\underset{\bullet\bullet}{V}-VCl_4 \longrightarrow ClAlR_2 + Cl_2\underset{\bullet\bullet}{V}-VCl_3R$$

2.60b

Stage 2:

$$ClAlR_2 \xrightarrow[\text{of Existence}]{\text{Decomposition State}} RClAl{\bullet}e + n{\bullet}R$$

$$RClAl{\bullet}e + Cl_2\underset{\bullet\bullet}{V}-VCl_3R \xrightarrow[\text{Transfer Species}]{\text{Abstraction of}} Cl_2AlR + Cl_2\underset{\bullet\bullet}{V}-\underset{\bullet e}{V}Cl_2R$$

$$Cl_2\underset{\bullet\bullet}{V}-\underset{\bullet e}{V}Cl_2R + R{\bullet}n \rightleftharpoons Cl_2\underset{\bullet\bullet}{V}-VCl_2R_2$$

2.61a

$$\underline{Overall\ equation}: AlR_3 + Cl_2\underset{\bullet\bullet}{V}-VCl_4 \longrightarrow RAlCl_2 + Cl_2\underset{\bullet\bullet}{V}-VCl_2R_2$$

2.61b

Stage 3:

$$RAlCl_2 \xrightarrow[\text{of Existence}]{\text{Decomposition State}} Cl_2Al{\bullet}e + n{\bullet}R$$

$$Cl_2Al{\bullet}e + Cl_2\underset{\bullet\bullet}{V}-VCl_2R_2 \xrightarrow[\text{Transfer Species}]{\text{Abstraction of}} Cl_3Al + Cl_2\underset{\bullet\bullet}{V}-\underset{\bullet e}{V}R_2Cl$$

$$Cl_2\underset{\bullet\bullet}{V}-\underset{\bullet e}{V}R_2Cl + R{\bullet}n \rightleftharpoons Cl_2\underset{\bullet\bullet}{V}-VClR_3$$

2.62a

$$\underline{Overall\ equation}: AlR_3 + Cl_2\underset{\bullet\bullet}{V}-VCl_4 \longrightarrow Cl_3Al + Cl_2\underset{\bullet\bullet}{V}-VR_3Cl$$

2.62b

Stage4:

$$AlR_3 \xrightarrow[\text{of Existence}]{\text{Decomposition State}} R_2Al{\bullet}e + n{\bullet}R$$

$$R_2Al{\bullet}e + Cl_2\underset{\bullet\bullet}{V}-VR_3Cl \xrightarrow[\text{Transfer Species}]{\text{Abstraction of}} R_2AlCl + Cl_2\underset{\bullet\bullet}{V}-\underset{\bullet e}{V}R_3$$

$$Cl_2\underset{\bullet\bullet}{V}-\underset{\bullet e}{V}R_3 + R{\bullet}n \rightleftharpoons \underset{(C)}{Cl_2\underset{\bullet\bullet}{V}-VR_4}$$

2.63*a*

$$\underline{Overall\ equation}: 2AlR_3 + Cl_2\underset{\bullet\bullet}{V}-VCl_4 \longrightarrow Cl_3Al + Cl_2\underset{\bullet\bullet}{V}-VR_4 + R_2AlCl$$

2.63b

Alkylation cannot go beyond this stage, because the equation will not be radically balanced. On the whole, two moles of AlR_3 were required to fully alkylate the vanadium salt (B) in order to produce (C), which can now be used to prepare the initiator. This is then followed by Equilibrium mechanism. To do this, there must be excess of AlR_3 in the system right from the onset.

Equilibrium mechanism

Stage 1:

$$AlR_3 \rightleftharpoons R{\bullet}e + n{\bullet}AlR_2$$

$$Cl_2\underset{\bullet\bullet}{V}-VR_4 \rightleftharpoons Cl_2\underset{\bullet\bullet}{V}-\underset{\bullet e}{V}R_3 + n{\bullet}R$$

$$R{\bullet}e + n{\bullet}R \rightleftharpoons R-R$$

$$R_2Al{\bullet}n + Cl_2\underset{\bullet\bullet}{V}-\underset{\bullet e}{V}R_3 \longrightarrow Cl_2\underset{\bullet\bullet}{V}-V^{\oplus}(R)_3(\diamond)_2 \ldots\ldots\ldots\ldots {}^{\ominus}Al(R)_2(\diamond)$$

(F) 2.64

$$\underline{\text{Overall overall equation}}: 2VCl_3 + 3AlR_3 \longrightarrow Cl_2\underset{\bullet\bullet}{V}(R_3)V^{\oplus}\ldots\ldots{}^{\ominus}AlR_2 + R-R + AlCl_3 + ClAlR_2 \qquad 2.65$$

(F) is the Z/N type of initiator all belonging to the family of initiators called **Covalently Charged-paired Initiators.** The two centers are active, one for Female and the other for Male identical to that of $AlR_3/TiCl_4$. It has syndiotactic placement for Males such as methyl acrylate, and isotactic placement for Females such as propene. Unlike the $AlR_3/TiCl_4$ combination, the molar ratio here for the catalyst to manifest itself is AlR_3/VCl_3 ratio of 3 to 2. Below this ratio, the initiator cannot be obtained, unless the $Cl_3V{:} \rightarrow VCl_3$ was used with a ratio of 2 to 2.

Like the $AlR_3/TiCl_4$ combination, the Electrostatically positively charged-paired initiator can also be obtained for this combination, since the followings are valid.

Stage 1:

$$R_2TiCl_2 \rightleftharpoons Cl_2RTi\bullet e + R\bullet n$$
$$Cl_2AlR \rightleftharpoons RClAl\bullet e + Cl\bullet nn$$
$$Cl_2RTi\bullet e + Cl\bullet nn \longrightarrow Cl_3TiR$$
$$RClAl\bullet e + R\bullet n \longrightarrow R_2AlCl$$
$$\underline{UNREACTIVE\ STAGE}$$

2.66

These are the products from the limited alkylation in $TiCl_4/AlR_3$ combination. They are unreactive. The R_2TiCl_2 was used in preparing the Z/N initiator in the presence of $RAlCl_2$ along with AlR_3. All were in Equilibrium state of existence. Even if $RAlCl_2$ was suppressed, it will still be unproductive with R_2TiCl_2 which is always in Equilibrium state of existence in the system. For the second combination, the followings are similarly obtained.

Stage 1:

$$Cl_2\underset{\bullet\bullet}{V}-VR_3Cl \rightleftharpoons Cl_2\underset{\bullet\bullet}{V}-\underset{\bullet e}{V}ClR + R\bullet n$$
$$AlCl_3 \rightleftharpoons Cl_2Al\bullet e + Cl\bullet nn$$
$$Cl_2\underset{\bullet\bullet}{V}-\underset{\bullet\bullet}{V}ClR + Cl\bullet nn \longrightarrow Cl_2\underset{\bullet\bullet}{V}-VCl_2R$$
$$Cl_2Al\bullet e + R\bullet n \longrightarrow Cl_2AlR$$
$$\underline{UNREACTIVE\ STAGE}$$

2.67

Stage 1:

$$Cl_2\underset{\bullet\bullet}{V}-VR_4 \rightleftharpoons Cl_2\underset{\bullet\bullet}{V}-\underset{\bullet e}{V}R_3 + R\bullet n$$
$$ClAlR_2 \rightleftharpoons R_2Al\bullet e + Cl\bullet nn$$
$$Cl_2\underset{\bullet\bullet}{V}-\underset{\bullet\bullet}{V}R_3 + Cl\bullet nn \longrightarrow Cl_2\underset{\bullet\bullet}{V}-VR_3Cl$$
$$R_2Al\bullet e + R\bullet n \longrightarrow AlR_3$$
$$\underline{UNREACTIVE\ STAGE}$$

2.68

In the first equation above, the products from alkylation after the first consumption of AlR_3 are unreactive. The same applies after full alkylation as shown by the last equation. Hence the second type of initiator is prepared as follows with excess VCl_3 in the system.

Equilibrium mechanism
Stage 1:

$$ClAlR_2 \rightleftharpoons R_2Al\bullet e + Cl\bullet nn$$

$$Cl\bullet nn + Cl_2\ddot{V}-VCl_4 \rightleftharpoons Cl_2\ddot{V}-VCl_4\bullet nn$$

(A)

$$R_2Al\bullet e + (A) \longrightarrow R_2Al^{\oplus}\ldots\ldots\ldots\ldots{}^{\ominus}VCl_5 - \ddot{V}Cl_2$$

(Electrostatically positively charged-paired Initiator) 2.69

$$\underline{Overall\ overall\ equation}: 2AlR_3 + 4VCl_3 \longrightarrow AlCl_3 + Cl_2\ddot{V}-VR_4 + R_2Al^{\oplus}\ldots\ldots{}^{\ominus}VCl_5 - \ddot{V}Cl_2$$

2.70

This can only be used to provide syndiotactic placement for Nucleophiles (Females), since there is only one active center. It cannot be used for Electrophiles (Males). For the initiator to be obtained, the ratio of the V/Al must be at least 2 to 1. While Zeigler and Natta by brilliance and accident, discovered this great family of initiators, they like the brilliant others, do not know what they discovered or understand how what they discovered works. Nothing in NATURE was created by accident, for if it was, then life and living will become meaningless. Though no two "combinations" are virtually the same in terms of the mechanisms involved in preparing the Initiators, the fundamental principles involved are the same.

2.3.4 **ZnR_2/H_2O, AlR_3/H_2O, and $FeCl_3$/Monomer Adduct Combinations**

The **isotactic** polymerization of propylene oxide (an epoxide) using catalyst systems such as ***zinc or aluminium alkyls in combination with water or alcohol***[24,25], a ***$FeCl_3$-propylene oxide adduct*** first reported in 1955[26], and ***even metal hydroxides and alkoxides under certain conditions***[27], is worthy of attention in this very little introduction into the New Frontiers. If isotactic placement as opposed to syndiotactic placement was obtained by the initiators generated from them, then paired media initiating system must be involved. Secondly, either there is only one reservoir present or one side of the horizontal plane of addition is sterically hindered. The monomer being Nucleophilic with transfer species, the natural route must either be electro-free-radical or positively charged.

$$N\bullet n \quad H_2C(-O-)CH-CH_3 \longrightarrow N\bullet n + e\bullet CH_2-CH(CH_3)-O\bullet nn \longrightarrow$$

$$NH + n\bullet CH_2 - C(\bullet e)H - CH_2 - O\bullet nn \quad \text{OR} \quad \ddot{N}H \;+\; CH_2{=}CH-CH_2-O\bullet nn$$

[Nucleo-free-radical] [Nucleo-non-free-radical] .2.71

Anionically or with free negative charge, the same as above is obtained, that is, no initiation is possible, the route being unnatural to it.

For the first combination-AlR_3/H_2O combination, the followings are obtained in identifying the initiator. Since two metallic elements are involved-H from water and Al from AlR_3, the process begins with Decomposition mechanism. Since Al is more electropositive than H, the first stage begins with aluminium trialkyl.

Decomposition mechanism

Stage 1: $AlR_3 \longrightarrow R\bullet n + e\bullet AlR_2$

$R_2Al\bullet e + H_2O \longrightarrow R_2Al-OH + H\bullet e$

$H\bullet e + n\bullet R \rightleftharpoons RH$ 2.72a

Overall Equation: $AlR_3 + H_2O \xrightarrow{HEAT} R_2AlOH + RH$ 2.72b

Equilibrium mechanism

Stage 1: $AlR_3 \rightleftharpoons R\bullet e + n\bullet AlR_2$

$R\bullet e + R_2AlOH \rightleftharpoons ROH + e\bullet AlR_2$

$R_2Al\bullet e + n\bullet AlR_2 \longrightarrow R_2Al^{\oplus}\Box - - {}^{\ominus}AlR_2\Box$

.(I) **THEINITIATOR** 2.73a

Overall Equation: $AlR_3 + R_2AlOH \longrightarrow ROH + (I)$ 2.73b

Overall overall Equation: $2AlR_3 + H_2O \longrightarrow ROH + (I) + RH$ 2.73c

This is almost in the same manner Z/N initiators were obtained from their combinations such as $Al_3R_3/TiCl_4$ combination. Note that the exact ratio of AlR_3/H_2O combination is 2:1. For the Z/N combination, for isotactic placement the ratio of Al to Ti is also 2:1. This is not just a matter of coincidence but Nature. Notice that the initiator **(Covalently charged-paired initiator)** is dual in character, that is, there are two active centers; one for a male and the other for a female. Notice the unique characters of some metals, wherein the covalent bond between two same metallic centers is chargedly paired, because of what is carried. The same initiator can also be used for propene, butenes, (vinyl chloride), and so many. When used for propylene oxide, the monomer stays all the time in the reservoir carried by the negative center being a Female to obtain the followings-

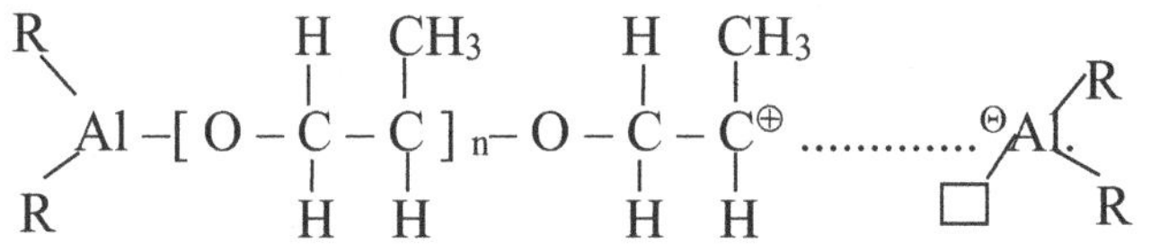

$$R_2Al-[O-\underset{H}{\overset{H}{C}}-\underset{H}{\overset{CH_3}{C}}]_n-O-\underset{H}{\overset{H}{C}}-\underset{H}{\overset{CH_3}{C^{\oplus}}}\cdots\cdots\cdots\cdots {}^{\ominus}Al(\square)R_2$$

Isotactic poly (propylene oxide) 2.74

For the AlR_3/R^1OH combination, the followings are similarly obtained.

Decomposition mechanism

Stage 1: $AlR_3 \longrightarrow R\bullet n + e\bullet AlR_2$

$$R_2Al\bullet e + R^1OH \longrightarrow R_2Al-OH + R^1\bullet e$$

$$R^1\bullet e + n\bullet R \rightleftharpoons RR^1 \qquad [R^1 < R] \qquad 2.75a$$

Overall Equation: $AlR_3 + R^1OH \xrightarrow{HEAT} R_2AlOH + RR^1$ 2.75b

Stage 2 is the same as above. The same initiator (I) is obtained. Al has remained trivalent.

For ZnR_2/H_2O combination, the situation is slightly different, based on the RADICAL Configuration (Not ELECTRONIC Configuration) of Zn, since electrons reside inside the nucleus of an atom. It is inside the Nucleus that we have Nucleonic or Electronic Configuration of atoms as will be shown downstream) in the Periodic Table. Zn is highly polar with six paired unbonded radicals and three vacant orbitals (in 4p) in the last shell. How can one reduce the three vacant orbitals to one and how can one remove the paired unbonded radicals so that zinc will now be made to carry electro-free or nucleo-free-radicals instead of electro-non-free or nucleo-non-free-radicals? One of the six paired unbonded radicals has been used for bonding to the two alkyl groups via excitation, leaving five. With the advantage offered by the presence of three vacant orbitals, one can begin to reduce the numbers of vacant orbitals.

Equilibrium Mechanism

Stage 1: $R_2Zn \rightleftharpoons R - Zn \bullet en + R \bullet n$

$$RZn \bullet en + R_2Zn \xrightleftharpoons{Activation} RZn - \underset{R}{\overset{R}{|}}\!\!Zn \bullet en \quad (A)$$

$$(A) + R \bullet n \longrightarrow RZn - ZnR_3 \quad (B) \qquad 2.76a$$

Overall Equation: $2ZnR_2 \longrightarrow RZn - ZnR_3$ 2.76b

A fraction of the ZnR_2 was kept in Equilibrium state of existence while another fraction was kept in Stable state of existence. Because of the presence of the vacant orbitals, the ZnR_2 in Stable state of existence was activated to give (A) which finally combined with R•n to give (B), reducing the number of vacant orbitals from 3 to 2 and the number of paired unbonded radicals from 5 to 4 on that stable Zn center. With focus still on that stable Zn center, the second stage follows as shown below.

Stage 2: $R_2Zn \rightleftharpoons R - Zn \bullet en + R \bullet n$

$$RZn \bullet en + (B) \xrightleftharpoons{Activation} RZn - Zn(R)(ZnR)(R)(R) \bullet en \quad (C)$$

$$(C) + R \bullet n \longrightarrow (RZn) - {}^{*}ZnR_4 - (ZnR) \quad (D) \qquad 2.77a$$

Overall Equation: $3ZnR_2 \longrightarrow (D)$ 2.77b

So far, notice that we have consumed 3 moles of ZnR_2. The stable central Zn element has been marked with asterisk above. It is the one carrying the load. It is the only one whose Nucleus can be used for radioactivity of the compound if the need arises. On that Zn center, there are three paired unbonded radicals left and one vacant orbital. The next step is to bond with the three paired unbonded radicals, so that Zn center can be made to carry free-radicals.

Stage 3: $ZnR_2 \rightleftharpoons R_2Zn\text{-}\square$

$R_2Zn\text{-}\square$ + (D) $\xrightarrow{\text{Dative Bonding}}$ (E) 2.78

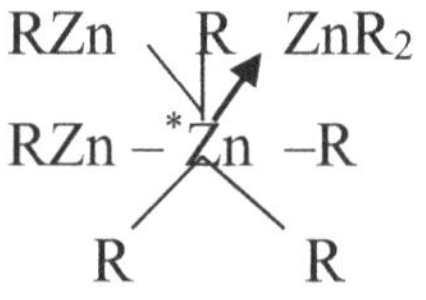

Note that no activation could further take place anymore despite the presence of one vacant orbital, because of the operating conditions. The ZnR_2 could no longer exist in Equilibrium state of existence being suppressed by the presence of (D). Instead, the (D) was datively bonded to ZnR_2 as shown above. With two paired unbonded radical now left, these are now bonded in the last two stages (Stages 4 and 5) to give the followings.

Stages 4 & 5:

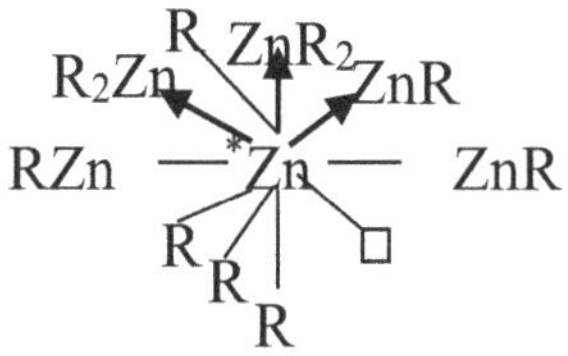

(F) 2.79

Another reason why this had to be done instead of forming additional bonds is because the last shell of Zn based on the Period she belongs to in the Periodic Table, cannot carry more than eighteen (18) radicals. Though the dative bonds formed in (F) above are chargeless, there are now 18 radicals in the last shell of Zn. It is full. With the presence of the dative bonds, Zn can now carry free-radicals and no more non-free radicals. Everything required to do the job are all in the system in the ratios desired as we shall very shortly show.

Overall Equations: $6ZnR_2 \longrightarrow$ (F) 2.80

Decomposition Mechanism

Stage 1:

(F) $\longrightarrow$ (G) + R•n

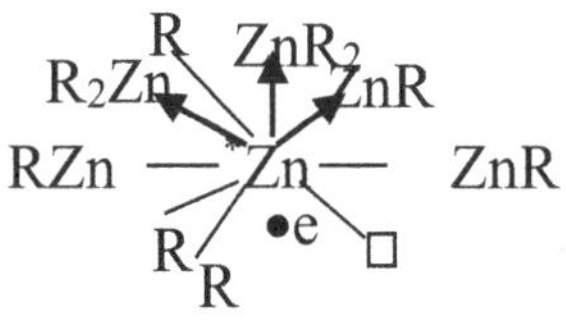

(G) + $H_2O \longrightarrow$ (G) – OH (H) + H •e

H•e + R •n $\rightleftharpoons$ RH 2.81a

Overall Equation: (F) + H_2O ⟶ RH + (H) 2.81b

Equilibrium Mechanism

Stage 1:

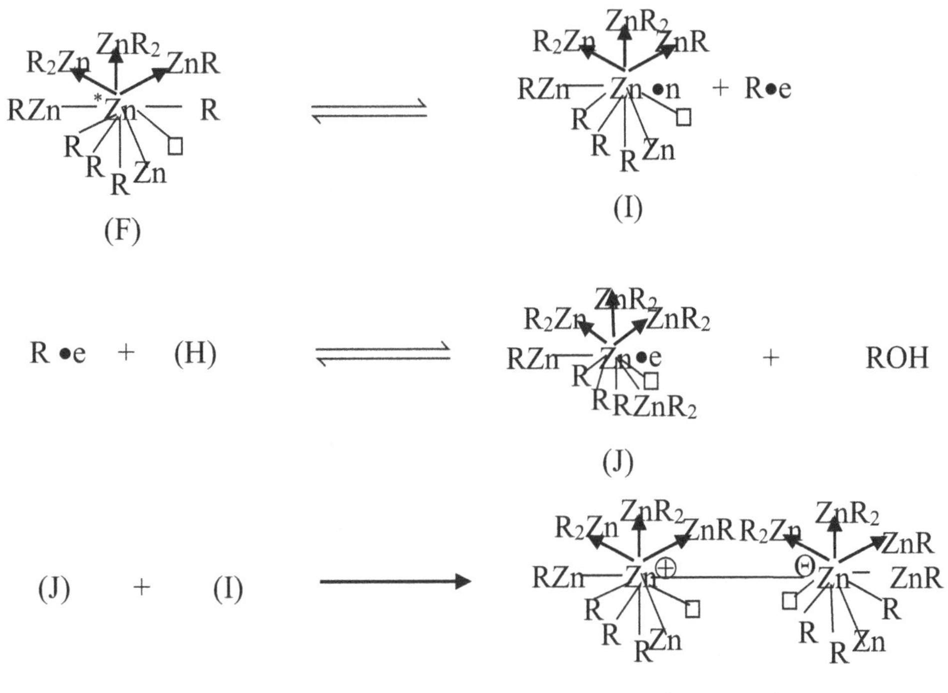

(F) ⇌ (I) + R•e

R•e + (H) ⇌ (J) + ROH

(J) + (I) ⟶ (II)

(II)THE INITIATOR

2.82

Overall overall Equation: $12ZnR_2 + H_2O \longrightarrow RH + ROH + (II)$ 2.83

(II) is the initiator of Z/N type. This is like the first case above- very unique set of initiators. Notice that the metals involved are Non-ionic metals. The initiator is a **Covalently charged-paired initiator**, since both centers are active- one for Males and the other for Females. Note the ratio of the combination-ZnR_2/H_2O used. It is 12 to 1. Without using this ratio, the initiator can never be obtained. Only isotactic placements can be obtained with these initiators. When used with propylene oxide, isotactic poly (propylene oxide) is formed. Zn which is naturally known to be divalent, can be observed to be also tetravalent and hexavalent under certain conditions without breaking the Boundary laws. Based on the mechanisms which we have been seeing so far, one is bound to ask wonderful questions. Why is there so much ORDERLINESS where no LAW is broken? Do these species "COMMUNICATE MATHEMATICALLY" between themselves? What is the essence of OPERATING CONDITIONS? Are these species LIVING and living in a different world of their own? The questions are too countless to list, but to end with WHAT IS THE FORCE BEHIND ALL THESE

GREAT WONDERS WHICH ARE INCOMPREHENSIBLE? That is the ONE AND ONLY MISTERY in humanity.

The greatest foundation for humanity is the PERIODIC TABLE even when what RADICALS are, had not been known. What are universally mistakenly called ELECTRONS are no electrons, but radicals. As already said and recalled again for conditioning purpose, the electron is one of the eight sub-atomic-particles in the Nucleus on the side of Matter, while POSITRON, its mirror image is on the side of Anti-matter. That electron in the Nucleus is still different from the universally so-called "electron" which does not exist. This has been the major missing link since antiquity.

The Equilibrium state of Existence of compounds is very important, because they are fixed and these are their FINGER-PRINTS, just like humans have finger-prints- the thumbs. One has already started preparing an encyclopaedia for the finger prints of compounds. For without knowing them, one cannot get the real mechanisms of all reactions, since before chemical reactions can take place between two components, one of them or both must exist in Equilibrium or Decomposition state of existence. How can a reaction take place when the two compounds are in Stable state of existence? ***To know these finger prints, many chemical reactions must be covered using literature data.***

Now, we will finally consider the case of the use of $FeCl_3$. Shown below is the Radical configuration of the last shell of Fe.

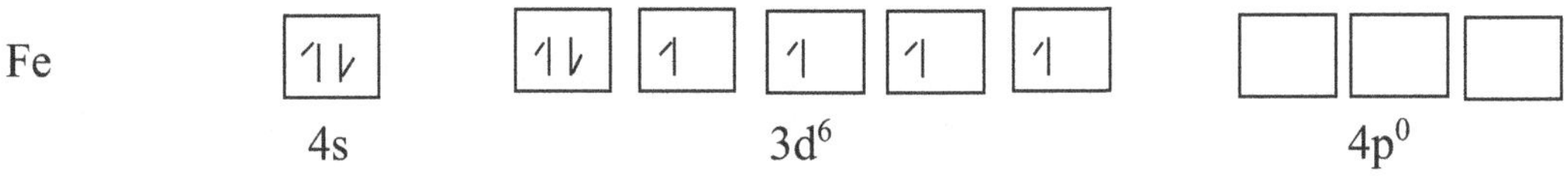

Fe	↿⇂	↿⇂	↿	↿	↿	↿			
	4s			$3d^6$				$4p^0$	

Radical Configuration of last shell of Fe 2.84

↿ is male and ⇂ is female of opposite spins. In the last shell after bonding with three chlorine atoms, there are two paired unbonded radicals, an electro-non-free-radical and three vacant orbitals left. ***With the presence of electro-non –free-radical, $FeCl_3$ can still be activated and can still exist in Equilibrium state of existence.***

Equilibrium mechanism

Stage 1: $FeCl_3 \rightleftharpoons Cl_2Fe \bullet en + Cl\bullet nn$

$$Cl_2Fe \bullet en + FeCl_3 \xrightleftharpoons{Activation} Cl_2Fe - \overset{*}{Fe}(Cl)(Cl)(Cl) \bullet en$$

(A)

(A) + Cl •nn ⟶ Cl_2~~Fe~~ --*$FeCl_4$.
(B) 2.85a

Overall Equation: $2FeCl_3$ ⟶ (B) 2.85b

Note that the central Fe atom has been highlighted. This was the fraction that was kept in Stable state of existence. On that Fe, the number of paired unbonded radicals has been reduced to one, leaving behind two vacant orbitals. That center is again activated in stage 2, noting that the last shell of Fe can accommodate only eighteen (18) radicals based on one of the Boundary laws already stated.

Stage 2: $FeCl_3$ ⇌ Cl_2Fe •en + Cl•nn

Cl_2Fe •en + (B) ⇌(Activation) Cl_2Fe – *Fe •e (bonded to Cl, $FeCl_2$, Cl, Cl, Cl)
(Fe*)

(Fe*) + H_2C—CH(CH_3)–O (propylene oxide) ⇌(Activation) (Fe*) – O – C(H)(H) – C(H)(CH_3) •e

⇌ (Fe*) – O – CH_2 – CH(•e)(•n) – CH_2 + H•e
(E)

H•e + Cl •nn ⇌ HCl

(E) ⟶(Deactivation) (Fe*) – O – CH_2 – CH = CH_2 + Heat
(F) 2.86a

Overall Equation: $3FeCl_3$ + H_3CCH_2CHO ⟶ HCl + (F) 2.86b

Note that the second step above is radically balanced, because the activation is taking place under equilibrium conditions (neither here nor there), the mechanism being Equilibrium mechanism. In the third step, propylene oxide was next activated and in the process transfer species of the first kind of the first type was released to form HCl and finally (F). At the end, the central Fe atom now has one vacant orbital and one electro-free-radical, since no paired unbonded radicals exist in the shell any more. The activation of the propylene oxide was necessitated by something else, as will shortly be explained.

Equilibrium Mechanism

Stage 3: HCl ⇌ H •e + Cl •nn

H•e + (F) ⇌ $HO - CH_2 - CH = CH_2$ + (Fe^*)

(Fe^*) ⇌ •e + $Cl_2Fe - Fe\oplus$ (with Cl, Cl, Cl, Cl, □, $FeCl_2$) ($Fe^\oplus$)

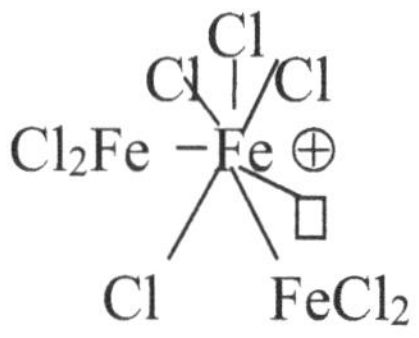

•e + nn• Cl ⇌ $Cl^\ominus$

$Cl^\ominus$ + (B) ⇌ $Cl_2Fe - Fe\,\theta$ (with Cl, Cl, Cl, Cl, Cl, □) (Fe^θ)

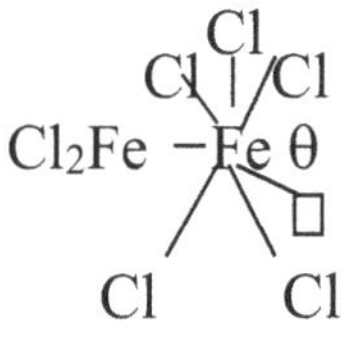

($Fe^\oplus$) + ($Fe^\ominus$) ⟶ $Cl_2Fe - Fe\oplus$ - ....$\ominus Fe - FeCl_2$ (with Cl, Cl, Cl, Cl, □, $FeCl_2$; □, Cl, Cl, Cl, Cl, Cl)

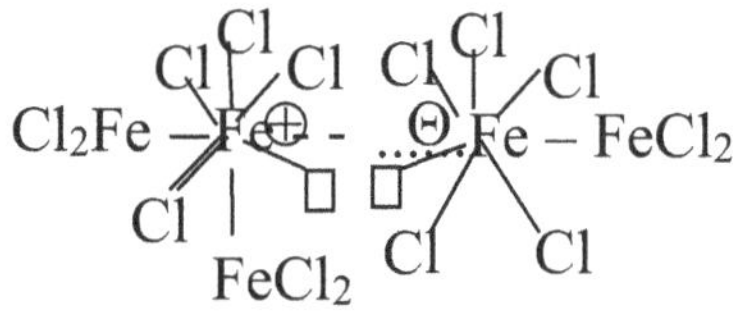

(III) THE INITIATOR

2.87a

Overall Equation: HCl + (F) + (B) ⟶ $HOCH_2CH{=}CH_2$ + (III)

2.87b

Overall overall Equation: $5FeCl_3$ ⟶ (III) 2.87c

The Equilibrium state of existence of (B) [$Cl_2Fe - FeCl_4$] and $Cl_2Fe - FeCl_5 - FeCl_2$ are as follows.

$Cl_2Fe - FeCl_4$ ⇌ (*Equilibrium State of Existence*) Cl •nn + en• Fe – Fe – Cl (with Cl, Cl, Cl, Cl)

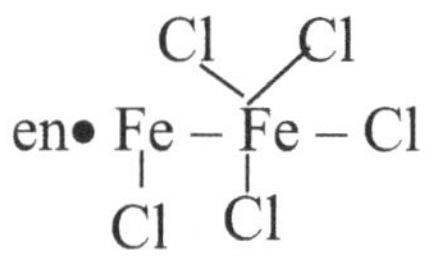

$2FeCl_3$ 2.88a

$Cl_2Fe - FeCl_5 - FeCl_2$ ⇌ (*Equilibrium State of Existence*) Cl •nn + $Cl_2Fe - Fe$ •e (with Cl, Cl, Cl, Cl, $FeCl_2$)

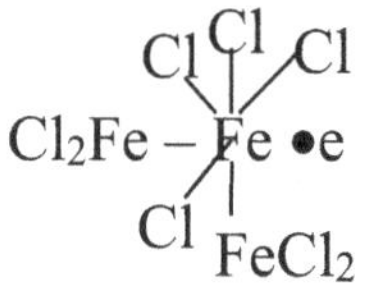

$3FeCl_3$

2.88b

It is no surprise therefore why the third step of Equation 2.85a is what it is, that is, that with a single right double headed arrow for Combination [Not Equilibrium or Decomposition States of existences]. It is also no surprise why propylene was involved in the third step of Stage 2 of Equation 2.86a, otherwise there would have been no stage. The stage would have been "stable, reactive and insoluble". That is, $FeCl_3$ and $Cl_2Fe - FeCl_4$ would be insoluble in themselves. (III) is the real initiator, which unlike the others is carrying an imaginary bond-electrostatic bond. The initiator is **Electro-statically positively charged-paired initiator,** because only the positive center is active and the positive charge is not ionic but free. The negative center is not active, i.e., it cannot be used. Note that, both centers carry reservoirs for monomers, of which the one on the negative end is only available for use. Notice first and foremost, that in ($Fe^{\oplus}$) and ($Fe^{\ominus}$) of the Initiator (III) of Equation 2.87a, and in Cl_2Fe-$FeCl_5$ and Cl_2Fe–$FeCl_5$–$FeCl_2$, the fourth electro-non-free-radical in the radical configuration of Fe in Equation 2.84, remained untouched throughout the stages to the end, but became an electro-free-radical. ***This clearly shows why Fe is not known to be tetravalent, despite the presence of that single radical in the last shell.*** Though in character, it is divalent and trivalent, it can be pentavalent and heptavalent, tetravalent, hexavalent or octavalent. But here, the even number states such as 4, 6, 8 have not been reflected, due to de-polarization of $FeCl_3$. Of these even numbered states, only valence states of 2 and 4 are known to exist.

Just from the paired unbonded radicals and vacant orbitals which we have since antiquity always neglected, *one can clearly observe the wonders of NATURE. All which have been highlighted above, are just but a grain of sand compared to what will be seen in The New Frontiers.*

2.3.5 **Hydrogenation of Some Cycloalkanes**

Not all cycloalkanes (All females) can be hydrogenated. Hydrogenation will depend on the size of the ring. The larger the size of the ring the more difficult it is to provide the MRSE to open the ring instantaneously. Cyclo-propane the smallest size in the family is the most strained and can therefore be readily opened. Cycloalkanes do not have Functional centers and can therefore only be opened instantaneously by providing the MRSE. As already said, they all contain the same amount of S.E. Hence, the order in which the rings are strained is as shown below.

Cyclopropane > Cyclobutane > Cyclopentane > Cyclohexane

Order of Strain in the rings

$$\text{Cycloalkanes (No functional center)} \qquad 2.89$$

Using Hydrogen catalyst or other chemical means, hydrogenation is carried out as follows starting with cyclopropane.

Stage1:

$$H_2 \underset{H_2Catalyst}{\overset{Pt}{\rightleftharpoons}} H\cdot e + H\cdot n$$

$$H\cdot e + \; H_2C{-}CH_2(CH_2) \underset{At\ 80^oC}{\overset{Unzipping\ of\ ring}{\rightleftharpoons}} H.e + n\cdot CH_2 - CH_2 - H_2C\cdot e$$

$$\rightleftharpoons \quad H - CH_2 - CH_2 - H_2C\cdot e$$

$$(A)$$

$$H\cdot n + (A) \xrightarrow{Combination\ State} H_3C - CH_2 - CH_3$$

$$(\mathrm{Pr}opane) \qquad 2.90a$$

$$\underline{Overall\ Equation}:\ H_2 + C_3H_6 \xrightarrow{Pt} C_3H_8 \qquad 2.90b$$

With the use of H_2 catalyst, it is a one-stage process. Hydrogen electro-free-radical does not have enough energy to open the ring instantaneously in the absence of paired unbonded radicals in the last shell of hydrogen. The energy required to instantaneously open the ring is provided by heat. The reaction above takes place at 80°C and above.[28] In the last step of the reaction above, worthy of note is one of the combination states of existence of the propane formed. Shown below are its Equilibrium states of existence (See Equation 2.49).

$$C_3H_8 \underset{at\ or\ below\ Melting\ point}{\overset{Equilibrium\ state\ of\ existence}{\rightleftharpoons}} H_3C\cdot e + n\cdot C_2H_5$$

$$(A)$$

$$\underset{at\ or\ below\ Boiling\ point}{\overset{Equilibrium\ state\ of\ existence}{\rightleftharpoons}} Part\ (A)\, above + \mathrm{Re}maining\ Part[H\cdot e + n\cdot C_3H_7]$$

$$\underset{at\ above\ Boiling\ point}{\overset{Equilibrium\ state\ of\ existence}{\rightleftharpoons}} H\cdot e + n\cdot C_3H_7 \qquad 2.91$$

None of the states above is carrying hydrogen nucleo-free-radical, that is, a hydride because Equilibrium state of existence is fixed and that in which hydrogen atom or electro-free- radical is held. Notice that the hydrogenation can only take place radically.

When cyclobutane is involved, in view of the increased size of ring, higher temperature of the order of 120°C is required to open the ring instantaneously. The larger the size of the ring, the higher the temperature required. If ethylene oxide was used in place of cyclopropane, the operating conditions for hydrogenation would be less than the 80°, because of the presence of paired unbonded radicals on the oxygen center (A functional center).

Stage 1:

$$H_2 \underset{\text{Hydrogen catalyst}}{\overset{Pt}{\rightleftharpoons}} H \cdot e + n \cdot H$$

$$H \cdot e + H_2C - CH_2 \text{ (ring with } \ddot{O}\text{)} \underset{\text{on the Functional center}}{\overset{\text{Attack of } H \cdot e}{\rightleftharpoons}} H - O - CH_2 - H_2C \cdot e \quad (A)$$

$$H \cdot n + (A) \xrightarrow{\text{Combination state}} HOC_2H_5 \qquad 2.92a$$

$$\underline{\textit{Overall Equation}}: H_2 + C_2H_4O \xrightarrow{Pt} C_2H_5OH \qquad 2.92b$$

The strain energy in ethylene oxide is smaller than that in cyclopropane, because of the presence of two paired unbonded radicals on the oxygen center. In the hydrogenation of ethylene oxide, notice that ethanol an alcohol is obtained with much ease. Notice that long chain of the component being hydrogenated cannot be obtained, because of the presence of both male and female parts of H_2 (H•e and H•n) in the system. In the presence of the opposite counterparts that are not paired, the chain can never grow. If some or all of the hydrogen atoms on the two monomers or compounds (ethylene oxide and cyclopropane) are replaced by other groups or components, the situation will become different in terms of strain energy contained or carried by the ring, the point of scission, and the type of initiator to use. However, in order to polymerize such monomers, one can use paired coordinated charged or radical initiators or just free-media radical initiators.

2.3.6 **Polymerization of Cycloalkanes**

Not all cycloalkanes can be polymerized. They are all Nucleophiles, i.e., Females. To polymerize cyclopropane, which is first member of the family, very mild operating conditions, are used, because of the small size of the ring. Their rings with no externally located activation center and no functional center can only be opened instantaneously, as has already been said. The larger the size of the ring, the higher the operating conditions (Temperature) needed for opening of their rings. Hence, only three-, four-, and five- membered rings can be opened instantaneously, based on the use of limited operating conditions. Herein, one will use coordination initiators such as BF_3/ROR combination or BF_3 alone[29, 30]. This begins for this case like most cases seen so far, with the Initiator preparation step. When initiators have to be prepared, it is better to be done first before the addition of monomer into the system.

Initiator preparation step

Stage 1:

$$ROR \underset{\text{State of Existence}}{\overset{\text{Equilibrium}}{\rightleftharpoons}} R\cdot e + nn\cdot OR$$

$$RO\cdot nn + BF_3 \underset{\text{Addition}}{\overset{\text{Electrostatic}}{\rightleftharpoons}} \underset{(A)}{n\cdot \overset{\overset{OR}{|}}{BF_3}}$$

$$R\cdot e + (A) \xrightarrow{\text{Combination State}} R^{\oplus}.......^{\odot}\overset{\overset{OR}{|}}{BF_3}$$

(*Electrostatically positively charged – paired Initiator*)
{*B*} 2.93

OR

Stage 1:

$$BF_3 \underset{\text{Existence}}{\overset{\text{Equilibrium State of}}{\rightleftharpoons}} F\cdot nn + e\cdot BF_2$$

$$F\cdot nn + BF_3 \underset{\text{Addition}}{\overset{\text{Electrostatic}}{\rightleftharpoons}} n\cdot BF_4$$

$$F_2B\cdot e + n\cdot BF_4 \xrightarrow{\text{Combination state}} F_2B^{\oplus}........^{\odot}BF_4$$

(*Electrostatically positively charged – paired Initiator*)
(*C*) 2.94

(B) and (C) are Electrostatically positively charged-paired (Not cationic coordination initiators as presently used in present-day Science) initiators obtained under mild but different operating conditions. The charges carried by the paired centers are not ionic (Real), but electrostatic (Imaginary) in character and they cannot be isolated. The electrostatic charges formed are strong enough to open a three-membered cyclopropane ring instantaneously as shown below using (B).

Stage 2: [Initiation Step]

$$R^{\oplus}.......^{\odot}BF_3(OR) + H_2C - CH_2 \xrightarrow[\text{Ring followed by Addition}]{\text{Instantaneous Opening of}} R - CH_2 - CH_2 - H_2C^{\oplus}......^{\odot}BF_3(OR) \quad (D)$$

(ring CH_2 bridging the two carbons)

2.95

The initiation above has taken place chargedly. This brings us to the end of the initiation step where the possibility of adding one monomer unit to the initiator exists. After the initiator is formed in Stage 1 by Equilibrium mechanism, this is then followed by addition of one monomer unit by Combination mechanism in Stage 2. If there is no monomer unit in the system, only (D) is formed, a product that is unstable, since it can terminate itself by releasing OR group to form an ether. The absence of CH_3 or groups of lesser capacity than R-CH_2-CH_2- group on the terminal carbon center carrying the positive charge makes termination by starvation impossible, that is, hydrogen cannot be released. Finally, in the initiators above [(B) and (C)], on the Boron center, there is no vacant orbital (a reservoir) where the monomer can reside for regular placements. Hence, these initiators with no reservoirs cannot be used for strong stereo-regular placements. In the absence of vacant orbital and steric limitations, the monomers can reside on any side of the horizontal axis to give irregular or amorphous polymer, non-crystalline in character. For the case of cyclopropane considered herein, a symmetric cyclic monomer, linear crystalline polyethylene is produced in the presence of more monomers in the system. Thus, after (D) is formed, propagation steps follow as shown below.

Propagation Step:
n Stages:

$$(D) + n(C_3H_6) \xrightarrow[\text{Steps}]{\text{Combination}} R - (CH_2)_3 - [C_3H_6]_{n-1} - CH_2 - CH_2 - H_2C^{\oplus}....^{\odot}BF_3(OR)$$

(Initiated Monomer) *(Ringed Monomer)* *(Living polymer Chain)* *(E)*

2.96

Additions of monomers continue until the optimum chain length dictated by the *Glass Transition temperature* of the polymer produced is reached in the absence of a terminating agent. The mechanism here is by Combination. A living polymer chain is eventually produced. How do we kill such a living polymer chain when hydrogen cannot be released as a cation and not as a free radical? To kill such living chain, like the case of polypropylene, the chain kills itself when no monomer is around or when the optimum chain length has been reached. But unlike the case of polypropylene, no transfer species can be released from the growing chain, but from the counter center as shown below. There is no need to use a foreign agent such an alcohol to kill the chain, unless the need arises.

Termination Step:

Stage 1:

$$(E) \xrightleftharpoons[\text{Existence}]{\text{Equilibrium State Of}} R-(CH_2)_3-(CH_2-CH_2-CH_2)_{n-1}-CH_2-CH_2-H_2C^{\oplus}.......{}^{\ominus}BF_3 \cdot e$$

$$(F) \quad + \quad nn\bullet OR$$

$$(F) \xrightleftharpoons[\text{From }(F)]{\text{Release Of }BF_3} R-(CH_2)_{3n}-CH_2-CH_2-H_2C\cdot e \; [-R^{//}\cdot e] + BF_3 + Energy$$

$$R^{//}\cdot e + nn\cdot OR \longrightarrow R^{//}OR \qquad 2.97$$

The dead polymer produced is ether which is dead. If it was an alcohol such as the use of water, then it will be living. For example, if an alcohol is used, the followings are obtained.

Stage1:

$$R^{/}OH \xrightleftharpoons[\text{of Existence}]{\text{Equilibrium State}} H\bullet e \quad + \quad nn\bullet OR^{/}$$

$$H\bullet e \quad + \quad (E) \xrightleftharpoons[\text{of OH}]{\text{Abtraction}} R-(CH_2)_3-(CH_2-CH_2-CH_2)_{n-1}-CH_2-CH_2-H_2C^{\oplus}....{}^{\ominus}BF_3 \cdot e$$

$$(F) \quad + \quad ROH$$

$$(F) \xrightleftharpoons[\text{From }(F)]{\text{Release Of }BF_3} R-(CH_2)_{3n}-CH_2-CH_2-H_2C\cdot e \; [-R^{//}\cdot e] + BF_3 + Energy$$

$$R^{//}\cdot e + nn\cdot OR \longrightarrow R^{//}OR \qquad 2.98$$

Like the case above, a dead polymer is produced along with a different alcohol. The dead polymer is ether. The polymeric ether is more stable than the polymeric alcohol. The Equilibrium state of existence of Ethers is as follows.

$$R^{//}OR \xrightleftharpoons[\text{of Existence}]{\text{Equilibrium State}} R\bullet e \quad + \quad nn\bullet OR^{//}$$

where $R^{//}$ is greater than that of R. $R^{//}$ and R are alkylane groups.

$$H_5C_2OCH_3 \rightleftharpoons H_3C\bullet e \; + \; nn\bullet OC_2H_5$$

$R^{//}$ is the polymer chain. 2.99

To produce a polyethylene, a hydride must be used. H_2 can be used only in the presence of H_2 catalyst as shown below.

Stage 1:

$$H_2 \underset{H_2\ Catalyst}{\overset{Pt}{\rightleftharpoons}} H\cdot e + n\cdot H$$

$$H\bullet e \ + \ (E) \underset{OR\ group}{\overset{Abstraction\ of}{\rightleftharpoons}} (F) + HOR$$

$$(F) \underset{From\ (F)}{\overset{Release\ Of\ BF_3}{\rightleftharpoons}} R^{//}\cdot e \ + \ BF_3$$

$$R^{//}\bullet e \ + \ n\bullet H \longrightarrow R-(CH_2)-[C_3H_6]_{n-1}-CH_2-CH_2-CH_3$$

$$(Polyethylene) \qquad 2.100a$$

$$\underline{OverallEquation}: H_2 + (E) \xrightarrow{Pt} H_2O + BF_3 + Polyethylene \qquad 2.100b$$

Note that when termination is from a growing polymer chain, dead terminal double or triple bonded polymers are produced. H cannot be released from any growing polymer chain as a hydride, i.e. as H•n, unless at very high operating conditions of the order of 1000^0C, such as used during polymerization of ethylene. This is said to be one of the sources of branching during the polymerization of ethylene. H can only be released from a growing polymer chain as an atom under very mild operating condition. When termination is from the counter center of a growing chain, living or dead polymers are obtained, depending on what the chain is carrying and the type of agent coming from the counter center. Termination from a growing polymer chain can give birth to another new chain, if there are monomers in the system. This is usually called ***"Transfer to monomer" step***. The transfer species released will begin another Initiation followed by propagation. There are different kinds and types of termination steps. Two growing polymer chains can add together and form a dead polymer only when the opposite of what the chain is carrying is not in the system (Termination by Combination). This can only be done radically, and indeed electro-free-radically.

Termination of a Nucleo-free- Radical Growing Polymer Chain of Ethylene as very high operating conditions- of the order of 1000⁰C

Stage1:

$$2N-(CH_2-CH_2)_{n-1}-CH_2-H_2C\cdot n \xrightarrow[transfer\ species\ at\ T>1000^0C]{Release\ of} 2N-(CH_2-CH_2)_{n-1}-CH=H_2C$$

$$2(A) \qquad\qquad 2(B)$$

$$(Dead\ terminal\ double\ bond\ polymer) + 2H\cdot n + Energy$$

$$2H\cdot n \xrightarrow{Combination\ State} H_2 \qquad 2.101a$$

$$\underline{Overall\ Equation}: 2(A)(A\ Radical) \xrightarrow[Mechanism]{Decomposition} H_2 + 2(B) + Energy \qquad 2.101b$$

The stage above has taken place via Decomposition mechanism. It cannot take place via Equilibrium mechanism. The N above is the nucleo-free-radical initiator. *It is possible that under such operating conditions, H_2 is indeed formed, along with 2H .n, ready to be used for re-initiation. With so much monomers, many living chains and H_2 in the system, the possibility of having branched chain can be made possible at such harsh operating conditions.*

Based on universal data, there are different types of polyethylene not with respect to density alone, but with respect to the source of the poly-ethylene. Those from ethylene (ethene) are called polyethylene (polyethene). Those from cyclopropane (cyclopropene) should be called poly(cyclo-propane) (polycyclopropene) and those from diazomethane should be called polymethylene. Based on the New Frontiers, cyclopropane is indeed cyclopropene, because of the presence of the invisible π-bond in the ring and the fact that it is an isomer to propene and not propane.

2.3.7 **Structures of Sydnones and Diazoalkanes**

There are countless numbers of compounds, whose true structures and even names are not known universally, ***because of lack of understanding of Boundary laws, full understanding of what Bonds and Charges are, full understanding of what Radicals and Electrons are, and much more.***

Sydnones were first prepared by Earl et al by the action of cold acetic anhydride on ***N-nitroso-N-phenylglycines***. Earl formulated the reaction as follow[31].

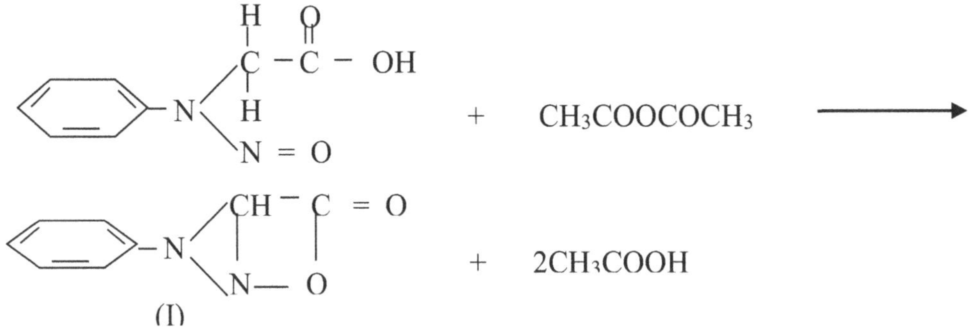

$$C_6H_5-N(N=O)-CH_2-C(=O)-OH \quad + \quad CH_3COOCOCH_3 \longrightarrow$$

$$C_6H_5-N\langle CH-C(=O)-O-N \rangle \ (I) \quad + \quad 2CH_3COOH \qquad 2.102$$

Sydnones are white or pale yellow crystalline compounds which are hydrolysed by hot five percent sodium hydroxide to the original N-nitroso-N-arylglycine, and by moderately concentrated hydrochloric acid to an acrylhydrazine, formic acid and carbon dioxide.

The structure (I) above proposed by Earl is said to be similar to that of β - lactone. But Baker et al offered a number of objections to this structure, one of the most important of which, is the fact that a system containing fused three- and four- membered rings would be highly strained, and consequently is unlikely to be produced by dehydration with acetic anhydride; β - lactones are not produced under these conditions[31]. Therefore, they proposed a five-membered ring which cannot be represented by any one purely covalent structure. They put forward a number of charged structures which are resonance stabilized.

Based on the current developments herein, the structure (I) is not the true structure, partly for the reasons already offered above. Very little of the Bakers et al structures are indeed correct. Nevertheless before proposing the true structure, the mechanism of the reaction of Equation 2.102 will first be provided.

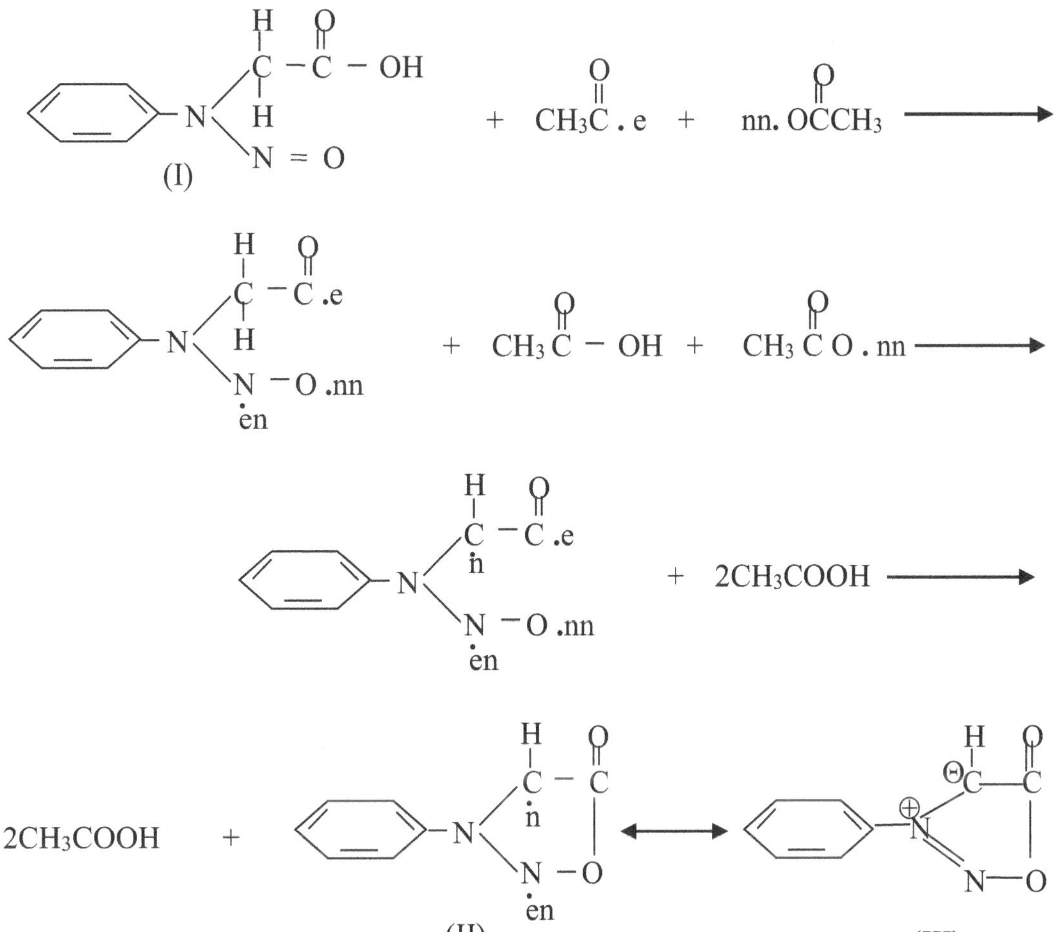

H O
C − C − OH
N H
N = O
(I)
\+ $CH_3\overset{O}{C}$. e + nn. $O\overset{O}{C}CH_3$ ⟶

H O
C − C .e
N H
N − O .nn
en
\+ $CH_3\overset{O}{C}$ − OH + $CH_3\overset{O}{C}$ O . nn ⟶

H O
C − C .e
n
N
N − O .nn
en
\+ $2CH_3COOH$ ⟶

$2CH_3COOH$ +
H O
C − C
N n
N − O
en
(II)
⟷
H O
⊖C — C
⊕N
N — O
(III)

2.103a

Shockingly enough, the reaction above, (I) down to (III), is just a one- or two-stage Equilibrium mechanism system and all seen so far chemically take place only radically. In the reaction above, the acetic anhydride was in Equilibrium state of existence to commence the reaction. Indeed, (I) should be the one to commence the reaction to first form acetic acid in one stage, since it is an acid. This has been used to illustrate some fundamental principles-order of States of existence of compounds, wherein all can be in Equilibrium state of existence at the same time or one is and the other is suppressed if for example they belong to different families. If they belong to the same family, it is only one at a time beginning with the first member of the family which is always the smallest or the last member of the family which is the largest in the family, depending on the type of family, and so on.

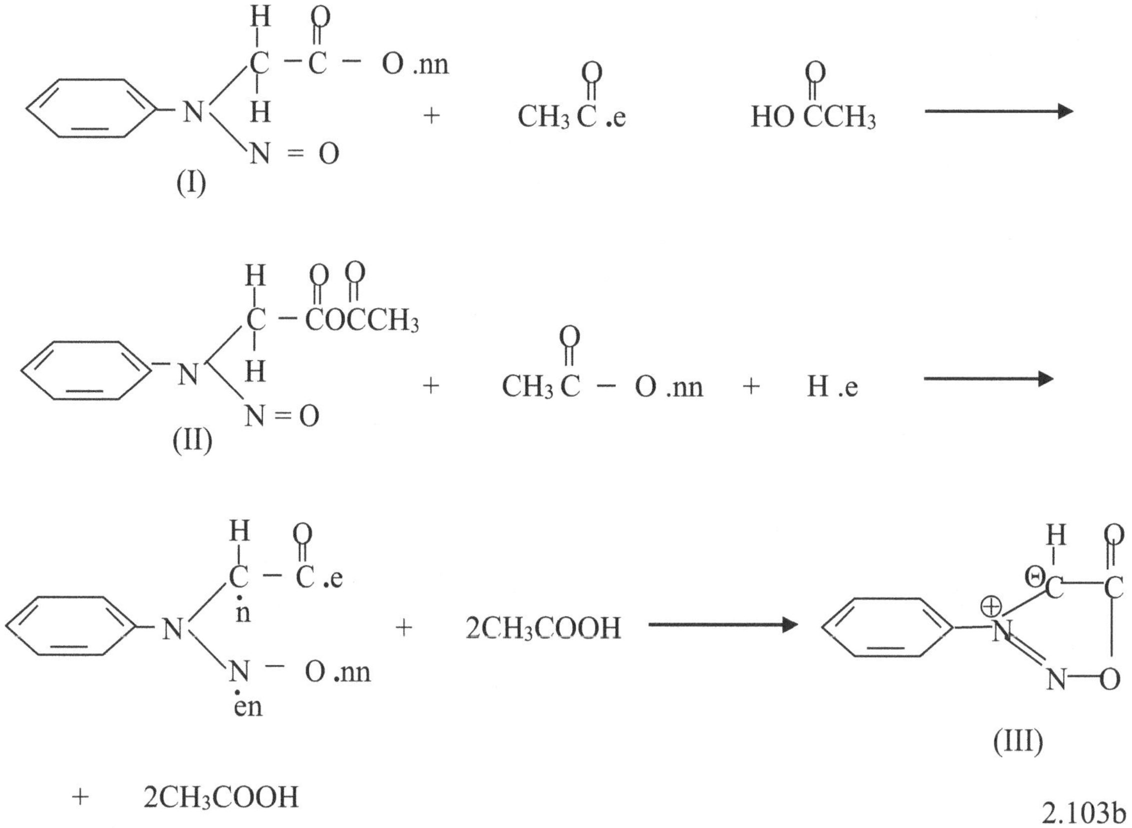

(I) + $CH_3\overset{O}{\overset{\|}{C}}$ .e HO$\overset{O}{\overset{\|}{C}}CH_3$ ⟶

(II) + $CH_3\overset{O}{\overset{\|}{C}}$ − O .nn + H .e ⟶

+ $2CH_3COOH$ ⟶ (III)

+ $2CH_3COOH$ 2.103b

It is Equation 2.103b that is favoured as opposed to Equation 2.103a. This is a two stage Equilibrium mechanism system. (III) is indeed what is obtained and that is the real or true structure of sydnone. (Note that when H is changed to R, then we have sydnones.) They are self-activated monomers or compounds. Thus, like diazoalkanes, it favours ***Polar/Radical resonance stabilization phenomenon***, but of the **closed-loop type and discrete in**

character. It is discrete, because based on the mechanism of resonance stabilization, it does not go beyond (III). It is the standing electro-radical that moves to grab a nucleo-free or non-free-radical either from an adjacently lo-

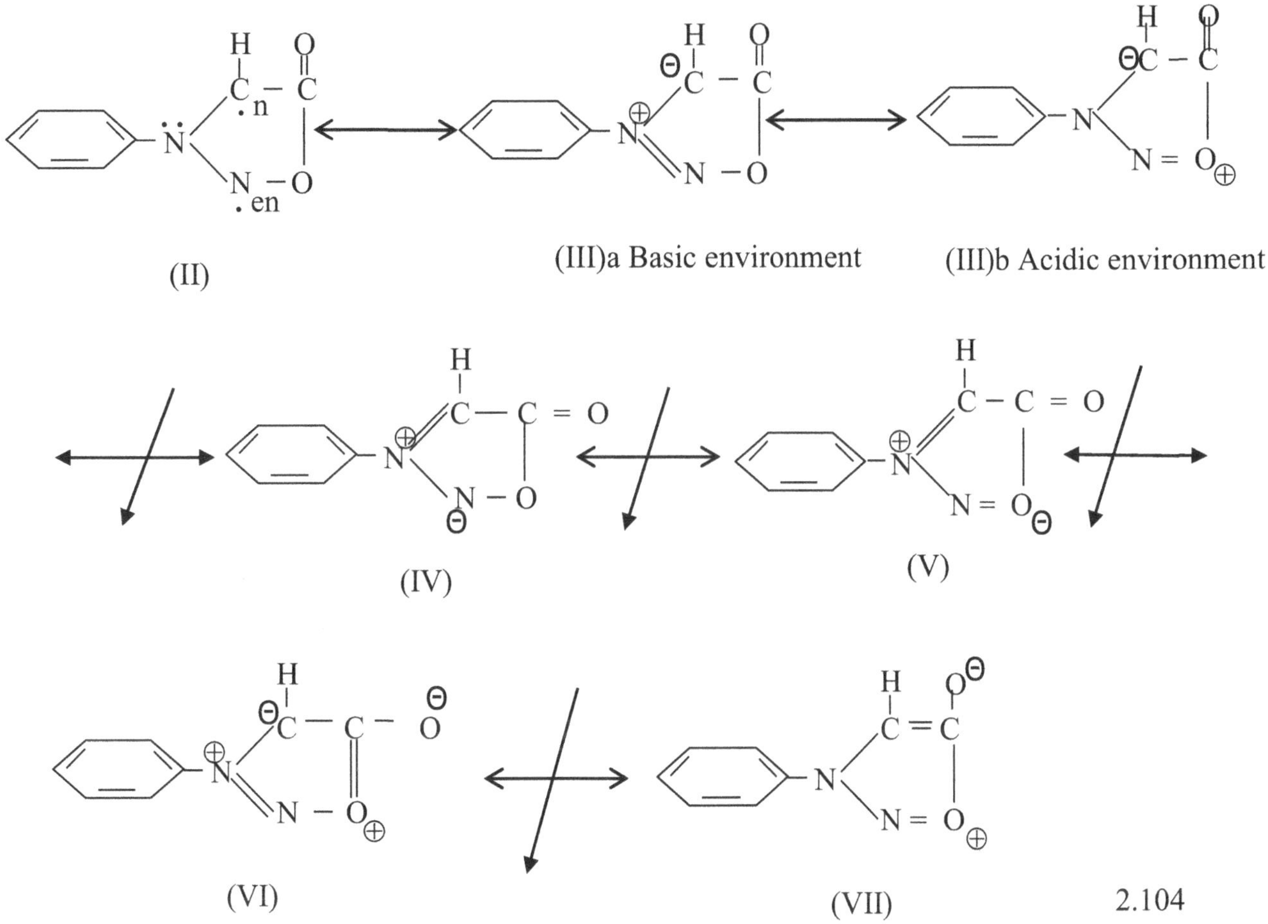

(II) (III)a Basic environment (III)b Acidic environment

(IV) (V)

(VI) (VII) 2.104

cated π-bond or from paired unbonded radicals, for which the existence of (IV), (V), (VI), and (VII) from (II) are not possible. (III)b is favoured, because the Electrophilic character of the monomer still remains. It is favoured in an acidic environment, while (III)a is favoured in a basic or neutral environment. In resonance stabilization, note that charges carried by their centers cannot be removed and moved. Nucleo-free- or non-free-radical can be removed from their carriers. Only electro-radicals can be moved from their centers or carriers to leave a positive charge or a hole behind depending on the receiving center or from paired unbonded radicals to leave a nucleo-radical behind. Recalled below is the case of diazomethane in the New Frontiers. Diazomethane looks like a three-membered ring too strained to exist as a ring.

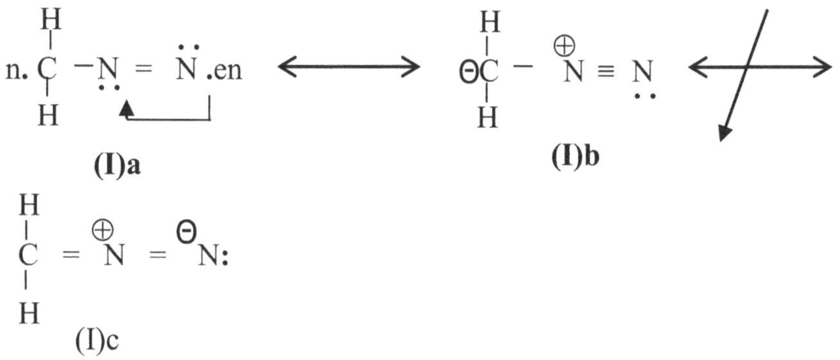

$$\text{n.}\overset{H}{\underset{H}{C}} - \underset{\cdot\cdot}{N} = \overset{\cdot\cdot}{N}\text{.en} \longleftrightarrow \ominus\overset{H}{\underset{H}{C}} - \overset{\oplus}{N} \equiv \underset{\cdot\cdot}{N} \not\longleftrightarrow$$

(I)a **(I)b**

$$\overset{H}{\underset{H}{C}} = \overset{\oplus}{N} = \overset{\ominus}{N}:$$

(I)c

<u>Opened - loop polar/radical resonance stabilization phenomenon</u>

2.105

This is also discrete in character. Unlike the case above, (I)c above is not one of resonance stabilized forms of diazoalkanes as will be seen in the New Frontiers. (I)a and (I)b are the two monoforms of diazomethane [Real name Diazomethylene to be seen in the New Frontiers]. (II), (III)a and (III)b of Equation 2.104 are the only forms of sydnone. The polar forms are the real and stable structures of Sydnones and Diazomethane or diazo-alkanes. When they are heated, the radical forms are obtained. This is the most important part of all the resonance stabilized forms. Note the name the New Frontiers has given to this kind of resonance stabilization phenomenon.

As can be seen above, in an attempt to provide the true structures of diazoalkanes and sydnones, one has been forced to introduce another kind of resonance stabilization phenomena with two of its type- Closed and Discrete, Opened and discrete. As has been said, based on what is contained in the New Frontiers, we have just only begun a long and endless journey. Yet, one cannot say that wonderful foundations have never been laid in the past, because without those foundations, what we have just begun to see would have been impossible, since the author is human.

2.4 Conclusions

For the first time, the existence of real and imaginary charges and bonds has been identified based on literature data, observations and the new concepts. In order words, for the first time the Real and Complex sides of Chemistry have been identified. From it, one has begun to show the origin of complex numbers in Mathematics. So also for the first one has begun to show the existences of male and female macromolecules. For the first time also, new theories on rings related to one of the subject areas of Feeling in Physics have been introduced. For the first time, one has begun to show what is INITIATION, and all the different types of Initiators in existence and so on. All these based on the New Science will find very strong applications to many disciplines. Why carbon monoxide is poisonous has been explained. Why vinyl chloride and vinyl acetate cannot be polymerized chargedly but only free-radically has been explained.

The mechanisms by which alternating and some other regular placements are obtained for Addition monomers have begun to be shown. How they are initiated, propagated and terminated have begun to be shown. The real structures of elements such as zinc dust and some other compounds have also begun to be shown. In all of them, notice the use of new forms of chemical equations providing explanations of things that could not clearly be explained in the past. Shocking to all is the realization of the fact that all chemical reactions take place only radically. Yet from what we have seen so far, charges are very important. Then one will begin to wonder what type of world we have been living in. It was only with polymeric reactions, we saw the involvement of charges only of the covalent, electrostatic, very little of the ionic and none yet of polar kinds. We have only just begun a long endless journey both in physical and meta-physical world in the real and imaginary domains.

2.5 *Foods for Thought*

The foods for thought are just enlightening questions, for there are many of us who cannot still believe what have been presented so far, particularly those who think they know so much, but know nothing. This is just the beginning of a New Science for which humanity cannot forget if and only if they have the ability to see with the eye of a needle.

PB2.1 Provide the new classifications for Bonds and Charges comparing it with present-day Knowledge.

PB2.2 Define the following terms as they apply to polymerization systems:

a) Initiators.
b) Initiation step.
c) Alkylation step.
d) Activation center with visible and invisible π-bonds.
e) Activation center with Functional center.
f) Activation center with paired unbonded radicals/Vacant orbitals.
g) Functional group.
h) Propagation step.
i) Termination step.
j) Transfer species of the first kind of the first type.
k) Strain energy (S E).
l) Minimum Required Strain Energy (Min. R S E).

PB2.3 Define the followings as they apply to chemical reactions on a micro-molecular scale.

a) Equilibrium stage.
b) Decomposition stage.
c) Hydrogenation stage.
d) Covalent bonds.
e) Polar bonds.
f) Electrostatic bonds.
g) Ionic bonds.
h) Dative bonds.
i) Isolations of charges and radicals with their carriers.
j) Electrodynamic forces of repulsion.
k) Isolation of charges and radicals from their carriers.
l) Diffusion of radicals and charges.

PB2.4 Explain the mechanisms for the production of Z/N and Electrostatically positively charge-paired initiators from $TiCl_4/AlR_3$ combinations. Compare the two types of initiators obtained from them.

PB2.5 Shown below are all known routes of polymerizations which include the use of the following initiators obtained from specific sources:

Free-media Initiators:

(i) Nucleo-free-radical initiators (In present-day Science, it is called Free-radical initiator and unknown),
(ii) Electro-free-radical initiators ("Unknown" in present-day Science)
(iii) Nucleo-non-free-radical initiators ("Unknown" in present-day Science)
(iv) Electro-non-free-radical initiators ("Unknown" in present-day Science)
(v) Anionic initiators ("Unknown" in present-day Science)

Paired-media Initiators:

(vi) Electrostatically positively charged-paired initiators (In present-day Science, it is called Cationic ion-paired initiator),
(vii) Covalently charged-paired initiators (In present-day Science, it is called Z/N initiator),
(viii) Ionically charged-paired initiators (In present-day Science, it is called Anionic Ion-paired initiator that which is acceptable),
(ix) Electrostatically negatively charged paired initiators ("Unknown" in present-day Science)
(x) Electrostatically cationic charged-paired initiators; exist but cannot be used for polymerization. ("Unknown" in present-day Science)
(xi) Electrostatically anionic charged-paired initiators ("Unknown" in present-day Science)
(xii) Free-radical-paired initiators ("Unknown" in present-day Science)
(xiii) Free-radical/Non-free-radical-paired initiators ("Unknown" in present-day Science)

All of them have been used in researches without the researchers knowing what is there.

Answer the following questions.

a. Give examples of each one of them.
b. Why is it that ethene (In present-day Science, it is called ethylene), favours the use of all the routes above, but propene (In present-day Science, it is called propylene) favours only routes which involve the use of initiators which are male in character?

c. Why is it that monomers such as vinyl chloride and methyl acrylate which are female in character favour free-radical polymerization routes and none of the charged polymerization routes?
d. Why is it that monomers such as vinyl ethers ($H_2C{=}CHOR$) favour the use of only specific types of initiators which are male in character? List the types of initiators.
e. Why is it that monomers such as acrylamide, acrylonitrile, acrolein, cumulenes, ketenes and so on favour the use of initiators which are female in character? List the types of initiators.
f. If SO_2 is to be polymerized, what types of initiators can be used for its homopolymerization.
g. List the driving forces favouring the existence of all the thirteen initiators.

PB2.6 Answer the following questions as briefly as possible.

a) Distinguish between an atom and its elements, using zinc, oxygen and carbon as examples.
b) What are driving forces favouring the existence of ionic bonds?
c) How can a ringed compound be opened? Explain. What of if the ringed compound has no point of scission, what do we do?
d) Carbon monoxide is known to be poisonous. Under what condition does this take place?
e) Are Deuterium and Tritium elements of hydrogen? Explain whether yes or no.
f) How are beautiful crystals obtained?

PB2.7 a) What are Alkenes? Name the first member of the family along with the next three members.

b) What are the alkylenes? If the first member is methylene, then name the next three members.
c) What are diazo-alkylenes? What are also diazo-alkanes? If they both exist, what are the possibilities of their being used as monomers? What is the origin of diazo-alkylenes which in present -day Science are called diazo-alkanes?

PB2.8 Describe the mechanisms for the polymerizations of:

a) Ethene (In present-day Science, it is called ethylene) starting from initiator preparation step.

b) Cyclo-propene (In present-day Science, it is called cyclopropane) starting from initiator preparation step.

c) Diazo-methylene (In present-day Science, it is called diazo -methane) starting from initiator preparation step.

PB2.9 a) Describe the mechanisms for the alternating copolymerization of sulfur dioxide with propene.

b) Under what conditions can alternating copolymers be obtained from a single monomer such as dimethylketene, that is, what types of initiators can be used to give alternating placements?

PB2.10 a) What is the origin of the existence of male and female compounds? Has it got to do with the existence of male and female chromosomes? Explain

b) What is Enzymatic engineering or indeed what are enzymes? For some of them that carry the genes as a backbone, then what is the so-called Genetic engineering? Can you show the structure of just one enzyme, such as fructase?

References

1. S. N. E. Omorodion, *"New Classifications for Radicals and Their Impacts"* Chapter 1 herein.
2. F. W. Billmeyer, "*Textbook of Polymer Science*", Interscience Publishers, Wiley Inc., New York, (1964).
3. F. W. Seymour, "*Introduction to Polymer Chemistry*", International Student Edition, McGraw-Hill, New York, (1971).
4. C. R. Noller, "*Textbook of Organic Chemistry*", W. B. Saunders Company, (1966), pg. 22.
5. E. D. Isaacs, A. Shukla, P.M. Platzman, D. R. Hamann, B. Barbiellini, and C. A. Tulk, *"Covalency of the Hydrogen Bond in Ice: A Direct X-Ray Measurement"*, Phys. Rev. Lett. 82, 1999, 600-603.
6. C. H. Cho, S. Singh and G. W. Robinson, *"Liquid water and biological systems: the most important problem in Science that hardly anyone wants to see solved"*, Faraday Discuss, 103 (1996), 19-27.
7. P. G. Kusalik and I. M. Svishchev, *"The Spatial Structure in Liquid Water",* Science, 265, 1994, 1219-1221.
8. S. Mashimo, *"Structure of water in pure liquid and biosystem"*, J. Non-crystalline Solids 172-174, 1994, 1117-1120.
9. R. C. Dougherty and L. N. Howard, *"Equilibrium structural model of liquid water. Evidence from Heat capacity, spectra, density and other properties",* J. Chem. Phys., 109, 1998, 7379-7393.
10. G. Odian, "*Principles of Polymerization*", McGraw-Hill, New York, (1970).
11. C. L. Fraser, A. P. Smith, *"Metal Complexes with Polymeric Ligands: Chelation and Metalloinitiation Approaches to Metal Tris (bipyridine) - Containing Materials"*, J. Polym. Sci., Part A: Polym. Chem. 2000, 38, 4704-4716.
12. J. L. Bender, K. W. Bothner, C. L. Fraser, M. Li, F. S. Richardson, Q. Shen, *"Luminescent, Polymeric Lanthanide Complexes for Block Copolymer Photonic Materials",* Polym. Prepr. (Am. Chem. Soc., Polym, Chem.) 2002, 43, 424.
13. A. P. Smith, C. L. Fraser, *"Ruthenium-Centered Heteroarm Stars by a Modular Coordination Approach Effect of Polymer Composition on Rate of Chelation",* Macromolecules, 36, 2003, 5520-5525.
14. D. J. Joubert, I. Tincul, J. R. Moss, Inorganic '99 Conf. Proceedings, 1999, p49.
15. W.C. Conner, E. L. Weist, Ito and F. Jacques, J. Phys. Chem., 93, 1989, 4138.
16. I.W.B. Innes (1968) Exp. Methods Catal. Res., 44.

17. F. Rodriquez, "*Principles of Polymer Systems*", McGraw-Hill, New York, (1970).
18. Chem. Eng. "*Gas-phase-High density Polyethylene process*", Chem. Eng. 72, Nov. 26, (1973).
19. C. R. Noller, "*Textbook of Organic Chemistry*", W. B. Saunders Company, 1966, pg. 380.
20. C. R. C. Handbook of Chemistry and Physics, 73rd edition, CRC Press, 1992.
21. John McMurry, Organic Chemistry, Fourth edition, Brooks/Cole, 1995.
22. C. R. Noller, " *Textbook of Organic Chemistry*", W. B. Saunders Company, 1966, pg. 63.
23. A. D. Dada, PhD thesis "Simulation of Polypropylene plant- NNPC, WRPC, Ekpan-Warri, Nigeria as case sturdy", 1993.
24 T. Tsuruta, S. Inoue, and K. Tsubaki, Macromol. Chem., 111:236 (1968).
25 T. Saegusa, H. Imai, and S. Matsumoto, J. Polymer Sci., A-1(6): 549 (1968).
26 M. E. Pruitt and J. M. Baggett, U. S. Pat. 2,706,182 to Dow Chemical Co. (April, 1955).
27 T. Tsuruta, Stereospecific Polymerization of Epoxides, A. D. Ketley (Ed), op. cit., vol. 2, chap. 4 of A. D. Ketley (ed.), "The Stereochemistry of Macromolecules", Marcel Dekker, Inc., New York, 1967.
28. C. R. Noller, "*Textbook of Organic Chemistry*", W. B. Saunders Company, 1966, pg. 638-658.
29. R. A. Patsiga, J. Macromol. Sci., - Revs. Macromol. Chem. C1 (2): 1967, 223.
30. G. Odian, *Principles of Polymerization*", McGraw-Hill, New York, 1970, pg. 503.
31. L. Finar, "Organic Chemistry", English Language Book Society and Longmans, Greens & Co Ltd., (1964), Vol. Two, pgs. 435 - 436.

Chapter 3

New Classification for Compounds and Their Impacts.

Herein contains new classification for compounds, based on the new concepts in last two chapters[1,2]. Therein, what Equilibrium, Decomposition, and Combination stages look like were begun to be shown. For the first time, using Equilibrium mechanism, Solubilisation and Insolubilisation have been clearly distinguished. Based on the new classification for compounds, Dissolution and Miscibilization have been well defined. How they take place have been clearly shown. The mechanisms of flow of electric current in fluids electrostatically have begun to be shown. Structures of interhalogens (in e.g. IF_7), and more elements of carbon atom have begun to be shown. How some compounds favour only transient existence and rearrange to give another compound via molecular rearrangements have begun to be introduced. Based on all these new concepts, the mechanisms of Cracking of hydrocarbons, the synthesis and decomposition of ammonia, and removal of so-called Green-house gases such as CO_2 have also been provided. The new concepts and other things introduced are too countless to list.

3.0 **Introduction**

Without introducing the new concepts on States of Existences, Stage-wise operations, new classification for mechanisms of systems, operating conditions, new classification for radicals[1], new classification for charges and bonds, activation centers, functional centers, ring theories and male/female macromolecules[2], the new classifications for compounds provided herein would not have been possible.

For years one has always wondered what the difference between solubility and dissolution is, since some authors use them interchangeably to mean the same; others use them differently and others like the author don't know what they are. The same too applies to dissolution and miscibilization, two different phenomena. In the same light, when a component cannot dissolve or miscibilize in a second component, it is believed by many schools of thought to mean Insolubilisation. What then are Solvents and Non-Solvent in this state of confusion? Of great shock is that all these terms exist; the only difference being their definitions and understanding. Insolubilisation is very different from Solubilisation and indeed not opposites as has started to be shown in Chapter2. While in Dissolution and Miscibilization, no chemical reaction takes place, in Solubilisation and Insolubilisation, chemical reaction takes place in one stage by Equilibrium mechanism with no products obtained. Understanding these phenomena will go a long way in bridging the gap between the Art and Science in for example Unit Separation Processes applied to chemical, petroleum, and environmental engineering processes, the pharmaceutical industries, applied medical sciences, corrosion studies and more and these will open new doors.

Present-day Science cannot explain how some compounds rearrange to give another compound such as vinyl alcohol to acetaldehyde[3], cis- and trans- 2-butene to 1-butene[4], allene to methyl acetylene[5] and more in an orderly manner. Based on the eye of the needle's use of universal data, there are several kinds and types of molecular rearrangements depending on the kind of movement (single and sequential) and the type of group delivering the component to be transferred (a situation of taking from the rich to the poor and never the reverse). These transfers do not take place indiscriminately like in all present-day chemical reactions.

Using new forms of chemical equations, one has begun to show what an Equilibrium stage is. A stage can be productive or non-productive (such as in solubilisation and insolubilisation) or unreactive.[2] Using the concept of stage-wise operations one has begun to show the mechanisms of flow of electric current in solutions. For the first time, why some compounds such as

IF_7 (heptafluoro-iodine) conduct electricity have begun to be shown through provision of their real structures.

3.1 New Classifications for Compounds

Based on the new concepts for charges, bonds, radicals and the use of the eye of the needle, there are three main kinds of compounds as shown in Figure 3.1.

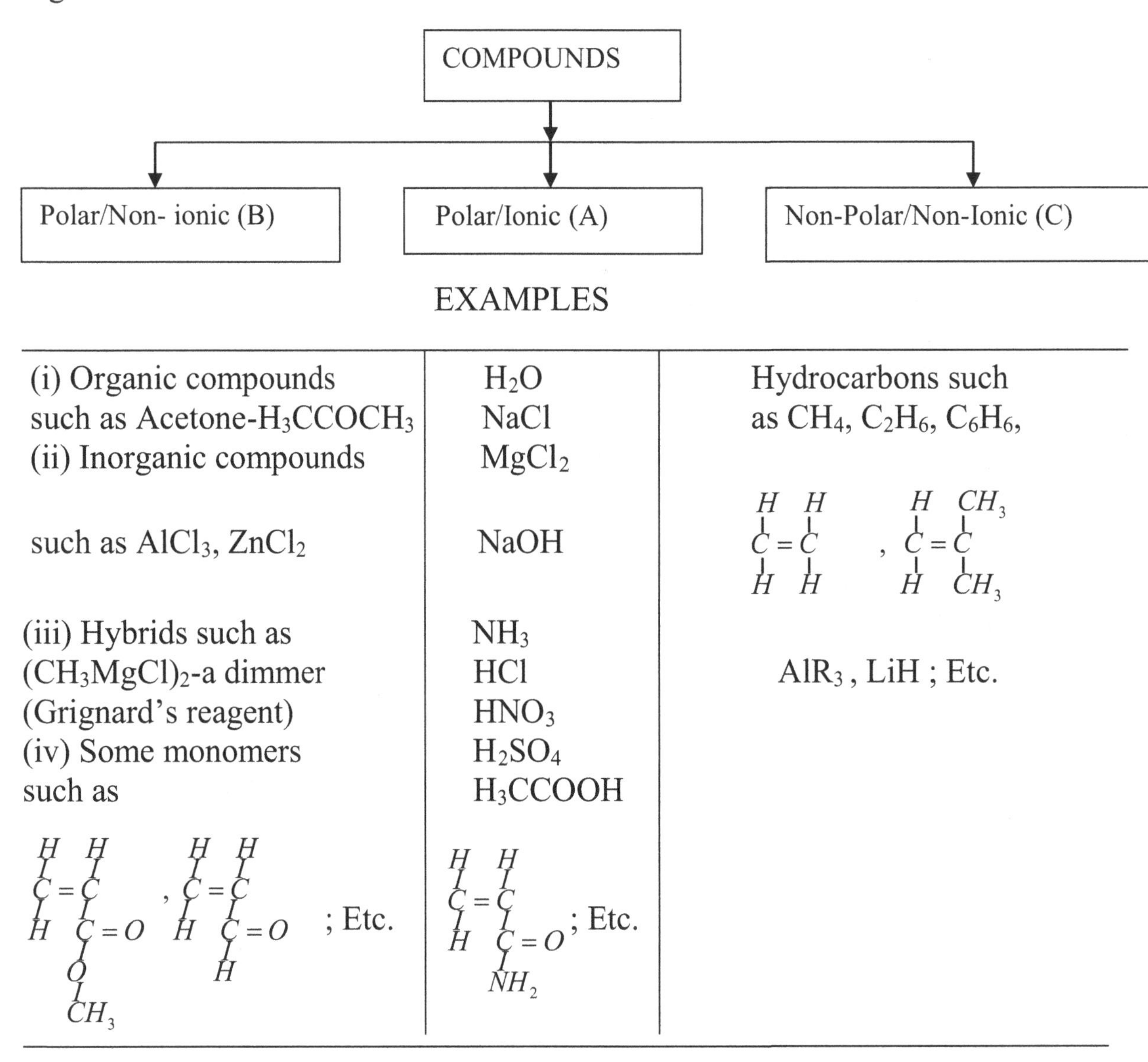

Figure 3.1 New classification for compounds

These main kinds are –

Polar/Ionic,
Polar/Non-ionic and
Non-polar/Non-ionic.

The fourth class based on Permutation and Combination theories in Mathematics, ***Non-polar/Ionic*** does not exist and does not arise, because ***before a compound can be ionic, it must first be polar.*** The word "polar" here refers to compounds, which either carry electronegative elements with centers of *paired unbonded radicals* in the last shell such as in O, Cl, N or electropositive elements with *paired unbonded radicals* in the last shell such as in Iron (Fe), Zinc (Zn) as already shown.[1] The three driving forces favouring the existence of ionic bonds and charges have already been stated[2] and briefly recalled again below-

(i) For the **cation** that is that carrying positive charge(s), the last shell, that is, the boundary must be **fully empty.**
(ii) For the **anion** that is that carrying the negative charge(s), the last shell, that is, the boundary must be **completely full and must be polar.**
(iii) Most importantly, **no hybridization** of the radicals in the sub-atomic orbital of the central atom takes place.

Shown below, are Equilibrium states of existence of the three members of compounds. Activated states of existence are also shown where possible. These states are the real ones and are fixed.

(A)Polar/Ionic compounds

The polar character is provided by the presence of paired unbonded radicals on some of the elements, and the ionic character is provided by the presence of ionic metals directly connected to a hetero atom.

(i)
$$H-\ddot{\underset{\cdot\cdot}{O}}-H \underset{}{\overset{Radically}{\rightleftharpoons}} H{\bullet}e + nn{\bullet}\ddot{\underset{\cdot\cdot}{O}}-H$$
$$Water \overset{Ionically}{\rightleftharpoons} H^{\oplus} + {}^{\ominus}{:}\ddot{\underset{\cdot\cdot}{O}}-H \qquad 3.1$$

(ii)
$$Na\ddot{\underset{\cdot\cdot}{Cl}}: \overset{Radically}{\rightleftharpoons} Na{\bullet}e + :\ddot{\underset{\cdot\cdot}{Cl}}{\bullet}nn$$
$$\text{Sodium chloride} \overset{Ionically}{\rightleftharpoons} Na^{\oplus} + :\ddot{\underset{\cdot\cdot}{Cl}}:^{\ominus} \qquad 3.2$$

(iii)
$$MgCl_2 \overset{Radically}{\rightleftharpoons} :\ddot{\underset{\cdot\cdot}{Cl}}{\bullet}nn + e{\bullet}MgCl$$
$$\text{Magnessium chloride} \overset{Ionically}{\rightleftharpoons} Mg^{+2} + 2\,{}^{\ominus}{:}\ddot{\underset{\cdot\cdot}{Cl}}: \qquad 3.3$$

(iv)

$$CH_3\overset{O}{\overset{||}{C}}OH \xrightleftharpoons{Radically} H\bullet e + nn\bullet\ddot{\underset{..}{O}}-\overset{:O:}{\overset{||}{C}}CH_3$$

$$\xrightleftharpoons[Male\ radical]{Activated\ by\ a} \begin{matrix} OH \\ | \\ e\bullet C-O\bullet nn \\ | \\ CH_3 \end{matrix} \quad ; \quad \begin{matrix} OH \\ | \\ \oplus C-O\ominus \\ | \\ CH_3 \end{matrix} \qquad 3.4$$

There is no molecular rearrangement in the last case above, since the same acid will be obtained with H as the transfer species when the CO center is activated noting that OH group is more radical-pushing than CH_3 group, that is, OH group is richer than CH_3 group [that is taking from the rich and not the poor]. The order of capacity of radical-pushing and radical-pulling groups have indeed been established in the New Frontiers in laying these foundations, for without it, the truth cannot emerge with so much confidence. {The universal data speak for themselves when the third eye is involved. The third eye is indeed not just the brain, but a gland closely located to the brain analogous to the BLACK BOX.}

(v)**Electrophile (Male) – Basic**

$$\underset{Acrylamide}{\begin{matrix} H \quad\ H \\ | \ Y \ | \\ C=C \\ | \quad\ | \ X \\ H \quad C=O \\ \quad | \\ \quad NH_2 \end{matrix}} \xrightleftharpoons[Female\ radical]{Activated\ by\ a} \begin{matrix} H \quad H \\ | \quad\ | \\ e\bullet C-C\bullet n \\ | \quad\ | \\ H \quad C=O \\ \quad | \\ \quad NH_2 \end{matrix}$$

$$\xrightleftharpoons[Male\ radical]{Activated\ by\ a} \begin{matrix} H \quad H \\ | \quad\ | \\ C=C \\ | \quad\ | \\ H \ e\bullet C-O\bullet nn \\ \quad | \\ \quad NH_2 \end{matrix} \qquad 3.5$$

$$\xrightleftharpoons[state\ of\ existence]{Radical\ Equilibrium} \begin{matrix} H \quad H \\ | \quad\ | \\ C=C \\ | \quad\ | \\ H \quad C=O \\ \quad | \\ \quad H\ddot{N}\bullet nn \end{matrix} + H\bullet e$$

$$\xrightleftharpoons[state\ of\ existence]{Ionic\ Equilibrium} \begin{matrix} H \quad H \\ | \quad\ | \\ C=C \\ | \quad\ | \\ H \quad C=O \\ \quad | \\ \quad H\ddot{N}:^{\ominus} \end{matrix} + H^{\oplus}$$

However, the transfer species of the first kind of the first type is the group NH_2 either removed as a radical or a charge when attacked by a male radical or male charged initiator respectively. Based on the operating conditions this can be made to undergo molecular rearrangement with NH_2 group as transfer species[6]. When NH_2 group is replaced with OH group, one also knows what to expect. These are **Polar/Ionic** from those groups as shown above for NH_2. When the two Hs on the NH_2 group and one H on OH groups are replaced with CH_3 types of groups (Alkylane groups which in present day Science are called alkyl groups), they become more stable and **Polar/Non-ionic**. When only one H on NH_2 is replaced with CH_3 types of groups, it is still **Polar/ Ionic.** However, they can still serve as transfer species for molecular rearrangements, since it is the entire group that is carried, unlike the next case below which is a Nucleophile.

(vi) **Nucleophile (Female)**

$$H(HO)C=CH_2 \xrightleftharpoons[\text{Activation}]{\text{Radical}} n\bullet C(H)(H)-C(H)(OH)\bullet e \xrightleftharpoons[\text{Rearrangement}]{\text{Molecular}} e\bullet C(H)(CH_3)-O\bullet nn \xrightarrow{\text{Deactivation}} H(CH_3)C=O$$

Cannot be readily isolated (Less Stable) — *(More Stable)*

Vinyl alcohol

Molecular Rearrangement of the 1st kind of the 1st type 3.6a

$$H(HO)C=CH_2 \xrightleftharpoons[\text{Activation}]{\text{Covalent}} \ominus C(H)(H)-C(H)(OH)\oplus \xrightleftharpoons[\text{Rearrangement}]{\text{Molecular}} \oplus C(H)(CH_3)-\ddot{O}:\ominus \xrightarrow{\text{Deactivation}} H(CH_3)C=O$$

Unstable — *(More Stable)*

Molecular Rearrangement of the 1st kind of the 1st type 3.6b

$$C(H)(H)=C(H)(OH) \xrightleftharpoons[\text{Operating Conditions}]{\text{Very low}} C(H)(H)=C(H)(O^{\ominus}) + H^{\oplus} \;;\; \xrightleftharpoons[\text{Operating Conditions}]{\text{Very low}} C(H)(H)=C(H)(O\bullet nn) + H\bullet e$$

(Ionically) — *(Radically)*

An organic Alcohol (Acid)

3.6c

Like the case above, it is Polar/Ionic. When activated, it molecularly rearranges to acetaldehyde which is Polar/Non-ionic and more stable. It is more stable, because C=O activation center is more Nucleophilic than C=C activation center, apart from fact that it is Polar/Non-ionic. Alcohols and Acids in general, have ionic and radical Equilibrium states of existence. ***Only Polar/Ionic compounds have both Ionic and Radical Equilibrium states of existence. Polar/Non-ionic compounds have only Radical Equilibrium, Decomposition and Combination states of existences. Polar/Ionic compound do not have Decomposition and Combination States of Existences expressed ionically, because ionically for example a compound which is electronegative such as O cannot be made to carry a cation (real). It can only be made to carry a positive charge of the electro-static type only. Only radically can such atoms be made to carry male radicals. Hence no chemical reaction can take place ionically. Polar/Non-ionic compounds do not have charged States of existences, because only ionic charges can be isolated. The same which apply to Polar/Non-ionic compounds also apply to Non-polar/Non-ionic compounds as will shortly become clear.*** When the H atom on OH group is changed to CH_3 types of groups, it no longer becomes an alcohol but ether. Also, it is now Polar/Non-ionic and more stable. Unlike the case shortly to be seen below and acrylamide above (v), both Electrophiles (Males), this Nucleophile (Female) cannot undergo molecular rearrangement as shown below only radically and not ionically or chargedly.

$$\begin{array}{c} H \quad H \\ I \quad I \\ C=C \\ I \quad I \\ H \quad O \\ I \\ CH_s \end{array} \xrightleftharpoons[\text{Activation}]{\text{Radical}} \begin{array}{c} H \quad H \\ I \quad I \\ n{\bullet}C-C{\bullet}e \\ I \quad I \\ H \quad O \\ I \\ CH_3 \end{array} \xrightleftharpoons[\text{Rearrangement}]{\text{Molecular}} \begin{array}{c} H \\ I \\ e{\bullet}C-O{\bullet}nn \\ I \\ C_2H_5 \end{array} \xrightarrow{\text{Deactivation}} \begin{array}{c} H \\ I \\ C=O \\ I \\ C_2H_5 \end{array}$$

(IMPOSSIBLE TRANSFER) (Cannot exist)

Methyl vinyl alcohol *Impossible Molecular Rearrangement* 3.7a

$$\begin{array}{c} CH_3H \\ I \quad I \\ C=C \\ I \quad I \\ CH_3O \\ I \\ CH_3 \end{array} \xrightleftharpoons[\text{Activation}]{\text{Radical}} \begin{array}{c} CH_3H \\ I \quad I \\ n{\bullet}C-C{\bullet}e \\ I \quad I \\ CH_3O \\ I \\ CH_3 \end{array} \xrightleftharpoons[\text{Rearrangement}]{\text{Molecular}} \begin{array}{c} H \\ I \\ e{\bullet}C-O{\bullet}nn \\ I \\ CCH_3 \\ H_3C \quad CH_3 \end{array} \xrightarrow{\text{Deactivation}} \begin{array}{c} H \\ I \\ C=O \\ I \\ CCH_3 \\ H_3C \quad CH_3 \end{array}$$

(Stable) (More Stable)

Molecular Rearrangement of the 1st kind of the 1st type 3.7b

In everything in life based on what an ATOM is, there are BOUNDARIES and DOMAINS, i.e., there are LIMITATIONS. The receiving C center above is carrying two H atoms. It cannot receive anything whose radical-pushing capacity is greater than what she is carrying. The radical-pushing capacity of CH_3 group is greater than that of H. The C center can only receive transfer species of the same or less radical-pushing capacity than H. In the second equation above, CH_3 moved because the receiving C center is carrying groups of the same capacity with the CH_3 group. Hence, the rearrangement was favoured. All the above apply to everything in life. ***Can one eat more than what the body requires for continued existence? Sow your Coat according to your size and so on.*** All rearrangements (Tautomerism) have their limitations. That is Nature. While vinyl alcohol can molecularly rearrange, methyl or ethyl or propyl etc., vinyl alcohol cannot molecularly rearrange, because the radical-pushing capacities of $C_3H_7 > C_2H_5 > CH_3 > CH_2F > CHF_2 > H > CF_3$. All these new concepts from CHEMISTRY are just a grain of sand compared to what is contained in the New Frontiers.

The instability of vinyl alcohol is so strong that, using it as a monomer for polymerization to obtain poly (vinyl alcohol) is impossible. Instead, the poly (vinyl acetate) is used to obtain the poly (vinyl alcohol) by hydrolysis as contained in the New Frontiers. One can see what the ionic characters of a monomer can do. In Step monomers, they are very significant, yet no chemical reaction takes place ionically. Even Step polymerization cannot take place ionically or covalently or electrostatically or polarly.

(vii) **Electrophile (Male) - Acidic**

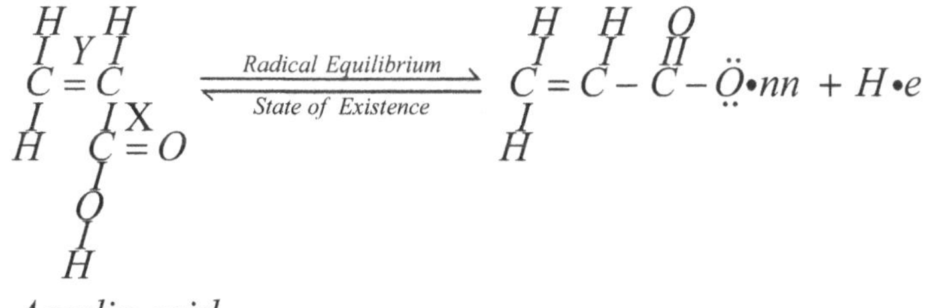

```
H   H                                        H   H   O
I Y I                                        I   I   II   ..
C = C        Radical Equilibrium             C = C - C - O•nn  +  H•e
I   I X      ⇌                               I               ..
H   C = O    State of Existence              H
    I
    O
    I
    H
```

Acrylic acid

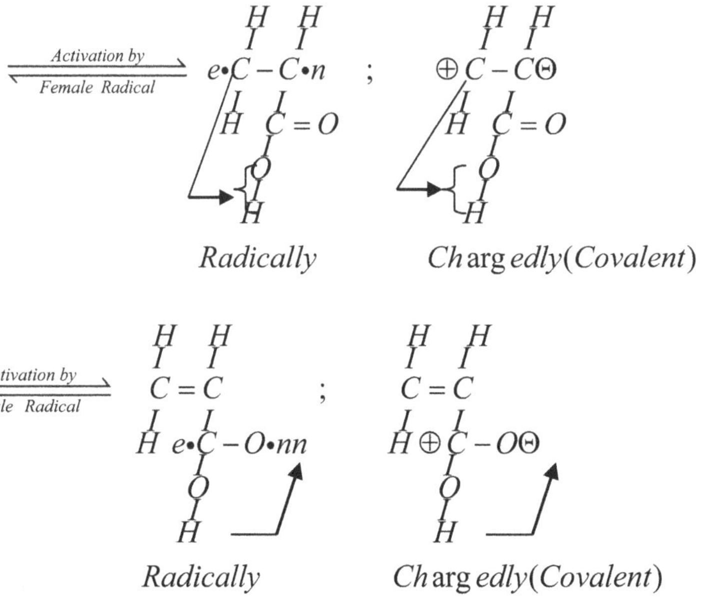

$$\xrightleftharpoons[\text{Female Radical}]{\text{Activation by}} \begin{matrix} H & H \\ | & | \\ e\bullet C & - C\bullet n \\ | & | \\ H & C = O \\ & | \\ & O \\ & | \\ & H \end{matrix} \quad ; \quad \begin{matrix} H & H \\ | & | \\ \oplus C & - C\ominus \\ | & | \\ H & C = O \\ & | \\ & O \\ & | \\ & H \end{matrix}$$

Radically   Chargedly(Covalent)

$$\xrightleftharpoons[\text{Male Radical}]{\text{Activation by}} \begin{matrix} H & H \\ | & | \\ C & = C \\ | & | \\ H & e\bullet C - O\bullet nn \\ & | \\ & O \\ & | \\ & H \end{matrix} \quad ; \quad \begin{matrix} H & H \\ | & | \\ C & = C \\ | & | \\ H & \oplus C - O\ominus \\ & | \\ & O \\ & | \\ & H \end{matrix}$$

Radically   Chargedly(Covalent)   3.8

Note that the transfer species of the first kind of the first type here like in acrylamide of (v) (which is NH_2 group) is OH group for the Y-center either in form of a radical or a charge which a male initiator when present grabs to prevent initiation, the monomer being a Male. When a female initiator is involved here, there is no transfer species involved for the Y-center which is natural to it. For the X-center, H is the transfer species for a female initiator and for molecular rearrangement, which when it takes place, the same product is obtained, unlike the case for the Y center. The C=O center is rarely involved when mildly activated, because its nucleophilic capacity is far stronger than that of C=C center. It is only when a strong male initiator is involved, that the C=O center gets activated along with the C=C center for it to be used either exclusively for homo-polymerization or for ***copolymerization with itself such as with acrolein.*** The C=O center can be activated at the same time with C=C center only if the initiator is strong in capacity for activation, ***since the monomer is not resonance stabilized.*** It is because the C=C center has transfer species to grab when attacked by a male initiator, hence copolymers cannot be obtained from the acrylic acid above, but can be obtained from acrolein positively or electro-free-radically. ***In acrolein another Electrophile,*** only an initiator such as that carried by sodium can be used to obtain homopolymers from acrolein via the C=O center as shown below. Nucleo-free-radically and with negatively charged-paired initiator, the main routes natural to the monomer, homopolymers are also only obtained via the Y center.

Initiator preparation step

Stage 1:

$$2NaC \equiv N \rightleftharpoons 2Na{\bullet}e + 2n{\bullet}C \equiv N$$

$$2n{\bullet}C \equiv N \longrightarrow N \equiv C - C \equiv N$$

Overall Equation: $2NaCN \longrightarrow 2Na{\bullet}e + NC - CN\,(Cyanogen)$ 3.9a

The mechanism above is Equilibrium mechanism that which better takes place via Decomposition mechanism, but for the operating conditions of temperature as low as -70^0C. It contains only two steps. The sodium carrying an electro-free-radical is just the atom of Na. No nucleo-free or non-free-radical is present in the system. Cyanogen gas is also a product.

Initiation Steps

(a) *With NaCN*

$$Na{\bullet}e + \underset{(I)}{(O{=}CH)HC{=}CH_2} \xrightarrow[Activation]{Y} NaH + e{\bullet}C(=O) - CH = CH_2$$

No Initiation 3.9b

(b) *With BF_3 / BF_3 Combination*

$$F_2B^{\oplus}......{}^{\odot}BF_4 + (I) \xrightarrow[Activation]{Y} F_2B - CH(CH{=}O) - CH_2^{\oplus}......{}^{\odot}BF_4$$

Initiation favoured 3.9c

(c) *With NaCN*

$$Na{\bullet}e + \underset{(I)}{(O{=}CH)HC{=}CH_2} \xrightarrow[Activation]{X} Na - O - CH{\bullet}e(HC{=}CH_2)$$

Initiation Step favoured 3.9d

(d) *With BF_3 / BF_3 Combination*

$$F_2B^{\oplus}........{}^{\odot}BF_4 + (I) \xrightarrow[Activation]{X} F_2B - O - CH^{\oplus}(HC{=}CH_2)........{}^{\odot}BF_4$$

Initiation Step favoured 3.9e

Thus, one can observe from the equations above why random copolymers can be obtained for acrolein with more of the C=O center along the chain than C=C center when positively charged centers are involved. There is more of C=O activation than C=C activation, because the route is natural to the C=O center, the X center. One can also observe why Na was able to abstract the H (Equation 3.9b), that which Boron (B) (Equation 3.9c) could not do, since there is no H^{θ}. On the other hand, Na and H belong to the same family and Na is far more electropositive than H and thirdly ***one of the best meals for Na is H.*** Based on the ways chemical and polymeric reactions have been observed to take place between families of compounds, using the EYE OF NEEDLE, one has begun to see that in general in Nature, ***Males or Electrophiles or electro-radicals are "POLYGAMISTS"*** **DEPENDING ON THE OPERATING CONDITIONS** ***; while Females or Nucleophiles or nucleo-radicals are "PROSTITUTES"*** **DEPENDING ON THE OPERATIONG CONDITIONS.** In these definitions, the most important is the section underlined above. This is to show the very great importance of OPERATING CONDITIONS in all operations in our world. Without it nothing takes place. In these definitions, one can see Monogamy, Sainthood, Polyandry, and Polygyny and so on. Imagine the situations where the females do not diffuse to grab a species to commence a reaction, but sit to be grabbed selectively and the situation where when "couples" are to be formed during alternating copolymerizations, the females are the ones that diffuse to grab a male of their choice. Imagine Na going for H in activated acrolein, in H_2O, in NaOH. Yet, Na and H belong to the same family as metals. NaH is a very unique compound used for hydrogenation in place of Hydrogen catalysts. Despite the great abilities of present-day Science in collection of data, none of them can be explained by them.

Thus, like in acrylamide where NH_2 is the transfer species coming from $CONH_2$ group, OH coming from COOH group in acrylic acid is the transfer species for the Y center. In vinyl alcohol, H coming from OH group is the transfer species for the X center (C = C). *When H in vinyl alcohol is changed to CH_3 coming from OCH_3 group, molecular rearrangement cannot take place, but CH_3 can still prevent initiation nucleo-free-radically as transfer species of the first kind of the first type.* With H, it is Polar/Ionic while with CH_3 it is Polar/Non-ionic. The same too applies to acrylic acid. ***One can observe the complex logical ordered network systems in NATURE.***

(B)Polar/Non-Ionic compounds

In all of them, the polar character is provided by paired unbonded radicals carried by some of the elements such as O, Cl, N, Fe, Zn and so on.

The non-ionic character is provided by the absence of ionic metals directly connected to a hetero atom.

(i) **Nucleophile (Female) – Formic acid**

$$\underset{\text{(Female)}}{\underset{\text{(Formic acid)}}{H\overset{\overset{O}{\|}X}{C}OH}} \xrightleftharpoons{\text{Radically}} H\bullet e + n\bullet\overset{\overset{O}{\|}X}{C}OH$$

$$\xrightleftharpoons{\text{Ionically}} \text{None (Because the C center is hybridized) and non-polar.} \quad 3.10$$

Note that unlike other acids, the hydrogen held in Equilibrium state of existence is not that attached to COOH group, but that attached to the carbon center based on universal experimental observations of its chemical reactions. This reason has been explained as shown in Equations 1.140 and 1.141 in Chapter1 during oxidation. Unlike other acids, there are two H atoms linked to the central C atom- one directly and the other via the O center. It is the H that is directly connected that can be seen and be used or held. ***Just like every first member of a family,*** its character is different from other members like acetic acid the next member in the family. ***Just as it is difficult to say that C_2H_5OH and all its members are alcohols, so also it is difficult to say that HCOOH is an acid. As it seems, just as H_2O is neither an acid or alcohol, CH_3OH and its members are acids. Just as HCOOH is neither an acid nor an alcohol, H_3CCOOH and its members are acids. It is only when CH_3OH and it members exist in Decomposition state of existence, that they can be said to be alcohols.*** While the other members of carboxylic acid family are Polar/Ionic, formic acid is Polar/Non-ionic.

With Alkanes, the first member is methane followed by ethane, propane, butane and so on. CH_4 is completely different from the other members, since alkenes cannot be obtained from methane. Ethene the first member of alkenes is not a real Nucleophile (Female). Propene the next member, followed by butene and so on are full Nucleophiles. All first members of any family of compounds are always different and unique from the other members in the same family, just as it is in Life. So far, one can imagine the roles H plays in humanity. It also plays major roles in the nucleus of the SUN, nucleus of the Planets such as ours (The Inner Core), and in the Nucleus of all atoms. Yet, as we shall see in the Appendix, it is the only atom that is indivisible amongst all atoms.

(ii) **Nucleophile (Female)** – Aldehydes (Acetaldehyde)

$$\underset{\substack{(Acetaldehyde)\\(Female)}}{CH_3\overset{\overset{O}{\|}X}{C}H} \xrightleftharpoons{Radically} H_3C\overset{\overset{O}{\|}X}{C}{\bullet}n + H{\bullet}e$$

$$\xrightleftharpoons{Ionically} None$$

$$\xrightleftharpoons{Activation} \underset{Radically}{e{\bullet}\overset{H}{\underset{CH_3}{C}}-O{\bullet}nn} \quad ; \quad \underset{Chargedly(Covalent)}{\oplus\overset{H}{\underset{CH_3}{C}}-\ddot{\underset{..}{O}}:\ominus}$$

3.11

(iii) **Nucleophile (Female)** – Ketones (Acetone)

$$\underset{\substack{Acetone\\(Female)}}{CH_3\overset{\overset{O}{\|}X}{C}CH_3} \xrightleftharpoons{Radically} H_3C\overset{\overset{O}{\|}X}{C}{\bullet}e + n{\bullet}CH_3$$

$$\xrightleftharpoons{Ionically} None$$

$$\xrightleftharpoons[Equilibrium]{Activated} H_3C\overset{\overset{.nn}{O}}{\underset{e}{C}} - \overset{H}{\underset{H}{C}}{\bullet}n \quad + \quad \underset{\text{(Transfer species of the first kind)}}{H{\bullet}e}$$

$$\xrightleftharpoons{Activation} \underset{Radically}{e{\bullet}\overset{CH_3}{\underset{CH_3}{C}}-O{\bullet}nn} \quad ; \quad \underset{Chargedly(Covalent)}{\oplus\overset{CH_3}{\underset{CH_3}{C}}-\ddot{\underset{..}{O}}:\ominus}$$

3.12

Aldehydes and ketones have in addition to other states of existence, Activated/Equilibrium state of existence. That of acetaldehyde has already been shown in Chapter 1. That for acetone has been shown above and note that the H held is transfer species of the first kind of the first type.

(iv) **Nucleophile (Female)** - Epoxide

$$H_2C - CH_2 \text{ (ring through } \ddot{O}:)$$

O:

X (Functional center with invisible π-bond inside ring)

$$\xrightleftharpoons[Instantaneously]{Radically} nn\bullet O-\underset{H}{\overset{H}{C}}-\underset{H}{\overset{H}{C}}\bullet e \;;\; \xrightleftharpoons{Chargedly} \Theta:\ddot{\underset{..}{O}}-\underset{H}{\overset{H}{C}}-\underset{H}{\overset{H}{C}}\oplus \qquad 3.13$$

It can be opened instantaneously when the MRSE is provided for the ring, and this can be done either by heating or through the functional center O using for example a male radical such as H·e or positively charged-paired-coordination initiators as already shown.[2]

(v) **Electrophile (Male)**

$H_2C - CH_2$
| |
H_2C O: X center
\ /
C
|| Y center
O
(Male)

$$\xrightleftharpoons{Radically} nn\cdot O-\underset{H}{\overset{H}{C}}-\underset{H}{\overset{H}{C}}-\underset{H}{\overset{H}{C}}-\overset{O}{\overset{||}{C}}\cdot e \qquad ;\; Ionically,\; None$$

$$\xrightleftharpoons[(Covalent)]{Chargedly} \Theta O-\underset{H}{\overset{H}{C}}-\underset{H}{\overset{H}{C}}-\underset{H}{\overset{H}{C}}-\overset{O}{\overset{||}{C}}\oplus \qquad 3.14$$

The ring can be opened instantaneously or through two centers placed on the ring. One center (X) a functional center is for the male initiator and the second center-π-bond activation center externally located (Y) is for the female initiator. Hence, the same monomer units are obtained when polymerized via both routes. Note the types of charges and radicals carried by the active centers. This will determine the types of initiators to use. Nucleo-free-radical and negatively charged centered initiators cannot be used, because of radical balancing. The natural routes are the nucleo-non-free-radical and anionic type of charge. For these families of Electrophiles, positively charged paired or electro-free-radical routes can almost be said to be natural to them. The only difference is that higher operating conditions (very far less than that of ethene a Nucleophile, nucleo-free-radically) will be required electro-free-radically than nucleo-non-free-radically, being an Electrophile. We can see that what are being observed with these monomers exist with humans except under very different operating conditions. For example, a nucleo-free-radical (a female radical will add to ethene (a female), only when an electro-free-radical is absent in that world and under very harsh operating conditions. This is not the case in our world. These families of monomers (Lactones) as has already been said do not have transfer species. While Lactams are Polar/Ionic from the NH center, Lactones are Polar/ Non-ionic.

(vi) **Nucleophile (Female)** - Formaldehyde

$$\begin{array}{c} O \\ \| \text{X} \\ H-C-H \end{array}$$

$$\xrightleftharpoons{Radically} H\cdot e + n\cdot \overset{\overset{O}{\|}}{C} - H \quad ;Ionically,\ None$$

$$\xrightleftharpoons{Activation} \underset{\underline{Radically}}{e\cdot \overset{H}{\underset{H}{\overset{|}{\underset{|}{C}}}} - O\cdot nn} \quad ;\underset{\underline{Chargedly}}{\oplus \overset{H}{\underset{H}{\overset{|}{\underset{|}{C}}}} - O\ominus} \qquad 3.15$$

This Nucleophile (Formaldehyde) which is Polar/Non-ionic, is the first member of the family of aldehydes. It can be stable or unstable, depending on what compound is in its neighbourhood. In the presence of warm water, it is stable, since it forms paraformaldehyde with it. In the presence of other members of the family and a ketone, it is unstable, because of the two H atoms connected to the central C atom.

Notice so far that all Polar/Non-ionic compounds can only show their States of existences radically and not chargedly or ionically. Based on all the considerations so far, the reasons are obvious and unquestionable. Can one remove a positive charge from a carrier, leaving the carrier behind as done in present-day Science? When a monomer is activated, can one remove the positive charge leaving the carrier with negative charge behind? Can one do the same thing for negative charges? Can one isolate $H_9C_4^{\ominus}$ in LiC_4H_9 leaving $Li^{\oplus}$ behind? Since NATURE does not allow a vacuum to exist, hence only ionic charges can be isolated. [Hence, there is no problem without a solution and hence there is no disease without a cure.] $Na^{\oplus}$ in NaCl can be isolated leaving $Cl^{\ominus}$, since it is completely ***full and POLAR*** while the other is completely ***empty and NON-POLAR***. Nothing is common between the two of them, just like when you mix a Polar/Ionic compound such as water with Non-polar/Non-ionic compound such as a hydrocarbon or fuel. Both are isolatedly placed with an interface. Since atoms which carry radicals can be isolated, all radicals can therefore be isolated.

(vii) **Interhalogen compounds (Females)**

$$(a)\ ICl \xrightleftharpoons{Radically} I\cdot en + nn\cdot Cl \quad ; \quad Ionically,\ None \qquad 3.16$$

$$(b)\ IF_7 \qquad \begin{array}{l} F^{\ominus}\ldots\ldots{}^{\oplus}F_{\ominus}-F \\ \quad\ \ \vdots \\ F-F^{\oplus\ominus}\ldots{}^{\oplus}F-F \\ \qquad\qquad\quad \uparrow \\ \qquad\qquad\quad I \end{array}$$

(A) *The real Structure*

$$IF_7 \xrightleftharpoons{Radically} \begin{array}{l} F^{\ominus}\ldots\ldots{}^{\oplus}F_{\ominus}-F \\ \quad\ \ \vdots \\ F-F^{\oplus\ominus}\ldots{}^{\oplus}F-F \\ \qquad\qquad\quad \dot{n}n \end{array} \quad + \quad I\cdot en \quad ; \quad Ionically,\ None$$

(B)

$(-+-+-+\ldots$*shows how current flows in the compound*$)$ 3.17

(B) above can be obtained either via one stage or many stages as will be shown in the New Frontiers. Note that IF_7 apart from the covalent bonds carried by it, also carries ***negative type of electrostatic bonds and charges*** and can therefore conduct electricity as will shortly be fully explained. However, notice how the + and – are alternatingly uniquely placed. One can observe that the atom held in Equilibrium state of existence is iodine a liquid metal in view of its electropositivity. It is far more electropositive than F. The Chemistry of Interhalogen compounds are very unique. So also, are halogenated carbons (Halocarbons and Halo-hydrocarbons). Because halogens are highly polar in character and electronegative with ability to be anionic, hence when present on Nucleophilic Alkenes, such as ethene, they make the alkene less nucleophilic (instead of more nucleophilic), such as in vinyl chloride. Its (Cl) strong nucleophilicity (i.e., female character) has suppressed the nucleophilicity of the C=C activation center. This is not the case with Interhalogen compounds which are strongly nucleophilic in character. ***Compounds with no activation centers that favour electrophilic attack are females.*** None favours nucleophilic attack when an unpaired or paired electrophile is present in the absence of activation centers. Invariably, all compounds without activation centers are Nucleophilic (Feminine) in character. From the beauty of the above, one can begin to see another origin of MOTHER NATURE and not Father Nature.

(viii)

$$CH_3Cl \xrightleftharpoons{Radically} Cl\cdot nn + e\cdot CH_3 \quad ; \quad \text{Ionically, None} \qquad 3.18$$

(Chloride)

(ix) $$CH_3I \xrightleftharpoons{Radically} I\cdot en + n\cdot CH_3 \quad ; \quad \text{Ionically, None} \qquad 3.19$$

[Note that C (Carbon atom), I (Iodine atom) and S (Sulfur atom) and not their elements are equi-electropositive at 2.5 and serve as the middle transition boundary atoms between Metals and Non-metals]. The equilibrium states of existence above were unquestionably established through their reactions from universal data. For example, in the halogenation of alkanes, there are marked differences between halogenation using F, Cl, and Br and from that using I. Alkanes, like hydrogen, can be made to react with fluorine, chlorine, or bromine.[7] It is said that H_2 gives two moles of hydrogen halide, whereas alkanes give one mole of hydrogen halide and one mole of halogenated alkane. Their reactions are believed to have "Initiation step and Propagation step"[7], something least to be expected and meaningless.

$$H_3C-H + F-F \longrightarrow R\bullet + HF + \bullet F \qquad Initiation$$

$$\left.\begin{array}{l} R\bullet + F_2 \longrightarrow RF + F\bullet \\ F\bullet + H_3C-H \longrightarrow FH + R\bullet \end{array}\right\} \text{Pr}\,opagation \qquad 3.20$$

$$Cl_2 \xrightarrow[hv]{Heat\ or} 2Cl\bullet \qquad Initiation$$

$$\left.\begin{array}{l} Cl\bullet + H-CH_3 \longrightarrow Cl-H + \bullet CH_3 \\ H_3C\bullet + Cl_2 \longrightarrow H_3C-Cl + Cl\bullet \end{array}\right\} \text{Pr}\,opagation \qquad 3.21$$

Note that, the R above is alkylane (alkyl) group such as CH_3. While F will fluorinate CH_4 in cycles one a time to give CH_3F, CH_2F_2, CHF_3 and CF_4 with great control needed in the absence of heat, Cl will chlorinate CH_4 only when heat is applied, for specific reasons as will shortly be explained.

$$CH_4 + Cl_2 \longrightarrow \underset{\substack{\text{Methyl chloride} \\ \text{(chloromethane)}}}{H_3CCl} + HCl \;;\; H_3CCl + Cl_2 \longrightarrow \underset{\substack{\text{Methylene chloride} \\ \text{(dichloromethane)}}}{H_2CCl_2} + HCl$$

$$H_2CCl_2 + Cl_2 \longrightarrow \underset{\substack{\text{Chloroform} \\ \text{(trichloromethane)}}}{CHCl_3} + HCl \;;\; CHCl_3 + Cl_2 \longrightarrow \underset{\substack{\text{Carbon tetrachloride} \\ \text{(tetrachloromethane)}}}{CCl_4} + HCl \qquad 3.22$$

Bromination of alkanes can also be carried out, except the fact that the products are believed to be of no practical use, that which is not true. Fluorination, Chlorination and bromination reactions are exothermic. Of the four halogens, Cl-Cl bond is the strongest (57 kcal/mole), followed by Br_2 (45 kcal/mole), then followed by F_2 (36 kcal/mole) and finally by I_2 (35 kcal/mole). The strength of their bond energies, that is energy required to dissociate them, has little or nothing to do with ability to exist in any of the states of existences. The Equilibrium state of existence of I_2 for example is stronger than that of Cl_2 which in turn is stronger than that of Br_2 which in turn is stronger than that of F_2. In the equations above nothing is known about states of existences, radicals, and too many things.

Looking at the equations above, one can see the merits and demerits in present-day Science. A look at Equations 3.20 and 3.21, just as it is everywhere, clearly shows that what radicals are, are unknown to humanity. Secondly, in chemical reactions (Micromolecules), there is nothing like Initiation and Propagation. These only exist in polymeric reactions (Macro-molecules). One can begin to imagine the state of confusion in our world. The equations as written look as if it is taking place stage-wisely; that which is not there. But they will say that it is there in self-defence; that which absolute nonsense is. How all the products are obtained are unknown to present-day Science. In fluorination of methane, the followings are obtained.

Fluorination

Stage 1:

$$H_4C \rightleftharpoons H\bullet e + n\bullet CH_3$$

$$H\bullet e + F_2 \rightleftharpoons HF + F\bullet en + Heat$$

$$F_2 \rightleftharpoons F\bullet en + F\bullet nn$$

$$F\bullet en + CH_4 \rightleftharpoons HF + e\bullet CH_3$$

$F{\bullet}en + n{\bullet}CH_3 \longrightarrow CH_3F + Heat$
(FAVOURED)

$H_3C{\bullet}e + nn{\bullet}F \longrightarrow CH_3F$
(Looks favoured, but not favoured)
NOT FAVOURED

$$\underline{Overall\ equation}: CH_4 + F_2 \longrightarrow H_3CF + HF + Energy \qquad 3.23a$$

Stage 2:

$H_3CF \rightleftharpoons H{\bullet}e + n{\bullet}CH_2F$

$H{\bullet}e + F_2 \rightleftharpoons HF + F{\bullet}en + Heat$

$F{\bullet}en + n{\bullet}CH_2F \longrightarrow CH_2F_2 + Heat$
(FAVOURED)

$F_2 \rightleftharpoons F{\bullet}en + F{\bullet}nn$

$F{\bullet}en + CH_3F \rightleftharpoons HF + e{\bullet}CH_2F$

$FH_2C{\bullet}e + nn{\bullet}F \longrightarrow CH_2F_2$
(Looks favoured, but not favoured)
NOT FAVOURED

$$\underline{Overall\ equation}: CH_3F + F_2 \longrightarrow H_2CF_2 + HF + Energy \qquad 3.23b$$

Stage 3:

$H_2CF_2 \rightleftharpoons H{\bullet}e + n{\bullet}CF_2H$

$H{\bullet}e + F_2 \rightleftharpoons HF + F{\bullet}en + Heat$

$F{\bullet}en + n{\bullet}CHF_2 \longrightarrow CHF_3 + Heat$

$$\underline{Overall\ equation}: CH_2F_2 + F_2 \longrightarrow HCF_3 + HF + Energy \qquad 3.23c$$

Stage 4:

$HCF_3 \rightleftharpoons H{\bullet}e + n{\bullet}CF_3$

$H{\bullet}e + F_2 \rightleftharpoons HF + F{\bullet}en + Heat$

$F{\bullet}en + n{\bullet}CF_3 \longrightarrow CF_4 + Heat$

$$\underline{Overall\ equation}: CHF_3 + F_2 \longrightarrow CF_4 + HF + Energy \qquad 3.23d$$

Notice why the reaction as observed is found to be violent and difficult to control. It is violent because the F_2 can be observed to be suppressed, that is, kept in stable state of existence in all the stages with release of heat. Its being suppressed is not as a result of the presence of the methane, but something else as will shortly be explained. In all the stages, large amounts of heat from two steps in each stage were generated because F was forced to carry an electro--free-radical which H was carrying to become an electro-non-free-radical. Can this take place ionically, wherein the F is made to carry the cation? For chlorination, the followings are obtained.

Chlorination

Stage 1:

$$H_4C \rightleftharpoons H\bullet e + n\bullet CH_3$$

$$H\bullet e + Cl_2 \rightleftharpoons HCl + Cl\bullet en + Heat$$

$$Cl\bullet en + n\bullet CH_3 \longrightarrow CH_3Cl + Heat$$

(LOOKS FAVOURED)

$$Cl_2 \rightleftharpoons Cl\bullet en + Cl\bullet nn$$

$$Cl\bullet en + CH_4 \rightleftharpoons HCl + e\bullet CH_3 + Heat$$

$$H_3C\bullet e + nn\bullet Cl \xrightarrow[by\ HCl]{Suppressed} CH_3Cl$$

FAVOURED

$$\underline{Overall\ equation}: CH_4 + Cl_2 \longrightarrow H_3CCl + HCl + Energy \qquad 3.24a$$

Stage 2:

$$H_3CCl \rightleftharpoons Cl\bullet nn + e\bullet CH_3$$

$$H_3C\bullet e + Cl_2 \rightleftharpoons H_3CCl + Cl\bullet en + Heat;$$

$$Cl\bullet en + nn\bullet Cl \rightleftharpoons Cl_2$$

(Stable, reactive and insoluble)

$$Cl_2 \rightleftharpoons Cl\bullet en + Cl\bullet nn$$

$$Cl\bullet en + CH_3Cl \rightleftharpoons HCl + e\bullet CH_2Cl + Heat$$

$$ClH_2C\bullet e + nn\bullet Cl \xrightarrow[by\ HCl]{Suppressed} CH_2Cl_2$$

FAVOURED

$$\underline{Overall\ equation}: ClCH_3 + Cl_2 \longrightarrow H_2CCl_2 + HCl + Energy \qquad 3.24b$$

Stage 3:

$$Cl_2 \rightleftharpoons Cl\bullet en + Cl\bullet nn$$

$$Cl\bullet en + CH_2Cl_2 \rightleftharpoons HCl + e\bullet CCl_2H + Heat$$

$$HCl_2C\bullet e + nn\bullet Cl \xrightarrow[by\ HCl]{Suppressed} CHCl_3$$

FAVOURED

$$\underline{Overall\ equation}: CH_2Cl_2 + Cl_2 \longrightarrow Cl_3CH + HCl + Energy \qquad 3.24c$$

Stage 4:

$$Cl_2 \rightleftharpoons Cl\bullet en + Cl\bullet nn$$

$$Cl\bullet en + CHCl_3 \rightleftharpoons HCl + e\bullet CCl_3 + Heat$$

$$Cl_3C\bullet e + nn\bullet Cl \xrightarrow[by\ HCl]{Suppressed} CCl_4$$

FAVOURED

$$\underline{Overall\ equation}: CHCl_3 + Cl_2 \longrightarrow CCl_4 + HCl + Energy \qquad 3.24d$$

Worthy of note is that the mechanisms for the fluorination of methane or alkanes in general, are quite different from the mechanisms for chlorination. It is important to observe that F_2 in view of its small size, ***[noting that electrostatic forces of repulsion between the two polar centers of F in F_2 or***

of Cl in Cl_2 or Br in Br_2 do not take place between the two centers, because of the sigma bond, i.e., the fact that the two elements are different-one a male (F•en) and the other a female(F•nn)], is very stable and therefore very difficult for it to exist in Equilibrium state of existence. So also is CF_4 known to be very stable and reactive. Secondly, notice the difference in Equilibrium state of existence of the fluorinated and chlorinated hydrocarbons.

$$
\begin{aligned}
&CH_3F \rightleftharpoons H\bullet e + n\bullet CH_2F \quad ; \quad CH_3Cl \rightleftharpoons H_3C\bullet e + Cl\bullet nn \\
&F_2CH_2 \rightleftharpoons H\bullet e + n\bullet CHF_2 \quad ; \quad Cl_2CH_2 \rightleftharpoons ClH_2C\bullet e + Cl\bullet nn \\
&F_3CH \rightleftharpoons H\bullet e + n\bullet CF_3 \quad ; \quad Cl_3CH \rightleftharpoons HCl_2C\bullet e + Cl\bullet nn \\
&CF_4 \overset{\text{Very stable}}{\rightleftharpoons} F_3C\bullet e + nn\bullet F \quad ; \quad CCl_4 \rightleftharpoons Cl_3C\bullet e + Cl\bullet nn \qquad 3.25
\end{aligned}
$$

Based on the Equilibrium state of existence of the fluorides, it is no surprise why CoF_3 (Cobaltic fluoride from cobaltous fluoride) was used as an alternative for fluorination of hydrocarbons[7]. Note that, Co a Transition metal cannot form hydrides with H, i.e., it cannot abstract H from the alkane.

The reason for the difference can be attributed mainly to the small size of the most electronegative atom. F (4.0) is far more electronegative than chlorine (3.0). The electronegativity of Cl (3.0) is close to that of Br (2.8). The potential difference between the electronegativities of C/H (0.4) is very small compared to that of C/F (1.4). The larger the potential difference, the greater the electrostatic forces of attraction. The potential difference for C/Cl is 0.5. One can see why the states of existence above are what they are and why CF_4 is very stable. When two atoms of the same electronegativity come together to form a molecule, a second reason is because the two atoms are two different elements of the same atom (Real). The first reason already given was the physical state of the atom (Real) and the third reason is an imaginary one. Radical do not repel or attract. In chlorination, in the first stage, it seemed as if the hydrocarbon could not allow the Cl_2 to exist in equilibrium state of existence. Indeed, when both are in Equilibrium state of existence, the same stage will be obtained. However, the chlorine was all the time kept in Equilibrium state of existence, all the time suppressing that of methane and its chlorinated ones, unlike fluorine. The same as apply to chlorination will also apply to bromination where the potential difference between the electronegativities of C/Br is 0.3. That for C/I is 0, noting however that iodine (I) looks more metallic in character than C, since it belongs to a Period where all the atoms are metallic. Higher operating conditions will be required for bromination than chlorination. With proper control based on use of exact molar ratios and a different type of reactor (not as used universally for halogenation), all the products can be obtained without the need for separation, since all the products come from different stages (and not in one stage).

For the iodination of methane, that which is said to be endothermic by about 14 kcal[7], the followings are obtained.

Iodination

Stage 1:

$$I_2 \rightleftharpoons I{\bullet}en + I{\bullet}nn$$
$$I{\bullet}en + CH_4 \rightleftharpoons HI + e{\bullet}CH_3$$
$$H_3C{\bullet}e + nn{\bullet}I \longrightarrow CH_3I$$
FAVOURED

$$CH_4 \rightleftharpoons H{\bullet}e + n{\bullet}CH_3$$
$$H{\bullet}e + I_2 \rightleftharpoons HI + I{\bullet}en$$
$$I{\bullet}en + n{\bullet}CH_3 \rightleftharpoons CH_3I$$
(*Stable, reactive and soluble*)
NOT FAVOURED

$$\underline{Overall\ equation}: CH_4 + I_2 \longrightarrow H_3CI + HI \qquad 3.26a$$

Stage 2:

$$I_2 \rightleftharpoons I{\bullet}en + I{\bullet}nn$$
$$I{\bullet}en + CH_3I \rightleftharpoons HI + e{\bullet}CH_2I$$
$$IH_2C{\bullet}e + nn{\bullet}I \longrightarrow CH_2I_2$$
FAVOURED

$$ICH_3 \rightleftharpoons I{\bullet}en + n{\bullet}CH_3$$
$$I{\bullet}en + I_2 \rightleftharpoons I_2 + I{\bullet}en$$
$$I{\bullet}en + n{\bullet}CH_3 \rightleftharpoons CH_3I$$
(*Stable, reactive and insoluble*)
NOT FAVOURED

$$\underline{Overall\ equation}: ICH_3 + I_2 \longrightarrow H_2CI_2 + HI \qquad 3.26b$$

Stage 3:

$$I_2 \rightleftharpoons I{\bullet}en + I{\bullet}nn$$
$$CH_2I_2 \rightleftharpoons I{\bullet}en + n{\bullet}CH_2I$$
$$IH_2C{\bullet}n + en{\bullet}I \rightleftharpoons CH_2I_2$$
$$I{\bullet}en + nn{\bullet}I \rightleftharpoons I_2$$
(*Stable, reactive and insoluble*)

$$I_2CH_2 \rightleftharpoons I{\bullet}en + n{\bullet}CH_2I$$
$$I{\bullet}en + I_2 \rightleftharpoons I_2 + I{\bullet}en$$
$$I{\bullet}en + n{\bullet}CH_2I \rightleftharpoons CH_2I_2$$
(*Stable, reactive and insoluble*)

$$\underline{Overall\ equation}: I_2CH_2 + I_2 \rightleftharpoons H_2CI_2 + I_2 \qquad 3.26c$$

As it seems, the iodination seems to stop in Stage 2, because the Equilibrium state of existence of the di-iodomethane cannot be suppressed by iodine, unlike during chlorination. This does not mean that iodoform[8] (CHI_3) and tetra-iodomethane (CI_4) do not exist. One can observe the very great significance of states of existences. All the halogens are uniquely different in characters despite the fact that they belong to the same family. If the Equilibrium state of existence of CH_2I_2 can be suppressed without suppressing that of iodine by a passive catalyst, then Stage 3 will be favoured. While energy was released in second step of the stages for fluorination and chlorination, because of the presence of halogens carrying electro-non-free-radicals in the presence of more electropositive atoms in the step, it could not be released in Step 2 of the stages for iodination, because C and I are equi-electropositive. It is no surprise therefore why for iodination to take place, more energy was required to start the reaction. That energy is that required to suppress the Equilibrium state of existence of methane in the presence of iodine. If it was not suppressed, the stage will be unproductive as shown below.

Stage 1:

$$I_2 \rightleftharpoons I\bullet en + I\bullet nn$$

$$CH_4 \rightleftharpoons H\bullet e + n\bullet CH_3$$

$$H\bullet e + nn\bullet I \rightleftharpoons HI$$

$$I\bullet en + n\bullet CH_3 \rightleftharpoons CH_3I$$

$(Stable, reactive\ and\ soluble)$

$$\underline{Overall\ equation}:\ CH_4 + I_2 \rightleftharpoons H_3CI + HI \qquad 3.27$$

Based on what has been highlighted above, one can observe why (viii) and (ix) of Equations 3.18 and 3.19 were chosen as examples for Polar/Non-ionic compounds.

(x) **Cuprous hydride**

$$\underset{\text{(Cuprous hydride)}}{CuH} \xrightleftharpoons{Radically} Cu{\cdot}en + H{\cdot}n \quad ; \quad \text{Ionically, None} \qquad 3.28$$

Copper has paired unbonded radicals in the last shell providing the polar character. Cupric hydride if it exists is also Polar/Non-ionic.

(C)Non-polar/Non-ionic compounds

These are compounds with no hetero atom or other atoms with paired unbonded radicals in their last shell. Though Sodium hydride (NaH) contains two ionic metals, it is Non-polar/Non-ionic. The bond in it is covalent of the elastic type. The same applies to all Ionic metallic hydrides.

(i) **Hydrocarbons (Methane)**

$$CH_4 \xrightleftharpoons{Radically} H\cdot e + n\cdot CH_3 \quad ; \; Ionically, \; None \qquad 3.29$$

(ii) **Hydrocarbons (Ethane)**

$$C_2H_6 \xrightleftharpoons[\text{At or bolow Melting point}]{Radically} H_3C\cdot e + n\cdot CH_3 \quad ; \; Ionically, \; None$$

$$\xrightleftharpoons[\text{at or above Boiling point}]{Radically} H\cdot e + n\cdot C_2H_5 \quad ; \; Ionically, \; None \qquad 3.30$$

(iii) **Hydrocarbon (Cyclopropane)**

This cannot exist in equilibrium state of existence, but can be activated.

$$H_2C-CH_2 \; (CH_2) \xrightleftharpoons{Radically} e\cdot \overset{H}{\underset{H}{C}} - \overset{H}{\underset{H}{C}} - \overset{H}{\underset{H}{C}}\cdot n \quad ; \; Ionically, \; None$$

$$\xrightleftharpoons{Chargedly} \oplus \overset{H}{\underset{H}{C}} - \overset{H}{\underset{H}{C}} - \overset{H}{\underset{H}{C}} \odot \qquad 3.31$$

(iv) **Hydrogen molecule**

$$H_2 \xrightleftharpoons[H_2\ Catalyst]{Radically} H\cdot e \; + \; n\cdot H \quad ; \text{Ionically,} \quad \text{None} \qquad 3.32$$

Covalently, there is none, ***since hydrogen like other ionic metals cannot carry a negative charge of the REAL type-Ionic or Covalent, and not even of the imaginary type-Electrostatic. Because it is univalent, it is not possible for it to carry negative Polar charge.***

(v) **Benzene (Aromatic hydrocarbon)**

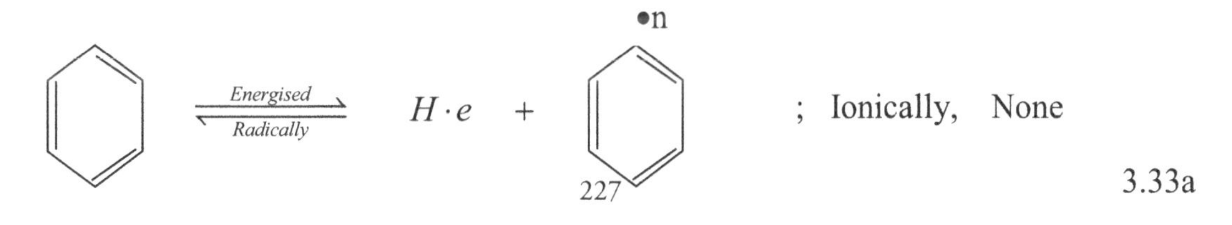

3.33a

(vi) **Toluene (Alipharomatic hydrocarbon)**

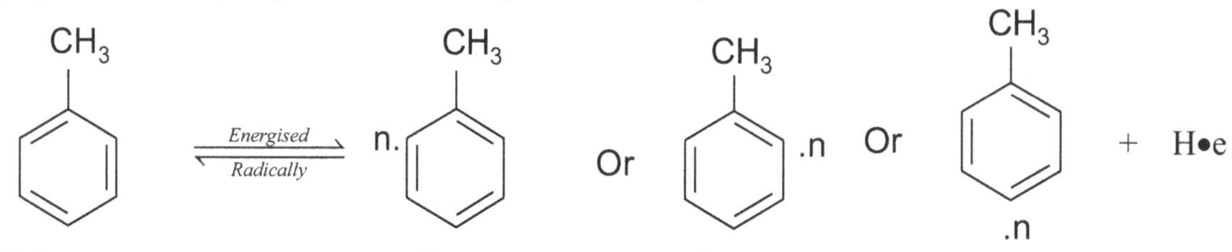

CH_3 Toluene $\underset{Radically}{\overset{Energised}{\rightleftharpoons}}$ n. CH_3 ortho- Or CH_3 .n ortho- Or CH_3 para- .n + H•e

; Ionically, None 3.33b

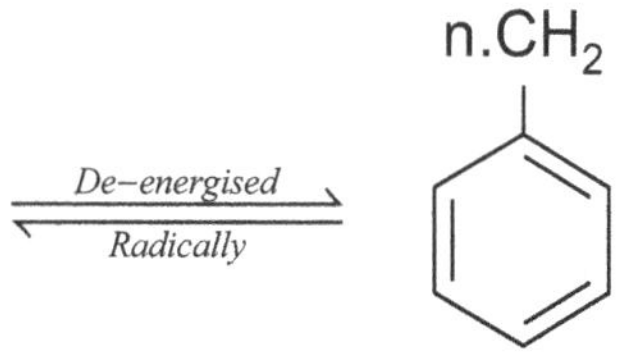

$\underset{Radically}{\overset{De-energised}{\rightleftharpoons}}$ n.CH_2 (benzyl) + H•e ; Ionically, None

3.34

Based on resonance stabilization phenomenon of some rings (Benzenes, Naphthalenes, Anthracenes and their likes), there are ***energized and de-energized rings*** based on the use of universal data. This has already been said in Chapter 1. None of the cases above has the ionic counterpart. For the aromatic ring, how the ortho-, meta-, and para- positions are obtained based on the new classification of groups and much more, have been found and partly shown herein and fully explained in the New Frontiers. This is not a game of Present-day Science. It is ORDERLINESS and the NEW SCIENCE. It has nothing to do with Kekule's formula or theory[9]. The theory may come up after laying this foundation. Though we know that one cannot put a cat behind a horse, that is what we have been doing by putting the electron outside the nucleus, ***thinking that the nucleus was positively charged while the electron outside the nucleus was negatively charged!!!***

(vii) **Lithium aluminium hydride and Grignard reagents (Polar/Non -ionic)**

$$Li^{\oplus}........^{\ominus}AlH_4 \underset{}{\overset{Radically}{\rightleftharpoons}} Li^{\oplus}......^{\ominus}\underset{e}{\dot{A}}lH_3 + H\cdot n\ ; \qquad Ionically, \quad None$$

Electrostatically cationically Charged – paired Bond
[Lithium aluminium hydride] 3.35*a*

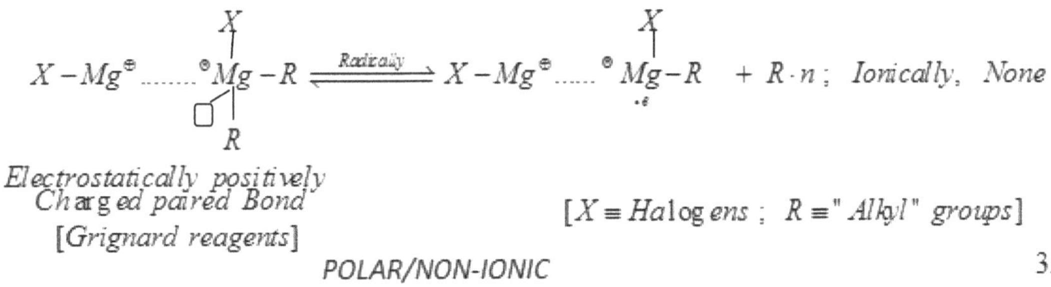

$$X-Mg^{\oplus}........{}^{\oplus}\overset{X}{\underset{R}{Mg}}-R \xrightarrow{Radically} X-Mg^{\oplus}......{}^{\oplus}\overset{X}{\underset{\cdot e}{Mg}}-R \quad + R\cdot n; \; Ionically, \; None$$

Electrostatically positively Charged paired Bond [Grignard reagents]

$[X \equiv Halogens; \; R \equiv "Alkyl" \; groups]$

POLAR/NON-IONIC 3.35b

Because lithium aluminium hydride and Grignard reagents are very unique compounds used for synthesis of many compounds, ***Grignard reagents which is Polar/Non-ionic*** has been placed in this sub-section. It is dimeric in character. The polar character is provided by the presence of X.

Our world has gone far in providing abundant data and great details, more than ninety-five percent of which unfortunately cannot be explained and properly used, because of the missing links. This is not only in the Sciences, Engineering and related disciplines, but also in other disciplines. Based on the new classifications for compounds above and the examples shown and new explanations for some phenomena such as the molecular rearrangement in Equations 3.6a and 3.6b, we have a long way to go. For the first time, the structures of Lithium aluminium hydride and Grignard reagents have been provided. Present-day Science knows fully well, based on what it does that the structure cannot be that of LiH and AlH_3. The same too applies to the structure of Grignard reagents which they know is dimeric. Based on what it does, they know that it cannot be RMgX where X is a halogen and R is an alkylane group such as CH_3. Based on the structures and their finger prints given above, one can observe how they do what they do. All the reactions are radical in character. While $LiAlH_4$ is an Electrostatically ***cationically*** charged-paired compound, Grignard reagents are Electrostatically ***positively*** charged-paired compounds, because the single positive charge on the Mg center is not ionic, but originally covalent. The electro-free-radical carried on the counter center- Al and Mg respectively is where the action begins as will be shown in the New Frontiers. ***The Al and Mg centers were able to respectively accommodate H and R, because of the presence of vacant orbitals in their last shell. Without the presence of vacant orbitals or paired unbonded radicals on the counter centers, electrostatic bonds can never be formed.*** Note that these cannot be used to conduct electricity. However, depending on the operating conditions, the Li case can very weakly be used as initiators only in Equilibrium state of existence. It cannot be used as initiators in Stable state of existence. The lithium center cannot be used cationically for polymerization and the electrostatic centers cannot be paired radically. The positive charge on the Mg center can be used positively for polymerization. They can only be used radically when they exist in Equilibrium state of existence.

(vi) The Butenes

$$\begin{array}{ccccccccc} H\ \ CH_3 & & H\ \ CH_3 & & H\ \ CH_3 & & H\ \ CH_3 & & H\ \ CH_3 \\ I\ \ \ I & & I\ \ \ I & & I\ \ \ I & & I\ \ \ I & & I\ \ \ I \\ C=C & \rightleftharpoons & n\cdot C-C\cdot e & \rightleftharpoons & H-C-C\cdot e & \rightleftharpoons & n\cdot C-C\cdot e & \rightleftharpoons & C=C \\ I\ \ \ I & & I\ \ \ I & & I\ \ \ I & & I\ \ \ I & & I\ \ \ I \\ H\ \ CH_3 & & H\ \ CH_3 & & H\ \ \cdot CH_2 & & H\ \ CH_3 & & H\ \ CH_3 \\ & & & & n & & & & \end{array} \qquad 3.36a$$

Methyl propene or isobutylene

$$\begin{array}{ccccccc} H\ \ H & & H\ \ H & & H\ \ H & & H\ \ H \\ I\ \ \ I & & I\ \ \ I & & I\ \ \ I & & I\ \ \ I \\ C=C & \rightleftharpoons & n\cdot C-C\cdot e & \rightleftharpoons & n\cdot C-C\cdot e & \longrightarrow & C=C \\ I\ \ \ I & & I\ \ \ I & & I\ \ \ I & & I\ \ \ I \\ CH_3\,CH_3 & & CH_3\,CH_3 & & H\ \ C_2H_5 & & H\ \ C_2H_5 \end{array} \qquad 3.36b$$

2 – *Butene* (*Cis*) 1 – *Butene*

Molecular rearrangement of 1st kind of the 1st type

$$\begin{array}{ccccccc} CH_3\ H & & CH_3H & & H\ \ H & & H\ \ H \\ I\ \ \ I & & I\ \ \ I & & I\ \ \ I & & I\ \ \ I \\ C=C & \rightleftharpoons & n\cdot C-C\cdot e & \rightleftharpoons & n\cdot C-C\cdot e & \longrightarrow & C=C \\ I\ \ \ I & & I\ \ \ I & & I\ \ \ I & & I\ \ \ I \\ H\ \ CH_3 & & H\ \ CH_3 & & H\ \ C_2H_5 & & H\ \ C_2H_5 \end{array} \qquad 3.36c$$

2 – *Butene* (*Trans*) 1 – *Butene*

Molecular rearrangement of 1st kind of the 1st type

Isobutylene rearranges to give the same isobutylene when activated. But 2-Butene (cis- or trans-) rearranges when activated to give the more stable 1-Butene with the transfer species of the first kind and of the first type being hydrogen electro-free radical. In present-day Science, it is believed that alkenes add to sulfuric acid according to ***Markovnikov rule*** (*which states that the reaction of a polar molecule with a carbon-carbon double bond results in the addition of the positive end of the polar molecule to the less alkyl-substituted end of the polarized double bond)* to give alkyl hydrogen sulfate.[4] Indeed, this is not the case as shown below. ***Secondly, present-day Science does not know what the word "Polar" is, based on the manner the word is used.*** There is nothing like a polarized double bond! Is it activated double bond? Even then, what "Activation" is, is unknown to present-day Science, since the word is sometimes used in so many different ways meaninglessly.

1-Butene

Stage 1:

$$H_2SO_4 \rightleftharpoons H\cdot e + nn\cdot OSO_3H$$

$$H\cdot e + \underset{H}{\overset{H}{\underset{|}{\overset{|}{C}}}} = \underset{C_2H_5}{\overset{H}{\underset{|}{\overset{|}{C}}}} \rightleftharpoons \underset{(A)}{H - \underset{H}{\overset{H}{\underset{|}{\overset{|}{C}}}} - \underset{C_2H_5}{\overset{H}{\underset{|}{\overset{|}{C}}}}\cdot e}$$

$$(A) + nn\cdot OSO_3H \longrightarrow CH_3CH_2\underset{OSO_3H}{\underset{|}{C}}HCH_3$$

$$s-Butyl\ hydrogen\ sulfate \qquad 3.37a$$

2-Butene

Stage 1:

$$H_2SO_4 \rightleftharpoons H\cdot e + nn\cdot OSO_3H$$

$$\underset{\substack{2-Butene\\(Trans)}}{\underset{H}{\overset{CH_3}{\underset{|}{\overset{|}{C}}}} = \underset{CH_3}{\overset{H}{\underset{|}{\overset{|}{C}}}}} \rightleftharpoons n\cdot\underset{H}{\overset{CH_3}{\underset{|}{\overset{|}{C}}}} - \underset{CH_3}{\overset{H}{\underset{|}{\overset{|}{C}}}}\cdot e \rightleftharpoons n\cdot\underset{H}{\overset{H}{\underset{|}{\overset{|}{C}}}} - \underset{C_2H_5}{\overset{H}{\underset{|}{\overset{|}{C}}}}\cdot e \quad [Molecular\ rearrangement]$$

$$H\cdot e + n\cdot\underset{H}{\overset{H}{\underset{|}{\overset{|}{C}}}} - \underset{C_2H_5}{\overset{H}{\underset{|}{\overset{|}{C}}}}\cdot e \rightleftharpoons \underset{(A)}{H - \underset{H}{\overset{H}{\underset{|}{\overset{|}{C}}}} - \underset{C_2H_5}{\overset{H}{\underset{|}{\overset{|}{C}}}}\cdot e}$$

$$(A) + nn\cdot OSO_3H \longrightarrow CH_3CH_2\underset{OSO_3H}{\underset{|}{C}}HCH_3$$

$$s-Butyl\ hydrogen\ sulfate \qquad 3.37b$$

2-Butene (Cis)

Stage 1:

$$H_2SO_4 \rightleftharpoons H\cdot e + nn\cdot OSO_3H$$

$$\begin{array}{c} CH_3 \quad CH_3 \\ | \quad\; | \\ C = C \\ | \quad\; | \\ H \quad\; H \end{array} \rightleftharpoons \begin{array}{c} CH_3\,CH_3 \\ | \quad\; | \\ n\cdot C - C\cdot e \\ | \quad\; | \\ H \quad\; H \end{array} \rightleftharpoons \begin{array}{c} H \quad\; H \\ | \quad\; | \\ n\cdot C - C\cdot e \\ | \quad\; | \\ H \quad\; C_2H_5 \end{array} \quad [Molecular\ rearrangement]$$

2 – *Butene*
(*Cis*)

$$H\cdot e + \begin{array}{c} H \quad\; H \\ | \quad\; | \\ n\cdot C - C\cdot e \\ | \quad\; | \\ H \quad\; C_2H_5 \end{array} \rightleftharpoons \begin{array}{c} H \quad\; H \\ | \quad\; | \\ H - C - C\cdot e \\ | \quad\; | \\ H \quad\; C_2H_5 \end{array}$$

(*A*)

$$(A) + nn\cdot OSO_3H \longrightarrow \begin{array}{c} CH_3CH_2CHCH_3 \\ | \\ OSO_3\,H \end{array}$$

s – *Butyl hydrogen sulfate* 3.37*c*

Isobutylene

Stage 1:

$$H_2SO_4 \rightleftharpoons H\cdot e + nn\cdot OSO_3H$$

$$H\cdot e + \begin{array}{c} H \quad\; CH_3 \\ | \quad\; | \\ C = C \\ | \quad\; | \\ H \quad\; CH_3 \end{array} \rightleftharpoons \begin{array}{c} H \quad\; CH_3 \\ | \quad\; | \\ H - C - C\cdot e \\ | \quad\; | \\ H \quad\; CH_3 \end{array}$$

(*B*)

$$(B) + nn\cdot OSO_3H \longrightarrow \begin{array}{c} (CH_3)_2CCH_3 \\ | \\ OSO_3\,H \end{array}$$

t – *Butyl hydrogen sulfate* 3.37*d*

Based on the mechanism provided so far, one can observe that the so-called Markovnikov rule is meaningless, just as so many countless rules in Chemistry. Even the present- day concept of a polar molecule is completely different from what has been seen so far. The mechanism provided above for the reactions is Equilibrium mechanism.

3.2 DISSOLUTION AND MISCIBILIZATION

Sodium chloride (NaCl) ***a solid,*** dissolves in water both being polar/ionic. For the same reasons, polyacrylamide, vinyl alcohol, acrylic acid, sodium hydroxide and more ***dissolve if solids or miscibilize if fluids (liquids, vapours and gases)*** in water. They are all **Polar/Ionic**. Because hydrocarbons such as kerosene, methane, etc. are **Non-polar/Non-ionic**, hence they can never dissolve if solid or miscibilize if fluid in for example water which is **Polar/Ionic**; neither will water dissolve if solid or miscibilize if fluid in them. They have nothing in common. The concepts of *Dissolution/ Miscibilization* are quite different from the concepts of *Solubilisation/ Insolubilisation* based on in-depth use of universal data.

In Miscibilization/Dissolution, the driving force is the similarity in the polar/ionic characters of the components. No stage-wise chemical reaction takes place whether the components are in STABLE or EQUILIBRIUM states of existence as will shortly be shown.

Polyacrylamide is a water-soluble polymer. Acrylamide and its polymer are Polar/Ionic. So also, is water and organic alcohol such as CH_3OH. One will recall the work on "Measurement of Molecular Weight Distribution of Polyacrylamide by Turbidimetric Titration"[10]. Methanol was then up to the present moment thought to be used as Non-solvent to precipitate polyacrylamide from its solution in water, according to size, the largest size being the first to precipitate. Therefore, when methanol the non-solvent was added continuously to an aqueous solution of polyacrylamide, another solution with increasing turbidity was obtained. One thought the methanol was pushing the polyacrylamide out of the solution. Unknown, was the fact that, the CH_3OH is not precipitating the polyacrylamide from solution but reacting with it to form another product which is Polar/Non-ionic as shown below.

Stage 1:

$$\underset{\text{[Polar/Ionic]}\ (A)}{\begin{array}{cc} H & H \\ | & | \\ C & = C \\ | & | \\ H & C = O \\ & | \\ & NH_2 \end{array}} \quad \underset{\text{of Existence}}{\overset{\text{Equilibrium State}}{\rightleftharpoons}} \quad \underset{(B)}{\begin{array}{cc} H & H \\ | & | \\ C & = C \\ | & | \\ H & C = O \\ & | \\ & nn\bullet NH \end{array}} \quad + \quad H\bullet e$$

$$H\bullet e \quad + \quad CH_3OH \quad \overset{\text{Abtraction}}{\rightleftharpoons} \quad H_2O \quad + \quad e\bullet CH_3$$

$$H_3C\bullet e \quad + \quad (B) \quad \longrightarrow \quad \underset{(C)}{H_2C = CH - CONH(CH_3)} \qquad 3.38a$$

Stage 2:

$$(C) \underset{\text{of Existence}}{\overset{\text{Equilibrium State}}{\rightleftharpoons}} H_2C = CH - CO\overset{CH_3}{\overset{|}{N}}\bullet nn \quad + H\bullet e$$
(D)

$$H\bullet e \quad + \quad CH_3OH \quad \overset{\text{Abtraction}}{\rightleftharpoons} \quad H_2O \quad + \quad e\bullet CH_3$$

$$H_3C\bullet e \quad + \quad (D) \quad \longrightarrow \quad H_2C = CH - CON(CH_3)_2$$
[Polar/Non-Ionic]
(E) 3.38b

Overall Equation: $(A) \quad + \quad 2CH_3OH \longrightarrow 2H_2O \quad + \quad (E)$ 3.38c

(A) above was just used in place of the monomer units of polyacrylamide. Note that the (E), the polymer chain has no double bond internally located. The formation of (E) for different sizes of polymer chains, is the source of its turbidity. (E) which is **Polar/Non-ionic**, was forced to partly precipitate from solution inside the water, because water is **Polar/Ionic.** The only thing common between the two of them is **"Polar"**. What is not common is more important than polarity. One is ionic, while the other is non-ionic. This is the source of the turbidity. As more of (E) is formed, the common character between the two phases increases in intensity, making the solution more turbid. The smallest chain is the first where the N center is alkylated and first to show turbidity. Invariable Methanol cannot be said to be a non-solvent, since WATER, ACRYLAMIDE, POLYACRYLAMIDE, and METHANOL are all **Polar/Ionic.** Methanol and water are SOLVENTS for Acrylamide and its polymer. Therefore, Acrylamide and its polymer being solids can readily DISSOLVE in water or methanol. ***While the acrylamide and polyacrylamide dissolve in water, they are both insoluble in water. While both also dissolve in methanol, they both react with methanol.*** The fact that two compounds may belong to the same family, such as Polar/Ionic, does not mean that productive chemical reactions cannot take place between them. In some, they do without a product such as sodium chloride and water. In another, they do and become productive such as above or hydrochloric acid and sodium hydroxide (the two products which should have been obtained from water and sodium chloride). To find a case where there is no reaction when they dissolve or misciblize is impossible, though no chemical reaction is involved during dissolution and miscibilization. The reaction could be productive or non-productive, but the fact is that there is a reaction when solubilisation or insolubilisation takes place.

3.3 SOLUBILISATION AND INSOLUBILISATION

In Solubilisation/Insolubilisation, chemical reactions take place ***in a single stage*** via Equilibrium mechanisms to lead to no products. Shown below are solubilisation of NaCl in H_2O (not Dissolution) and the insolubility of NaOH in H_2O (also not Dissolution). NaCl and NaOH both dissolve in water, since they all belong to the same family (Polar/Ionic).

3.3.1 SOLUBILITY

(a)Stage1:[FAVOURED]

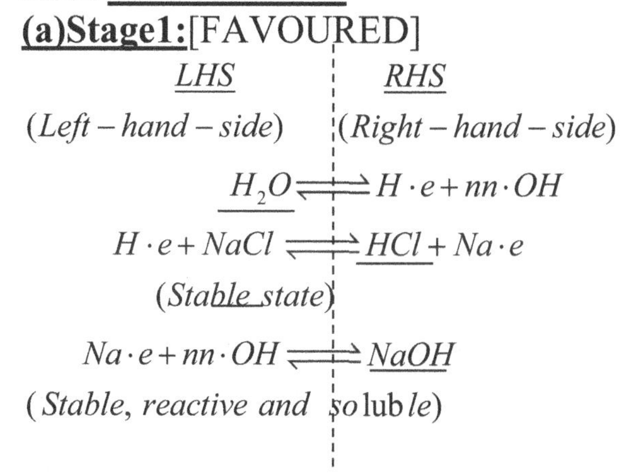

$$\underline{LHS} \quad (Left-hand-side) \quad \vdots \quad \underline{RHS} \quad (Right-hand-side)$$

$$\underline{H_2O} \rightleftharpoons H\cdot e + nn\cdot OH$$

$$H\cdot e + NaCl \rightleftharpoons \underline{HCl} + Na\cdot e$$

$$(Stable\ state)$$

$$Na\cdot e + nn\cdot OH \rightleftharpoons \underline{NaOH}$$

$$(Stable,\ reactive\ and\ soluble) \qquad 3.39a$$

LHS and RHS above are left-hand-side and right-hand-side respectively. Notice that above, the Equilibrium state of existence of sodium chloride is suppressed by the presence of water. Note that HCl produced in the third step could not suppress the Equilibrium state of existence of NaOH. If NaCl had suppressed the Equilibrium state of existence of water, heat is released in the process, while the stage will still remain unproductive as shown below.

(b) Stage 1: [NOT FAVOURED]

$$\underline{LHS} \quad (Left-hand-side) \quad \vdots \quad \underline{RHS} \quad (Right-hand-side)$$

$$\underline{NaCl} \rightleftharpoons Na\cdot e + nn\cdot Cl$$

$$Na\cdot e + \underline{H_2O} \rightleftharpoons \underline{NaH} + HO\cdot en + Heat$$

$$HO.en \rightleftharpoons \underline{H.e} + en.O.nn + Heat$$

$$nn.O.en + NaH \rightleftharpoons Na.e + nn.OH$$

$$\rightleftharpoons NaOH$$

$$H\cdot e + Cl\cdot nn \rightleftharpoons HCl$$

$$(Stable,\ reactive\ and\ soluble) \qquad 3.40a$$

$$\underline{Overall\ Equation}: H_2O + NaCl \rightleftharpoons HCl + NaOH + Heat \qquad 3.40b$$

When both are in Equilibrium state of existence, the followings are obtained.

(c) Stage 1: [NOT FAVOURED]

$$\underline{LHS} \quad \vdots \quad \underline{RHS}$$

$$(Left-hand-side) \quad (Right-hand-side)$$

$$\underline{NaCl} \rightleftharpoons Na\cdot e + nn\cdot Cl$$

$$\underline{H_2O} \rightleftharpoons H\cdot e + nn\cdot OH$$

$$H\cdot e \; + \; nn\cdot Cl \rightleftharpoons \underline{HCl}$$

$$Na\cdot e \; + \; nn\cdot OH \rightleftharpoons \underline{NaOH}$$

$$(Stable,\ reactive\ and\ so\text{lub}le) \qquad 3.41a$$

$$\underline{Overall\ Equation}:\ H_2O + NaCl \longrightarrow HCl + NaOH \qquad 3.41b$$

Like the first case above (a), no heat is released for this last case, since when salt is added to a beaker of water, the beaker is not warmer than what it was. As will shortly become obvious, the sodium chloride is the one suppressed (a), being a solid. Notice that, on the LHS of the vertically drawn line, we have NaCl and water, that which can be seen, but on the RHS, we have HCl and NaOH, that which cannot be seen. But, they are there imaginarily, because we can see them above, but not in the real part of our complex world. ***Since both sides have different compounds, the Sodium chloride is said to be soluble in water.*** Sodium chloride dissolves in water and also soluble in water. What of what we see on the LHS is the same as what we see on the RHS? What does this mean? In some cases, a reaction takes place to give the same products as reactants **not held in equilibrium,** while in others, reaction takes place to give the same product as the reactants, **held in Equilibrium.** One can see the very important need of establishing the order of capacities of so many things, just as we do in our world, but differently.

3.3.2 INSOLUBILITY

(a) Stage 1: [FAVOURED]

$$\underline{LHS} \quad \vdots \quad \underline{RHS}$$

$$\underline{NaOH} \rightleftharpoons Na\cdot e + nn\cdot OH$$

$$Na\cdot e + \underline{H_2O} \rightleftharpoons \underline{NaH} + HO\cdot en + Heat$$

$$HO.en \rightleftharpoons \underline{nn.O.en} + H.e + Heat$$

$$nn.O.en + NaH \rightleftharpoons Na.e + nn.OH + Heat$$

$$\rightleftharpoons NaOH$$

$$H\cdot e + nn\cdot OH \rightleftharpoons H_2O$$

$$(Stable,\ reactive\ and\ insoluble) \qquad 3.42a$$

$$\underline{Overall\ Equation}:\ NaOH + H_2O \rightleftharpoons NaOH + H_2O + Heat \qquad 3.42b$$

(b) Stage 1: [NOT FAVOURED]

$$\underline{LHS} \quad \vdots \quad \underline{RHS}$$

$$\underline{H_2O} \rightleftharpoons H\cdot e + nn\cdot OH$$

$$H\cdot e + \underline{NaOH} \rightleftharpoons NaO\cdot en + \underline{H_2} + Heat$$

$$NaO\cdot en \rightleftharpoons Na\cdot e + en\cdot O\cdot nn + Heat$$

$$nn\cdot O\cdot en + \underline{H_2} \rightleftharpoons HO\cdot nn + H\cdot e + Heat$$

$$\rightleftharpoons \underline{H_2O}$$

$$Na\cdot e + nn\cdot OH \rightleftharpoons \underline{NaOH}$$

$$(Stable,\ reactive\ and\ insoluble) \qquad 3.43a$$

$$\underline{Overall\ Equation}:\ NaOH + H_2O \rightleftharpoons NaOH + H_2O + Heat \qquad 3.43b$$

(c) Stage 1: [NOT FAVOURED]

$$\underline{LHS} \quad \vdots \quad \underline{RHS}$$

$$\underline{H_2O} \rightleftharpoons H\cdot e + nn\cdot OH$$

$$\underline{NaOH} \rightleftharpoons Na\cdot e + nn\cdot OH$$

$$H\cdot e + nn\cdot OH \rightleftharpoons \underline{H_2O}$$

$$Na\cdot e + nn\cdot OH \rightleftharpoons \underline{NaOH}$$

$$(Stable,\ reactive\ and\ insoluble) \qquad 3.44a$$

$$\underline{Overall\ Equation}:\ NaOH + H_2O \rightleftharpoons NaOH + H_2O \qquad 3.44b$$

Whether the last step is equilibrium or a single double headed arrow is shown for water as the last step, the situation remains the same.

Notice how transfer species are abstracted. Instead of H•e abstracting OH from NaOH, as it did in Equation 3. 38a and 3.38b where it abstracted OH from CH_3OH, in Equation 3. 38a and 3.38b, it abstracted H from the NaOH, ***because the component abstracted is not the component usually held in Equilibrium state of existence, unless there is only one state of existence for the compound such NaCl or the same component held is not removed as it was held, such as in H_2O. In addition, only ionic metal or metals more electropositive than H can displace H from a compound electro-free-radically.*** One can see the great dominance of ionic metals amongst all metals, being the first members of every Period in the Periodic Table. OH was removed from CH_3OH, because that is not the component held in Equilibrium state of existence of CH_3OH. OH was not the component removed in NaOH, because that is the component held in Equilibrium state of existence of NaOH. H was the component removed from water as a hydride (Not as H.e an atom) by Na, because Na an ionic metal is more electropositive than H which is also an ionic metal. If OH was removed from H_2O by Na, then we move from more stable state (Na.e) to less stable state (H.e), that which is against one of the laws of Nature. If H had been removed in Equation 3.38a or 3.38b, the stage would have been unproductive. On the other hand while NaOH is basic, CH_3OH is not. CH_3OH is an alcohol when it decomposes, but an acid when in Equilibrium state of existence. From the last three equations above, it is obvious that sodium hydroxide and water cannot both be in Equilibrium state of existence, since no heat is released. One of them must be suppressed and that one is water and not the pellets. For so many years, one has always wondered why when pellets of NaOH are put into a beaker of water, so much heat is released. Based on the equations above, one can begin to see the source of the heat in (a) where NaOH was the component held in Equilibrium state of existence.

Notice that on the LHS and RHS of the vertical line, we have the same NaOH and H_2O. That on the LHS can be seen, but not that on the RHS. ***Since both sides are the same, NaOH is said to be insoluble in water.*** Thus, sodium hydroxide dissolves in water, but insoluble in water, just like the case of acrylamide/H_2O. Zinc chloride ($ZnCl_2$) is insoluble in chlorine (Cl_2), but dissolves in Cl_2. One can begin to see the origin of turbidity/precipitation, particularly with respect to productive chemical reaction being involved as a stage and Ionic/Non-ionic characters.

3.3.3 PRODUCTIVE STAGE

The reaction between sodium hydroxide (Basic) and hydrochloric acid (Acid) is irreversible, because whatever the type of operating conditions applied, the products from the reaction cannot be reversed. The products are sodium chloride and water as shown below.

(a)Stage 1: [NOT FAVOURED]

$$NaOH \rightleftharpoons Na \cdot e + nn \cdot OH$$
$$HCl \rightleftharpoons H \cdot e + Cl \cdot nn$$
$$H \cdot e + nn \cdot OH \rightleftharpoons H_2O$$
$$Na \cdot e + nn \cdot Cl \longrightarrow NaCl \qquad 3.45a$$
$$\underline{Overall\ Equation}: NaOH + HCl \longrightarrow NaCl + H_2O \qquad 3.45b$$

Herein, NaOH and HCl are both in Equilibrium state of existence. Under such conditions, no heat is released, clear indication that this cannot be the real mechanism, though the stage is productive.

(b) Stage 1: [NOT FAVOURED]

$$NaOH \rightleftharpoons Na \cdot e + nn \cdot OH$$
$$Na \cdot e + HCl \rightleftharpoons NaCl + H \cdot e$$
$$H \cdot e + nn \cdot OH \rightleftharpoons H_2O$$
$$(Stable,\ reactive,\ and\ soluble) \qquad 3.46a$$
$$\underline{Overall\ Equation}: NaOH + HCl \rightleftharpoons NaCl + H_2O \qquad 3.46b$$

In the second step of the stage above, it is Cl that is abstracted at the end of which the stage is unproductive, because NaCl was suppressed by water. Probably the wrong component must have been abstracted.

(c)Stage 1: [NOT FAVOURED]

$$NaOH \rightleftharpoons Na \cdot e + nn \cdot OH$$
$$** Na \cdot e + HCl \rightleftharpoons NaH + Cl \cdot en + Heat$$
$$Cl \cdot en + nn \cdot OH \longrightarrow HOCl \qquad 3.47a$$
$$\underline{Overall\ Equation}: NaOH + HCl \longrightarrow NaH + HOCl + Heat \qquad 3.47b$$

The reaction continues in the next stage, because the products above may be reactive.

Stage 2:

$$HOCl \rightleftharpoons H \cdot e + nn \cdot OCl$$

$$ClO \cdot nn \rightleftharpoons nn \cdot O \cdot en + Cl \cdot nn + Heat$$

$$nn \cdot O \cdot en + NaH \rightleftharpoons HO \cdot nn + Na \cdot e$$

$$H \cdot e + nn \cdot OH \rightleftharpoons H_2O$$

$$Na \cdot e + nn \cdot Cl \longrightarrow NaCl \qquad 3.47c$$

$$\underline{Overall\ Equation}: NaOH + HCl \longrightarrow NaCl + H_2O + Heat(2) \qquad 3.47d$$

Though the right component has been abstracted above, one cannot say that this is the mechanism, because the two stages are productive. How components are abstracted is very important to know, because these are not done indiscriminately. Hence, the example above has been shown.

(d) Stage 1: [FAVOURED]

$$HCl \rightleftharpoons H \cdot e + nn \cdot Cl$$

$$H \cdot e + NaOH \rightleftharpoons H_2 + NaO \cdot en + Heat$$

$$NaO \cdot en \rightleftharpoons Na \cdot e + nn \cdot O \bullet en + Heat$$

$$nn \cdot O \cdot en + H_2 \rightleftharpoons HO \cdot nn + e \cdot H$$

$$\rightleftharpoons H_2O$$

$$Na \cdot e + nn \cdot Cl \longrightarrow NaCl \qquad 3.48a$$

$$\underline{Overall\ Equation}: NaOH + HCl \longrightarrow NaCl + H_2O + Heat(2) \qquad 3.48b$$

$$HCl \; > \; NaOH \; > \; H_2O \; > \; NaCl$$

$$\underline{ORDER\ OF\ EQUILIBRIUM\ STATE\ OF\ EXISTENCE} \qquad 3.48c$$

Based on (a) of Equation 3.42a, it is (d) that is favoured, with heat released from two steps in the stage. If NaOH could suppress the Equilibrium state of existence of H_2O and NaCl, then why can acid not suppress them when water cannot exist in Equilibrium state of existence in the presence of inorganic acids? Though (a) and (c) are all productive, only one is the real mechanism and that one is (d). (b) informs us that NaOH cannot suppress the Equilibrium state of existence of inorganic acids. In NATURE, there is nothing like an ALTERNATIVE route or way or mechanism. Everything depends on the OPERATING CONDITIONS and types of compounds involved.

In solubilisation, the stage is ***fully in equilibrium*** since there is no combination (single right-handed full-headed arrow) in the last step of the stage. In it NaCl was suppressed by water. The same was observed in Productivity in (d) of Equation 3. 48a- the last two steps. In that stage, HCl is held in Equilibrium state of existence suppressing that of NaOH. In the process of abstraction, energy was released followed by a second one in the next step at the end of which H_2O and NaCl were formed with water being the first to be formed. In solubilisation, the stage is also ***fully in equilibrium*** since there is no combination in the last step of the stage, with products same as reactants held in equilibrium. It could also ***not be fully in equilibrium*** as the case of NaOH/H_2O (Equation 3.43a) but productive, with products same as the reactants. It is also a non-productive stage. In a productive stage in Equilibrium mechanisms, the last step in a stage is always a Combination state step (one single right-handed full-headed arrow), otherwise the rate of product formation cannot be expressed or represented mathematically. Without the one single arrow, no product can be obtained. ***If there are two or more single arrows at the end of a stage, the stage is unreactive as has been shown already and will be shown downstream.*** In a stage, it is the electro-radical that initiates attack all the time for all mechanisms. As will be shown in the New Frontiers with respect to Orderliness, based on the use of the EYE OF THE NEEDLE above, the order of the Equilibrium states of existence of the compounds used in NaCl/H_2O/HCl/NaOH systems have been shown in Equation 3.48a above. This is how they will operate wherever two or three of them co-exist in Nature. *All chemical reactions take place only radically, while all polymeric reactions take place radically and chargedly.* The charges that appear in some stages are non-ionic in character, as already shown and shown again below with provision of how the real structure of ammonium hydroxide was obtained.

3.3.4 OBTAINING REAL STRUCTURE OF AMMONIUM HYDROXIDE

Its structure has already been shown (See Figure 2.2), but how obtained has not yet been shown.

(i) Stage1:

$$H_2O \rightleftharpoons H\cdot e + nn\cdot OH$$

$$H\cdot e + \ddot{N}H_3 \rightleftharpoons e\cdot NH_4$$

$$HO\cdot nn + e\cdot NH_4 \longrightarrow HO^{\ominus}.......^{\oplus}NH_4$$

$$(Ammonium\ \ hydroxide) \qquad 3.49a$$

$$\underline{Overall\, equation:}\; H_2O \;+\; NH_3 \;\longrightarrow\; NH_4OH \qquad 3.49b$$

Note that, e• NH_4 cannot exist based on Boundary laws, because N center is carrying nine radicals instead of eight. It is put in there for simplicity. It is $H_4N^{\oplus}$ that is actually formed by transferring the radical to nn• OH to give $HO^{\ominus}$ as shown in the last step. The same applies to all other cases where electrostatic bonds are in general formed.

$$HO^{\ominus}.......{}^{\oplus}NH_4 \underset{Of\ Existence}{\overset{Equilibrium\ State}{\rightleftharpoons}} HO^{\ominus}.....{}^{\oplus}\underset{n}{\dot{N}}H_3 + H\cdot e \qquad 3.49c$$

$$H_2O \;>\; NH_3 \;>\; NH_2R \;>\; NHR_3$$

$$\underline{ORDER\ OF\ EQUILIBRIUM\ STATE\ OF\ EXISTENCE} \qquad 3.50$$

In the same manner as above apart from other means, was how the real structures of many compounds such as Lithium aluminium hydride, Grignard and many other reagents and more were obtained. The reason why they do it is nothing else other than the presence of paired unbonded radicals and vacant orbitals carried by the CENTRAL ATOMS. *In every compound, there is always a CENTRAL ATOM. In water, the central atom is O; in ammonia, the central atom is N. What they carry and use to work with are very important, since they determine their characters. In ethene, the central atoms are the two C atoms (any of them can carry an electro-free-radical when activated); in propene, the central atom is C, the one carrying a group (CH_3). In Lithium aluminium hydride, the central atom is Al (not Li which is an ionic metal); in Grignard reagent, the central atom is only one Mg, the one carrying the imaginary negative charge. It is not the one carrying the positive charge which is a hole. In NaCl, the central atom is Na and not Cl, the one carrying paired unbonded radicals. The paired unbonded radicals on the Cl center may attract some electropositive non-polar atoms with vacant orbitals and become shielded. When Na is replaced with another atom such as H, the compound no longer becomes what it was- a salt, but the central atom still remains H. NaF, NaBr are all different types of salt in which none is edible. The central atom in H_2 is the one carrying the atom, the one carrying the electro-free-radical. The central atom in Cl_2 is the one carrying the electro-non-free-radical, and not the atom. The central atom in O_2 is the one carrying two electro-non-free-radicals ($en{\bullet}\ddot{\underset{..}{O}}{\bullet}en$) an element of oxygen yet to be discovered and not the oxygen atom. It is not oxidizing oxygen element. In benzene, the central atom is any of the six C atoms. In toluene, the central atom is the one carrying CH_3 group. The CENTRAL ATOM is the HEAD. It is the one carrying the LOAD. It is the one that dictates where action begins. It is the one carrying or indirectly supplying* *the electro-radical. It is therefore also a SLAVE. The CENTRAL ATOM does not necessarily have to be the largest or the smallest of all the atoms in a compound. It must have far more than what the others have. That is the way it is supposed to be in our world, that which is almost the case.*

Notice that the Equilibrium state of existence of water is stronger than that of ammonia as shown in Equation 3.49a. The charges in the ammonium hydroxide are electrostatic charges and the bond is ***electrostatic bond of the negative type***, just like in the second example shown below.

3.3.5 OBTAINING REAL STRUCTURE OF AMMONIUM CHLORIDE

Stage 1:

$$HCl \underset{\text{Existence of HCl}}{\overset{\text{Equilibrium State of}}{\rightleftharpoons}} H\cdot e \; + \; nn\cdot Cl$$

$$H\cdot e + NH_3 \underset{\text{Ammonia}}{\overset{\text{Sitting on Stable}}{\rightleftharpoons}} e\cdot NH_4$$

$$Cl\cdot nn + e\cdot NH_4 \xrightarrow[\text{State}]{\text{Combination}} Cl^{\ominus}\ldots\ldots{}^{\oplus}NH_4$$

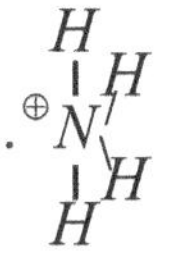

$$(A) - Ammonium\ Chloride \qquad 3.51a$$

$$\underline{Overall\ Equation}: HCl + NH_3 \longrightarrow H_4NCl \qquad 3.51b$$

$$NH_4Cl \underset{\text{Existence of Ammonium Chloride}}{\overset{\text{Equilibrium State of}}{\rightleftharpoons}} H\cdot e \; + Cl^{\ominus}\ldots\ldots{}^{\oplus}N(H)(H)(H)\cdot nn \qquad 3.51c$$

Notice that their Equilibrium state of existence is that in which H can only be held radically and not chargedly as $H^{\oplus}$, because if held as a cation, ammonia will be repelled away by the negative charge on Cl and the pairing will cease to exist. Also, notice that the Equilibrium state of existence of the ammonia which is a base is also suppressed by the inorganic acid -HCl. Water was able to suppress its Equilibrium state of existence, because water is both a weak acid and a weak alcohol. The Equilibrium state of existence of a compound is not necessarily determined by the number of H atoms carried by a central atom, but by the family to which the compound belongs to. Not all members of a family can suppress the Equilibrium state of existence of ammonia. For example, while the Equilibrium state of existence of ammonia cannot be suppressed by F_2, it can be suppressed by Cl_2 as shown below. The mechanisms of how all chemical and polymeric reactions take place have since antiquity been unknown. With respect to chlorination of ammonia the laboratory data usually reported in the literature and textbooks have been wonderful guides[11,12]. ***The whole journey so far looks as if the world data most of which have questionable explanations were put in place for one to use to lay new foundations for humanity. One did not need to go to the laboratory anymore to start doing what, when data were in abundance to start the job- an endless journey. It is endless, because given one billion years of continuous writing, the job will never be complete. But however, the need to lay the foundations is the most important, since it marks the beginning of a new era.***

3.3.6 CHLORINATION AND FLUORINATION OF AMMONIA

Chlorination

Stage 1a:

$$Cl_2 \xrightleftharpoons[\text{Chlorine}]{\text{Equilibrium State of}} Cl\cdot en + nn\cdot Cl$$

$$Cl\cdot en + NH_3 \xrightleftharpoons{\text{Abstraction}} HCl + en\cdot NH_2$$

$$Cl\cdot nn + en\cdot NH_2 \xrightarrow[\text{State}]{\text{Combination}} Cl-NH_2 \qquad 3.52a$$

$$\underline{Overall\ Equation}: NH_3 + Cl_2 \longrightarrow HCl + Cl-NH_2 \qquad 3.52b$$

Chlorine electro-non-free radical is always very thirsty for hydrogen hydride. Hence, it abstracts hydrogen from ammonia. It cannot sit on the ammonia because the central atom N does not have a vacant orbital for her to use. Instead the N center has paired unbonded radicals to fill empty holes such as that carried by $H^{\oplus}$. On other hand in the second step above, the H removed as a nucleo-free-radical is not the same H held as an electro-free-radical when ammonia is in Equilibrium state of existence. It cannot remove NH_2 as a nucleo-non-free-radical because that is how it is held in its Equilibrium state of existence. To continue further chlorination or retain the chloramine in the system, the HCl must be removed as shown below.

(Neutralization of the Acid)
Stage 2:

$$HCl \rightleftharpoons H\cdot e + nn\cdot Cl$$

$$H\cdot e + NaOH \rightleftharpoons H_2 + NaO\cdot en + Heat$$

$$NaO\cdot en \rightleftharpoons Na\cdot e + nn\cdot O\bullet en + Heat$$

$$nn\cdot O\cdot en + H_2 \rightleftharpoons HO\cdot nn + e\cdot H$$

$$\rightleftharpoons H_2O$$

$$Na\cdot e + nn\cdot Cl \longrightarrow NaCl \qquad 3.53a$$

$$\underline{Overall\ Equation}: NaOH + HCl \longrightarrow NaCl + H_2O + Heat(2) \qquad 3.53b$$

$$\underline{Overall\ overall\ equation}: NH_3 + Cl_2 + NaOH \longrightarrow H_2NCl + NaCl + H_2O + Heat \qquad 3.54$$

Instead of HCl reacting with H_2NCl in the presence of NaOH, it reacted with the NaOH, because NaOH is less stable than H_2NCl.

If the HCl is not removed, that is, no NaOH in the system, then the followings are obtained.

Stage 2a:

$$HCl \rightleftharpoons H\cdot e + Cl\cdot nn$$

$$H\cdot e + H_2NCl \rightleftharpoons H_2 + en\cdot NHCl + Heat$$

$$ClHN\cdot en + Cl\cdot nn \longrightarrow NHCl_2$$

$$\underline{Overall\ equation}: HCl + H_2NCl \longrightarrow H_2 + NHCl_2 \qquad 3.55a$$

$$\underline{Overall\ overall\ equation}: Cl_2 + NH_3 \longrightarrow H_2 + NHCl_2 + Heat \qquad 3.55b$$

Thus, with one mole of Cl_2, dichloramine is obtained along with release of H_2 and energy. In two stages, dichloramine was obtained. Notice what was removed in the second step of the stage. That is component held as an electro-free-radical in its Equilibrium state of existence. In the presence of more Cl_2 in the system, the followings are obtained

Stage 3a:

$$Cl_2 \xrightleftharpoons[\text{Chlorine}]{\text{Equilibrium State of}} Cl\cdot en + nn\cdot Cl$$

$$Cl\cdot en + NCl_2H \xrightleftharpoons{\text{Abstraction}} HCl + en\cdot NCl_2$$

$$Cl\cdot nn + en\cdot NCl_2 \xrightarrow[\text{State}]{\text{Combination}} Cl - NCl_2 \qquad 3.56a$$

$$\underline{Overall\ Equation}: NH_3 + 2Cl_2 \longrightarrow HCl + NCl_3 + H_2 + Heat \qquad 3.56b$$

With more chlorine in the system, one should expect that hydrogen will combine with it to form HCl. As shown below, this is not possible. This can only be done in the presence of H_2 catalyst as also shown below in Stage 4a.

Stage 4:

$$Cl_2 \xrightleftharpoons[\text{Chlorine}]{\text{Equilibrium State of}} Cl\cdot en + nn\cdot Cl$$

$$Cl\cdot en + H_2 \xrightleftharpoons{\text{Abstraction}} HCl + e\cdot H$$

$$Cl\cdot nn + e\cdot H \rightleftharpoons HCl$$

$$(Stable,\ reactive\ and\ soluble) \qquad 3.57a$$

$$\underline{Overall\ Equation}: H_2 + Cl_2 \rightleftharpoons 2HCl \qquad 3.57b$$

Stage 4a:

$$H_2 \xrightleftharpoons[\text{Catalyst}]{H_2} H\cdot e + n\cdot H$$

$$H\cdot e + Cl_2 \xrightleftharpoons{\text{Abstraction}} HCl + en\cdot Cl + Heat$$

$$Cl\cdot en + n\cdot H \longrightarrow HCl \qquad 3.58a$$

$$\underline{Overall\ Equation}: H_2 + Cl_2 \xrightarrow[Cat.]{H_2} 2HCl \qquad 3.58b$$

$$\underline{Overall\ overall\ equation}: NH_3 + 3Cl_2 \xrightarrow[Cat.]{H_2} NCl_3 + 3HCl \qquad 3.59$$

One can observe how these reactions take place in stages. This calls to attention, the manners by which chemical reactions are carried out in laboratories in continuous systems. Components are added indiscriminately without taking into considerations the great significance of MOLAR RATIOS. When the one and only one mechanisms are provided, one sees exactly what should be done and not be done. One should know when to add and when not to add. One should know whether the reaction is going to be endothermic or exothermic. One should know what types of operating conditions to use. There are still lots to be revealed about the GREAT WONDERS of NATURE too much to comprehend.

How can present-day Science, say that a compound such as nitrogen trichloride is a powerful oxidizing agent, when the compound is not carrying OXYGEN!? What is doing the job, when NCl_3 is used as a bleaching agent is Cl using the molecular oxygen of the air. All these can be very misleading for which one can imagine the state of Confusion the world has ever been.

Stage 1:

$$NCl_3 \rightleftharpoons Cl_2N\cdot nn + Cl\cdot en$$

$$\underset{}{Cl\cdot en} + \underset{(AIR)}{O=O} \rightleftharpoons \underset{(A)}{Cl-O-O\cdot en}$$

$$(A) + Cl_2N\cdot nn \longrightarrow \underset{(B)}{Cl-O-O-NCl_2} \qquad 3.60a$$

$$\underline{Overall\ equation}: NCl_3 + O_2 \longrightarrow Cl-O-O-NCl_2 \qquad 3.60b$$

Stage 2: [NOT FAVOURED]

$$2Cl-O-O-NCl_2 \rightleftharpoons 2Cl-O\cdot en + 2nn\cdot O-NCl_2$$

$$2Cl-O\cdot en \rightleftharpoons 2Cl\cdot en + 2nn\cdot O\cdot en + 2Heat$$

$$2Cl_2N-O\cdot nn \rightleftharpoons 2Cl\cdot nn + \underset{(C)}{2Cl-\overset{\cdot en}{\underset{\cdot\cdot}{N}}-O\cdot nn}$$

$$2Cl\cdot en + 2Cl\cdot nn \rightleftharpoons 2Cl_2$$

$$(C) \xrightarrow[Release\ of\ Energy]{Deactivation} 2Cl-N=O + Heat \qquad 3.61a$$

$$\underline{Overall\ equation}: 2Cl-O-O-NCl_2 \longrightarrow 2Cl-N=O + 2Cl_2 + \underline{\underline{O_2}} + Heat \qquad 3.61b$$

$$\underline{Overall\ overall\ equation}: 2NCl_3 + 2O_2 \longrightarrow 2Cl-N=O + 2Cl_2 + \underline{\underline{O_2}} + Heat \qquad 3.61c$$

One can observe the great importance of Molar ratios. The NCl_3 through Cl converted the molecular oxygen of the AIR to Oxidizing oxygen molecule. NCl_3 under such conditions cannot be said to be an oxidizing agent. It is a ***Molecular oxygen converter***. It is not even a catalyst. It is different from $KMnO_4$ or HOOH which are carrying the oxidizing oxygen. These are Oxidizing agent. Imagine if the Equilibrium state of existence of Cl-O-O-

NCl_2 used in Equation 3.61a had been differently represented, then the followings would have been obtained.

Stage2:[FAVOURED]

$$Cl-O-O-NCl_2 \rightleftharpoons Cl-O\cdot nn + en\cdot O-NCl_2$$

$$Cl_2N-O\cdot en \rightleftharpoons Cl\cdot en + Cl-N=O + Heat$$

$$Cl\cdot en + Cl-O\cdot nn \longrightarrow Cl-O-Cl \qquad 3.62a$$

$$\underline{Overall\ equation}: 2Cl-O-O-NCl_2 \longrightarrow 2Cl-N=O + 2Cl-O-Cl + Heat \qquad 3.62b$$

Stage3:

$$2Cl-O-Cl \rightleftharpoons 2Cl-O\cdot en + 2nn\cdot Cl$$

$$2Cl-O\cdot en \rightleftharpoons 2Cl\cdot en + \underline{\underline{O_2}} + Heat$$

$$2Cl\cdot en + 2Cl\cdot nn \xrightarrow[\underline{O_2}]{Suppressed\ by} 2Cl_2 \qquad 3.62c$$

$$\underline{Overall\ equation}: 2Cl-O-Cl \longrightarrow \underline{\underline{O_2}} + 2Cl_2 + Heat \qquad 3.62d$$

With this Equilibrium state of existence, by coincidence the stage is productive and in the third stage, the same final products in Equation 3.61b are obtained after decomposition of Cl-O-Cl in that Stage 3. *The coincidence is as a result of the fact that Cl and N are equi-electronegative (3.0).* However, the state in the first step of the stage above is the correct Equilibrium state of existence of the peroxide. Note how heat one of the forms of energy was released in the second steps of the stages. Oxygen being more electronegative than Cl, cannot be the one carrying an electro-non-free-radical. Hence heat was released. Thus, when nitrogen trichloride is used as a bleaching agent, three stages are involved before the bleaching action begins. Worthy of note is the Equilibrium state of existence of NCl_3.

Without knowing the Equilibrium state of existence of compounds, one cannot provide the mechanisms for chemical reactions. The same applies to all systems with or without life, not necessarily through the use of chemical equations, but something else. Since establishing the Equilibrium state of Existence of compounds is not an easy task, hence there is need to provide ***handbooks or encyclopaedia for States of Existences*** for not only all compounds but also for other systems in other disciplines. As simple as these new concepts may be, because of the lack of them, hence today, many things cannot be done, such as our inability to cure diseases in biological systems. Unknown to humanity, is the simple fact that all diseases have cure, for the solutions to all problems are right in front of our nose.

During chlorination of NH_3 a strong base in Stage 1a of Equation 3.52a, one moved to chloramine (NH_2Cl) thereby becoming less basic because of the presence of one Cl atom instead of former H atom. The H atom on NH_3 has been replaced with one Cl atom. When that H was connected to N, it was basic, but when it was connected to X (Cl), it started becoming acidic (where X is a halogen atom). The same applies with O. Why? It looks as if when a central atom such as N is singly bonded to one H atom, it is basic, just as when an atom such as Cl is connected to a central atom H, it is acidic. When the central atom (N) is made to carry two or more H atoms, it becomes more basic. When the same central atom (N) is made to carry one Cl atom, it is acidic, but when the same central atom is made to carry two or more Cl atoms, it becomes more acidic. If this was the case, then why is NR_3 (Tertiary amine) found to be more basic than R_2NH (Secondary amine) which in turn is more basic than RNH_2 (Primary amine) which in turn is more basic than ammonia (NH_3)? Unless either the reverse is the case, or the effect of R groups (Alkylane groups such as CH_3) on the N center is different from that of Cl. Indeed, they are both different since R groups are RADI-CAL- PUSHING GROUPS of greater capacity than H, while Cl, F, etc. are RADICAL – PULLING groups. Therefore, the reverse is not the case, clear indication that the universal observations and data like in many but few cases are in order as will be further shown in the New Frontiers. One can see the need for the use of the EYE OF THE NEEDLE. From chloramine, we moved next to dichloramine ($NHCl_2$) with basicity decreasing because of the presence of one additional Cl atom. Finally, with NCl_3, which contains no H, it is no longer basic, but fully acidic. Hence the Equilibrium state of this last case is different from the others. One can see what a CENTRAL ATOM means. Present-day Science sometimes makes allusion to CENTRL ATOM without knowing what it actually means.

$$NH_3 \rightleftharpoons H\cdot e + nn\cdot NH_2$$
$$ClNH_2 \rightleftharpoons H\cdot e + nn\cdot NHCl$$
$$Cl_2NH \rightleftharpoons H\cdot e + nn\cdot NCl_2$$
$$Cl_3N \rightleftharpoons Cl\cdot en + nn\cdot NCl_2 \qquad 3.63$$

For the fluorination of NH_3, the situation is completely different, since as has been shown, F_2 is very stable.

Fluorination

Stage 1:

$$NH_3 \xrightleftharpoons[\text{Ammonia}]{\text{Equilibrium State of}} H\cdot e + nn\cdot NH_2$$

$$H\cdot e + F_2 \xrightleftharpoons{\text{Abstraction}} HF + en\cdot F + Energy$$

$$F\cdot en + nn\cdot NH_2 \xrightarrow[\text{State}]{\text{Combination}} F - NH_2 \qquad 3.64a$$

$$\underline{Overall\ Equation}: NH_3 + F_2 \longrightarrow HF + F - NH_2 + Energy \qquad 3.64b$$

HF unlike HCl is too weak to exist in Equilibrium state of existence in the presence of the second product.

Stage 2:

$$FNH_2 \xrightleftharpoons[\text{compound}]{\text{Equilibrium State of}} H\cdot e + nn\cdot NHF$$

$$H\cdot e + F_2 \xrightleftharpoons{\text{Abstraction}} HF + en\cdot F + Energy$$

$$F\cdot en + nn\cdot NHF \xrightarrow[\text{State}]{\text{Combination}} H - NF_2 \qquad 3.65a$$

$$\underline{Overall\ Equation}: NH_2F + F_2 \longrightarrow HF + H - NF_2 + Energy \qquad 3.65b$$

Stage 3:

$$HNF_2 \xrightleftharpoons[\text{compound}]{\text{Equilibrium State of}} H\cdot e + nn\cdot NF_2$$

$$H\cdot e + F_2 \xrightleftharpoons{\text{Abstraction}} HF + en\cdot F + Energy$$

$$F\cdot en + nn\cdot NF_2 \longrightarrow NF_3 \qquad 3.66a$$

$$\underline{Overall\ Equation}: NF_2H + F_2 \longrightarrow HF + NF_3 + Energy \qquad 3.66b$$

$$\underline{Overall\ overall\ equation}: NH_3 + 3F_2 \longrightarrow NF_3 + 3HF + Energy \qquad 3.66c$$

It should also be noted that in the second step of the stages, heat is released. NF_3 a colourless gas, differs widely from all other nitrogen halides since as can be seen compared to NCl_3, it is very stable and known not to be explosive. However, the fluorination reaction looks more explosive than the chlorination reaction, since more heat is released. Hence, it is much more conveniently prepared by the electrolysis of fused ammonium hydrogen bifluoride, NH_4HF_2, a crude gas said to contain hydrogen fluoride, hydrogen, nitrogen, nitrous oxide, oxygen and ozone and small amounts of NHF_2. It does not react with water or caustic alkali, although if sparked with water vapor, it decomposes into HF and oxides of nitrogen[12].

$$3NF_3 \quad + \quad 3H_2O \xrightarrow[(Heat\ from\ Water\ vapor)]{Sparked} \quad 6HF \quad + \quad N_2O_3 \qquad 3.67$$

This reaction is being used to show how stable fluorinated compounds are in general including the acid.

Stage1:

$$H_2O \underset{(Water\ vapor)}{\overset{High\ Temperature}{\rightleftharpoons}} H\cdot e \ + \ nn\cdot OH$$

$$H\cdot e + NF_3 \rightleftharpoons HF \ + \ en\cdot NF_2 \ + \ Heat$$

$$F_2N\cdot en \ + nn\cdot OH \longrightarrow F_2N-OH \qquad 3.67a$$

$$\underline{Overall\ equation}: NF_3 \ + \ H_2O \longrightarrow HF \ + \ F_2N-OH \quad + \quad Heat \qquad 3.67b$$

OR

Stage 1a:

$$H_2O \rightleftharpoons H\cdot e \ + \ nn\cdot OH$$

$$H\cdot e \quad + \quad NF_3 \rightleftharpoons HNF_2 \ + \quad en\cdot F \ + \ Heat$$

$$F\cdot en \quad + \quad nn\cdot OH \longrightarrow HOF \qquad 3.68a$$

$$\underline{Overall\ equation}: H_2O \ + \ NF_3 \longrightarrow HNF_2 \ + \ HOF \ + \ Heat \qquad 3.68b$$

Stage 1aa

$$HOF \rightleftharpoons H\cdot e \ + \ nn\cdot OF$$

$$FO\cdot nn \rightleftharpoons F\cdot nn \ + \ en\cdot O\cdot nn \ + \ Heat$$

$$H\cdot e \ + \ nn\cdot F \rightleftharpoons HF$$

$$nn\cdot O\cdot en + HNF_2 \rightleftharpoons HO\cdot nn + en\cdot NF_2$$

$$\longrightarrow HONF_2 \qquad 3.68c$$

$$\underline{Overall\ equation}: \ H_2O \ + \ NF_3 \longrightarrow HF \ + \ HONF_2 \ + \ Heat \qquad 3.68d$$

Stage 2:

$$F_2N-OH \rightleftharpoons F_2N-O\cdot nn \ + \ H\cdot e$$

$$H\cdot e + NF_3 \rightleftharpoons HF + en\cdot NF_2 \ + \ Heat$$

$$F_2N-O\cdot nn + \ en\cdot NF_2 \longrightarrow F_2N-O-NF_2 \qquad 3.69a$$

$$\underline{Overall\ equation}: F_2N-OH + NF_3 \longrightarrow HF \ + \ F_2N-O-NF_2 \ + \ Heat \qquad 3.69b$$

Stage 2a:

$$HONF_2 \rightleftharpoons H\cdot e \ + \ nn\cdot ONF_2$$

$$H\cdot e \ + \ NF_3 \rightleftharpoons HNF_2 \ + \ en\cdot F \ + \ Heat$$

$$F\cdot en \ + \ nn\cdot ONF_2 \longrightarrow FONF_2 \qquad 3.70a$$

$$\underline{Overall\ equation:}\ HONF_2 \ + \ NF_3 \longrightarrow HNF_2 \ + \ FONF_2 \ + \ Heat \qquad 3.70b$$

AND/OR

Stage 2aa:

$$HNF_2 \rightleftharpoons H\cdot e \ + \ nn\cdot NF_2$$

$$H\cdot e \ + \ FONF_2 \rightleftharpoons HF \ + \ en\cdot ONF_2 \ + \ Heat$$

$$F_2N\cdot nn \ + \ en\cdot ONF_2 \longrightarrow F_2NONF_2 \qquad 3.70c$$

$$\underline{Overall\ equation:}\ HNF_2 \ + \ FONOF_2 \longrightarrow HF \ + \ F_2NONF_2 \ + \ Heat \qquad 3.70d$$

In Nature as already said, there is nothing like an alternative mechanism even when productive stages are obtained. In one of two alternatives, something must be wrong. Something is wrong with Stages 1 and 2 of Equations 3.67a and 3.69a. In the second step of those stages, F was abstracted and that is the component held in Equilibrium state of existence of NF_3 as a nucleo-non-free-radical. This is against the Laws of Nature with respect to abstraction of components. It is NF_2 which is held as an electro-non-free-radical in its state of existence that should be abstracted as a nucleo-non-free-radical. Hence in place of Stage 1, we now have Stages 1a and 1aa. In place of Stage 2, we have Stages 2a and 2aa. These are now four stages instead of two stages.

Stage 5:

$$H_2O \rightleftharpoons H\cdot e \ + \ nn\cdot OH$$

$$H\cdot e + F_2N-O-NF_2 \rightleftharpoons HF \ + \ F\overset{\bullet en}{N}-O-NF_2 \ + \ Heat$$

$$F\overset{\cdot en}{N}-O-NF_2 + nn\cdot OH \longrightarrow F-\overset{\overset{OH}{|}}{N}-O-NF_2$$

$$(A) \qquad 3.71a$$

$$\underline{Overall\ equation:} F_2N-O-NF_2 \ + \ H_2O \longrightarrow F(OH)N-O-NF_2 \ + \ HF \ + \ Heat$$

$$3.71b$$

Stage 6:

$$F-\overset{\overset{OH}{|}}{N}-O-NF_2 \rightleftharpoons H\cdot e + F-\overset{\overset{O\cdot nn}{|}}{N}-O-NF_2 \quad (B)$$

$$(B) \rightleftharpoons F\cdot nn + en\cdot\overset{\overset{O\cdot nn}{|}}{\ddot{N}}-O-NF_2 \quad (C)$$

$$H\cdot e + F\cdot nn \rightleftharpoons HF$$

$$(C) \xrightarrow[\text{Release of Heat}]{\text{Deactivation}} O=N-O-NF_2 \qquad 3.72a$$

$$\underline{\text{Overall equation}}: F(OH)N-O-NF_2 \longrightarrow HF + O=N-O-NF_2 \qquad 3.72b$$

Even in the presence of water vapor, the HF could not exist in Equilibrium state of existence. It has remained in Stable state of existence so far. So also is NF_3. One can imagine all the shocking revelations based on understanding how NATURE operates, made possible by universal data.

Stage 7:

$$H_2O \rightleftharpoons H\cdot e + nn\cdot OH$$

$$H\cdot e + F_2N-O-N=O \rightleftharpoons HF + F\overset{\bullet en}{N}-O-N=O$$

$$F\overset{\cdot en}{N}-O-NF_2 + nn\cdot OH \longrightarrow F-\overset{\overset{OH}{|}}{N}-O-N=O \quad (D) \qquad 3.73a$$

$$\underline{\text{Overall equation}}: F_2N-O-N=O + H_2O \longrightarrow F(OH)N-O-N=O + HF \qquad 3.73b$$

Stage 8:

$$F-\overset{\overset{OH}{|}}{N}-O-N=O \rightleftharpoons H\cdot e + F-\overset{\overset{O\cdot nn}{|}}{N}-O-N=O \quad (E)$$

$$(E) \rightleftharpoons F\cdot nn + en\cdot \underset{..}{\overset{O\cdot nn}{\overset{|}{N}}}-O-N=O$$

$$(F)$$

$$H\cdot e + F\cdot nn \rightleftharpoons HF$$

$$(F) \xrightarrow[\text{Release of Heat}]{\text{Deactivation}} O=N-O-N=O + Heat \qquad 3.74a$$

$$\underline{\text{Overall equation}}: F(OH)N-O-N=O \longrightarrow HF + O=N-O-N=O + Heat \qquad 3.74b$$

$$\underline{\text{Overall overall equation}}: 2NF_3 + 3H_2O \longrightarrow 6HF + O=N-O-N=O \qquad 3.75$$

In eight stages, the mechanisms of the reaction of Equation 3.68 were provided. Note the order of existence of the Equilibrium states of existence of the compounds. For example, the Equilibrium state of existence of F(OH)N-O-NF_2 or F(OH)N-O-N=O is stronger than that of H_2O which in turn is greater than HF, with all of them less than that of HCl. Though compounds may belong to the same family such as the Halogens, all the members which still have something common with them, have very different "personalities", i.e., properties which cannot be changed. What manifest them are the types of neighbours around and operating conditions.

What one has always been doing, is collection of universal data for countless cases and provide their mechanisms without breaking any of the LAWS of NATURE, with the realization of the fact that LAWS OF NATURE, unlike MATERIAL laws used universally, have NO EXCEPTION. One has started provision of these LAWS in the New Frontiers. For example, the first law of NATURE is TO BE FREE AND NOT TO BE FREE (See Figure 1.3). This gave birth to all other first Laws used in all disciplines. In addition, this first law in CHEMISTRY gave birth to all the other laws in CHEMISTRY. So also, it is in other disciplines. In Religion, when all the Laws of Moses (Ten Commandments) and all the AVARTARS are put together, they are embracement of this first law in CHEMISTRY. As you sow your seed, so shall thou reap it or Do unto others as you want them to do unto you and so on. In Engineering this first law is called THE FUNDAMENTAL LAWS OF CONSERVATION IN ENGINEERING-with respect to Materials, Energy and Momentum. All these can be found inside CHEMISTRY. All four founding fathers of Engineering apply them differently. In Physics, for every Action, there is always an equal and opposite reaction. In Mathematics, NOTHING IS THERE FOR NOTHING- the Equality and Inequality signs. Note that one did not say "NOT ANYTHING IS THERE FOR NOTHING" or the reverse as some put it. All these can be found in Chemistry. In all disciplines including RELIGION, this first law has never been fully understood and properly applied. That has been the major problem universally and the missing Link-FREE AND NON-FREE RADICALS.

Nitrogen tribromide itself has never been obtained, but its hexamine, **$NBr_3.6NH_3$,** is formed by the action of bromine on ammonia at 100^0C and a pressure of 1-2 mm, the reaction products being rapidly **cooled** to -95^0C[12]. When iodine is added to aqueous ammonia, a very dark coloured precipitate is formed which is often called "nitrogen iodide". This product, however, always contains hydrogen, and after drying approximates to the composition **NI_3NH_3.** Most of the ammonia can be removed by washing with a suitable solvent[12]. Based on what has been seen so far, under well controlled conditions, similar type of products obtained during fluorination and chlorination can also be obtained during bromination and iodination, via different mechanisms. Of particular interest here, are the structures of the types of compounds highlighted above, since in general universally, the structures are unknown. The N center in $NBr_3.6NH_3$ does not have vacant orbitals for the paired unbonded radicals on N to use to form Dative bonds. Therefore, their structures are as shown below.

$$NBr_3 \rightleftharpoons Br_2N \cdot nn \quad + \quad Br \cdot en \qquad 3.76a$$

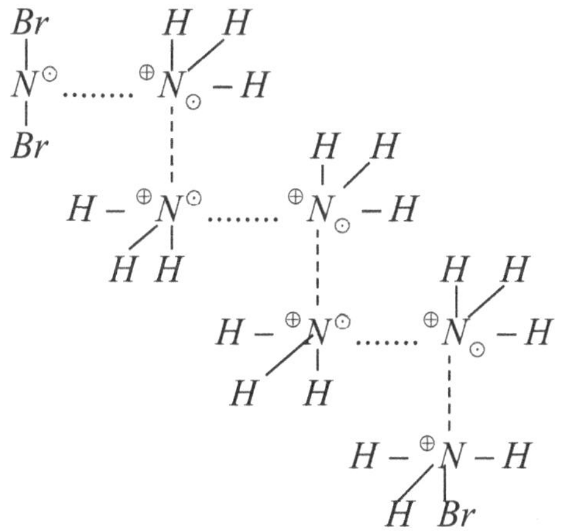

Br
|
N⊙ .... ⊕N⊙(H)(H) – H
|
Br
H – ⊕N⊙(H)(H) .... ⊕N⊙(H)(H) – H
H – ⊕N⊙(H)(H) .... ⊕N⊙(H)(H) – H
H – ⊕N(H)(Br) – H

$\underline{NBr_3.6NH_3}$ 3.76b

$$NI_3 \rightleftharpoons I_2N\cdot nn + I\cdot en \qquad 3.77a$$

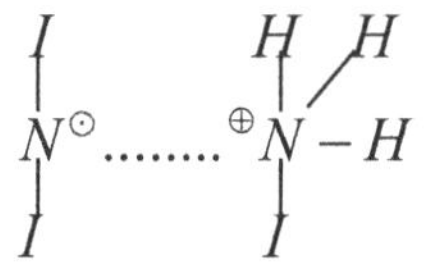

$$\begin{matrix} I & & H\ \ H \\ | & & |\ / \\ N^{\ominus} & \cdots\cdots\cdots & {}^{\oplus}N - H \\ | & & | \\ I & & I \end{matrix}$$

$$\underline{NI_3NH_3} \qquad 3.77b$$

One can observe the great significance of all these new countless numbers of concepts. These as can be observed can be made to conduct electricity.

3.3.7 MECHANISM OF FLOW OF CURRENT IN FLUIDS

How current flows in fluids has never been known, though many will claim that it is known. Unknown is the fact that of all Engineering disciplines, only Electrical Engineering is hundred percent imaginary in character. Not until we are able to build buildings hanging in space, just as the Planets, Stars, Moon, Sun, and other bodies are held together in space in our Solar system by electrostatic forces which are real and imaginary, Civil Engineering is hundred percent real. What was just said above is impossible at this point in time for Civil Engineers to do, because as will be shown in the Appendix, those bodies referred to including our planet Earth are weightless bodies with respect to the SUN. When a plane is flying in our air space, it becomes a weightless body for specific reasons, but cannot yet be made to hang in space. Yet some birds can do it. Whether our Technology has developed to this level or not, it is just still an ART and not yet a SCIENCE. Chemical and Mechanical are both hybrids of the real and imaginary. These are four founding fathers of Engineering which have given birth to so many children- their sub-disciplines. In showing how current flows in fluids, water is going to be used, since the same mechanism will apply to all of them, noting that water is Polar/Ionic. The case of IF_7 which has been shown and like the cases above are Polar/Non-ionic.

Stage1:

$$H_2O \rightleftharpoons H\cdot e + nn\cdot OH$$

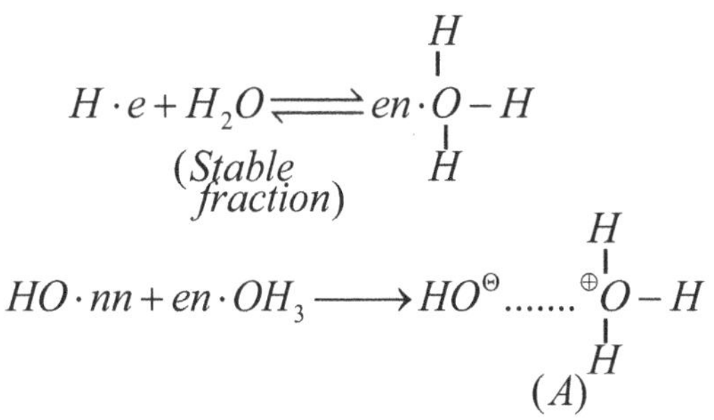

$$H\cdot e + H_2O \rightleftharpoons en\cdot \overset{\displaystyle H}{\underset{\displaystyle H}{O}} - H$$

(Stable fraction)

$$HO\cdot nn + en\cdot OH_3 \longrightarrow HO^{\ominus}\cdots\cdots {}^{\oplus}\overset{\displaystyle H}{\underset{\displaystyle H}{O}} - H \qquad 3.78a$$

$$(A)$$

Stage2:

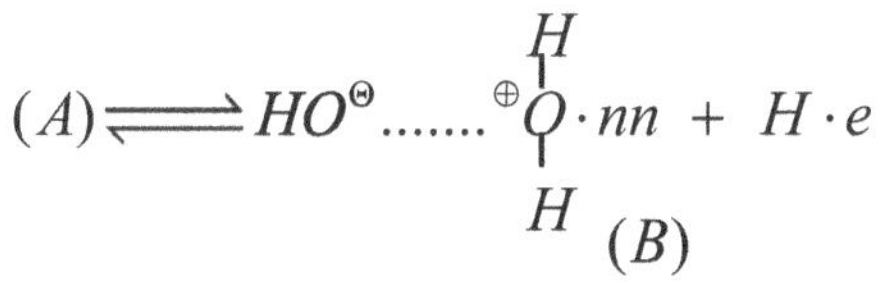

$$(A) \rightleftharpoons HO^{\ominus} \ldots\ldots {}^{\oplus}\underset{H}{\overset{H}{O}} \cdot nn \;+\; H \cdot e \quad (B)$$

$$H \cdot e + H_2O \rightleftharpoons en \cdot OH_3$$

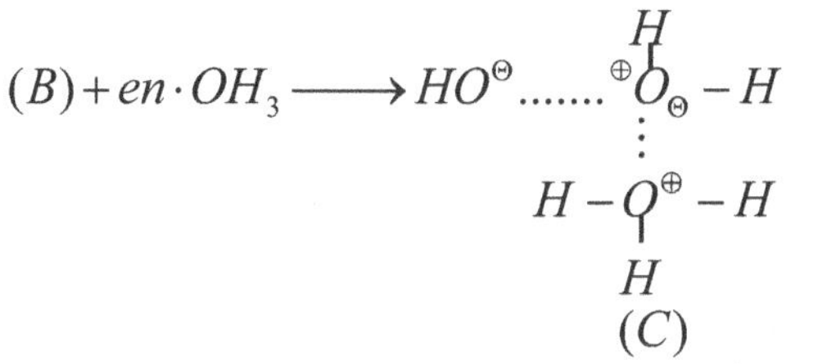

$$(B) + en \cdot OH_3 \longrightarrow HO^{\ominus} \ldots\ldots {}^{\oplus}\overset{H}{O}_{\ominus} - H \;\vdots\; H - \underset{H}{O^{\oplus}} - H \quad (C) \qquad 3.78b$$

Stage3:

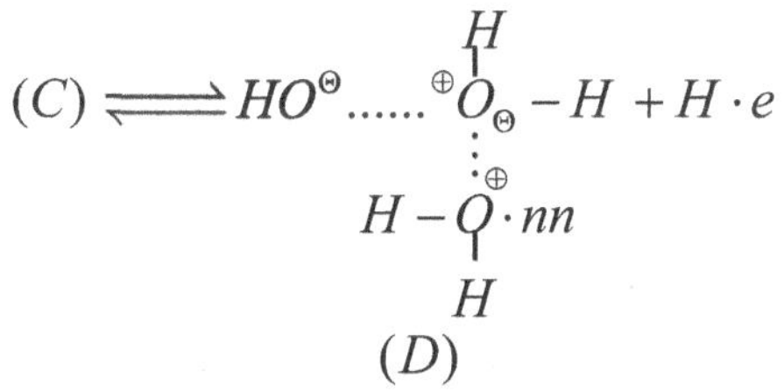

$$(C) \rightleftharpoons HO^{\ominus} \ldots\ldots {}^{\oplus}\overset{H}{O}_{\ominus} - H + H \cdot e \;\vdots^{\oplus}\; H - \underset{H}{O} \cdot nn \quad (D)$$

$$H \cdot e + H_2O \rightleftharpoons en \cdot OH_3$$

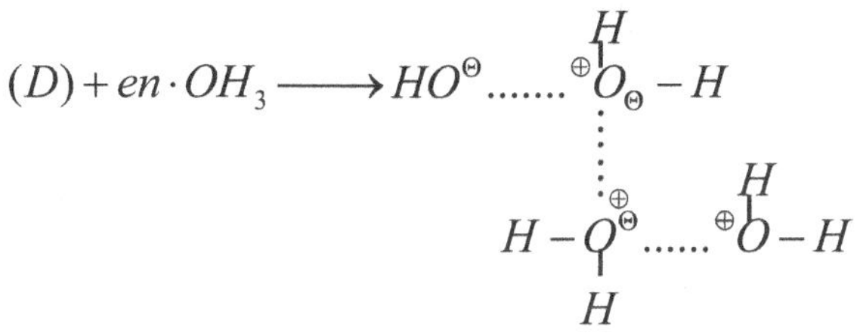

$$(D) + en \cdot OH_3 \longrightarrow HO^{\ominus} \ldots\ldots {}^{\oplus}\overset{H}{O}_{\ominus} - H \;\vdots^{\oplus}\; H - \underset{H}{O^{\ominus}} \ldots\ldots {}^{\oplus}\overset{H}{O} - H \qquad 3.78c$$

Addition of water molecules continues until consumed with a giant molecule through which electric current flows is obtained. There are two major types of electrostatic bonds – **the positive and negative types**. The types used above are the negative types just as in IF_7. The type shown in Equations 3.35a and 3.35b are of the positive type. Not all Polar/Non-ionic compounds can conduct electricity. All Polar/Ionic compounds can conduct electricity. All Non-Polar/Non-ionic compounds cannot be made to conduct electricity. One can observe the very great importance of all the new countless numbers of concepts that have been introduced.

3.4 Additional applications of the New Concepts.

3.4.1 Definition of STEP monomers

(a) STEP MONOMER OF THE FIRST KIND

Based on what has been seen so far, for the first time, one started showing what an ADDITION monomer is, unlike what has been known in Present-day Science. We saw the three kinds of ADDITION monomers-

i) Those with visible and invisible π-bonds,
ii) Those with functional centers found only in some ringed compounds,
iii) Those with vacant orbital/paired unbounded radicals such in CO.

All these were called Activation centers. In addition, there are different kinds and types of ADDITION POLYMERIZATION SYSTEMS. The four kinds include-

i) Bulk polymerization systems,
ii) Solution polymerization systems,
iii) Suspension polymerization systems,
iv) Emulsion polymerizations systems.

For the first time as will be shown in the New Frontiers is the existence of mini-reactors inside macro-reactors as largely exist in NATURE-Suspension and Emulsion systems. Though Present-day Science think that Addition monomers add by Combination forwardly alone, for the first time as one has started showing herein in Chapter 2 (See Equations 2.34 to 2.36) and will be fully shown down-stream in the New Frontiers, some ringed Addition monomers with functional centers largely undergo polymerization via Equilibrium mechanism only electro-free-radically with the chain growing backwardly instead of forwardly.

Though the definition of a STEP monomer seems to be known by Present-day Science, its definition is very far from completeness. Today it is believed that only compounds with two or more functional groups (not functional centers in rings as additionally believed in present-day Science) can be said to be STEP monomers, without the realization that there are different kinds and types of functional groups and different kinds and types of STEP POLYMERIZATION SYSTEMS. As will shortly be shown, not all compounds with functional groups can be made to undergo STEP polymerization and in addition a compound does not necessarily have to have functional groups before it can be said to be a STEP monomer. There are cases where as will shortly be shown, the compound without functional groups is polymerized using STEP polymerization kinetics. Whatever the kind of STEP monomer, unlike ADDITION POLYMERIZATION SYS-

TEMS, small bye-molecular products must be released for every addition of STEP monomers in STEP POLYMERIZATION SYSTEMS. For one kind of STEP monomer, that is, those with functional groups, ***the functional groups of one of the two monomers must first and foremost be Polar/Ionic.*** If all the monomers are Polar/Non-ionic or Non-polar/Non-ionic, some with aromaticity in particular can still be used as a STEP monomer, because they satisfy the mechanisms involved during STEP polymerization. The mechanisms involved in all these systems are unknown in Present-day Science. Yet, they will claim that they (The so-called EXPERTS) know, that which is very painful. None of the definitions given for all disciplines are in place, because if they were in place, the developments of the disciplines would have been a very different scenario. The most important discipline in humanity as has been said is CHEMISTRY and most important area of discipline inside CHEMISTRY is POLYMER CHEMISTRY, because it is a complete embracement of LIFE and LIVING.

Shown below are two compounds with functional groups.

$$\underline{(H_3C)_2N} - (CH_2)_n - \underline{N(CH_3)_2} \quad ; \quad \underline{H_2N} - (CH_2)_n - \underline{NH_2}$$
$$(A) - Polar\ /\ Non-ionic \qquad (B) - Polar\ /\ Ionic \qquad 3.79$$

The functional groups have been underlined above. Only (B) can be used as a STEP monomer. One of the types of this kind of STEP POLYMERIZATION SYSTEMS is that in which only ALTERNATING placement of monomer units are obtained along the chain. Therefore, two different types of STEP monomers must be used as shown below for the polymerization of an organic di-acid with a di-amine both in the same solvent such as water. The di-acid, di-amine and water belong to the same family.

Stage1:

$$H_2N(CH_2)_6NH_2 \rightleftharpoons H\cdot e + nn\cdot NH(CH_2)_6NH_2 \ (A)$$

$$H\cdot e + HOOC(CH_2)_4COOH \rightleftharpoons H_2O + e\cdot \overset{O}{\overset{\|}{C}}(CH_2)_4COOH \ (B)$$

$$(A) + (B) \longrightarrow H_2N(CH_2)_6\overset{H}{\overset{|}{N}} - \overset{O}{\overset{\|}{C}}(CH_2)_4COOH \quad (C) \qquad 3.80a$$

$$\underline{Overall\ equation}: HOOC(CH_2)_4COOH + H_2N(CH_2)_6NH_2 \longrightarrow H_2O + (C) \qquad 3.80b$$

Stage2:

$$H_2N(CH_2)_6NH_2 \rightleftharpoons H\cdot e \ + \ nn\cdot NH(CH_2)_6NH_2 \ (A)$$

$$H\cdot e \ + \ (C) \rightleftharpoons H_2O + e\cdot \overset{O}{\overset{\|}{C}}(CH_2)_4\overset{O}{\overset{\|}{C}}-\overset{H}{\overset{|}{N}}(CH_2)_6NH_2 \ (D)$$

$$(A) \ + \ (D) \longrightarrow H_2N(CH_2)_6\overset{H}{\overset{|}{N}}-\overset{O}{\overset{\|}{C}}(CH_2)_4\overset{O}{\overset{\|}{C}}-\overset{H}{\overset{|}{N}}(CH_2)_6NH_2 \quad (E) \qquad 3.81a$$

$$\underline{Overall\ equation}: \ HOOC(CH_2)_4COOH \ + \ 2H_2N(CH_2)_6NH_2 \longrightarrow 2H_2O \ + \ (E) \quad 3.81b$$

This takes place when the di-amine is in excess, after many moles of (C) have been formed.

Stage 3:

$$H_2N(CH_2)_6\overset{H}{\overset{|}{N}}-\overset{O}{\overset{\|}{C}}(CH_2)_4\overset{O}{\overset{\|}{C}}-\overset{H}{\overset{|}{N}}(CH_2)_6NH_2 \ (E) \rightleftharpoons H\cdot e + nn\cdot \overset{H}{\overset{|}{N}}(CH_2)_6\overset{H}{\overset{|}{N}}-\overset{O}{\overset{\|}{C}}(CH_2)_4\overset{O}{\overset{\|}{C}}-\overset{H}{\overset{|}{N}}(CH_2)_6NH_2 \ (F)$$

$$H\cdot e + HOOC(CH_2)_4COOH \rightleftharpoons H_2O \ + \ e\cdot \overset{O}{\overset{\|}{C}}(CH_2)_4COOH \ (B)$$

$$(F) + (B) \longrightarrow HOOC(CH_2)_4\overset{O}{\overset{\|}{C}}-\overset{H}{\overset{|}{N}}(CH_2)_6\overset{H}{\overset{|}{N}}-\overset{O}{\overset{\|}{C}}(CH_2)_4\overset{O}{\overset{\|}{C}}-\overset{H}{\overset{|}{N}}(CH_2)_6NH_2 \quad (G) \qquad 3.82a$$

$$\underline{Overall\ equation}: \ 2HOOC(CH_2)_4COOH \ + \ 2H_2N(CH_2)_6NH_2 \longrightarrow 3H_2O \ + \ (G) \quad 3.82b$$

This takes place only when one mole of the di-acid is added after (C) is formed. For the first time, one can observe the one and only mechanism for STEP POLYMERIZATION SYSTEMS. More will be seen in the New Frontiers. They are all via Equilibrium mechanism. Worthy of note is that while the two monomers are Polar/Ionic, the reactions can only take place radically, because of the presence of covalent charges which cannot be isolatedly placed in some of the steps in the Stages. On the other hand, all states of existences can indeed only be expressed radically. One can see how the small bye-molecular products are released in every stage. In this case, it is water. One can also imagine the number of stages that will be involved to produce high molecular weight alternating copolymers. Because the mechanism is via Equilibrium mechanism, hence far longer polymerization times of the order of hours have been the observations for these systems over the years compared to short polymerization times of the order of second and minutes via Combination mechanism. Because for this case, the system is entirely Polar/Ionic, hence with respect to the problems of Solubility,

Insolubility and Immiscibility, the water must be removed continuously from the system by operating at fairly high temperatures, otherwise the growth of the polymer chain will be limited. Why this has been the experience industrially over the years could not be explained by Present-day Science. On the whole, it can be seen that the entire exercise in the past has been an ART and not a SCIENCE just as in almost all cases. Throughout the stages, notice that it is the di-amine (and not even the mono-amine) that was always kept in Equilibrium state of existence all the time, otherwise, ammonia would have appeared as the small molecular bye-products.

Now considering the case of two monomers in which only one is Polar/Ionic, such as polymerization of a di-amine in water and a di-acyl chloride in an organic solvent.

$$\underset{\substack{Hexamethylene\ diamine \\ in\ Water \\ (Polar/Ionic)}}{n(H_2N(CH_2)_6NH_2)} \quad + \quad \underset{\substack{Sebacoyl\ chloride \\ in\ Cl_2C=CCl_2 \\ (Polar/Non-ionic)}}{n(Cl-\overset{O}{\overset{\|}{C}}-(CH_2)_8-\overset{O}{\overset{\|}{C}}-Cl)} \longrightarrow$$

$$\underset{Nylon\,6-10}{H_2N(CH_2)_6\overset{H}{\overset{|}{N}}-[\overset{O}{\overset{\|}{C}}(CH_2)_8\overset{O}{\overset{\|}{C}}-\overset{H}{\overset{|}{N}}(CH_2)_6\overset{H}{\overset{|}{N}}]_{n-1}-\overset{O}{\overset{\|}{C}}(CH_2)_8\overset{O}{\overset{\|}{C}}Cl} \quad + \quad (n-1)HCl \qquad 3.83$$

The same mechanisms as above also apply here, with the small bye-molecular products now being HCl. However, notice that at low temperatures unlike the case above, there is no chemical reaction whatsoever between the two solvents. They are both in Stable state of existence. Both being far apart in polarity cannot make them miscibilize. Hence, two phases are created in which organic solvent stays at the bottom being denser than water. Eventually, polymerization takes place at the interface. Hence this in present –day Science has been called INTERFACIAL CONDENSATION. Indeed, it is not a Condensation process. The real name is INTERFACIAL STEP POLYMERIZATION. The two solvents being immiscible, polymerization is forced to take place at the interface with only the electro-free-radical carrying species doing the movements. Water being stable cannot sit on the di-amine to form hydroxides. It can only stand and remain stable unlike the di-amine. The di-amine is insoluble in the water, but miscibilizes in it since they belong to the same family of Compounds. The di-acyl chloride is also insoluble in its solvent, but miscibilizes in it because they belong to the same family of Compounds. At the operating conditions (Room temperature), the di-amine cannot activate the organic solvent which has a π-bond type of activation center, because of the operating conditions. Both of them do not however belong to the same family of compounds. Nothing is taking place between the di-acyl chloride and the water since both are indeed in Stable state of existence and since both belong to two different families of Compounds. With all these in place, new doors are opened and clear understanding become so visible and unquestionable and the exercise is no longer an ART but a SCIENCE. Eradication of the ART (with no understanding) is essentially the ESSENCE of the NEW FOUNDATIONS being laid for humanity and continuity. Like the case above where water had to be continuously evaporated (not condensed), so that polymers can continuously be produced, the same applies here with the presence of so much HCl in the system. Unlike water which is stable, the acid is not and as such, it can readily sit on the di-amine and make it inactive. Secondly, the polymer produced must be continuously removed from the interface, if the contact between the monomers in the system must be maintained. So far, we are beginning to see the true definition of a STEP monomer.

(b) STEP MONOMERS OF THE SECOND KIND

i) Use of Benzene as STEP monomer

In another kind of STEP monomer, i.e., those with no functional types of groups, small bye-molecular products are still released in every stage. Unknown to Present-day Science, is that benzene a very unique compound, can both be used as a STEP and ADDITION monomer. It is preferably easy to use as an Addition monomer when it is carrying one radical-pushing group such as CH_3.

Benzene accommodated within interlayer space of graphite-alkali metal (Potassium and rubidium) intercalation compounds was found to be polymerized not only to biphenyl, but also to terphenyl and quarterphenyl, while only biphenyl was formed by the action of potassium or rubidium metal alone under the same conditions[13]. The formation of higher oligomers of benzene in the interlayer space in graphite intercalation compounds was ascribed to the amphoteric nature of the compounds which were said to be capable of both accepting and donating electrons[13]. In the presence of ionic metals such as Na, K, Rb, benzene ring can easily be kept in a de-energized state of existence. Hence the followings take place when benzene is with K alone.

Stage 1:

2K ⇌ 2 K •e

2 K •e + 2 (benzene) ⇌ 2 (•e benzene) + 2 KH

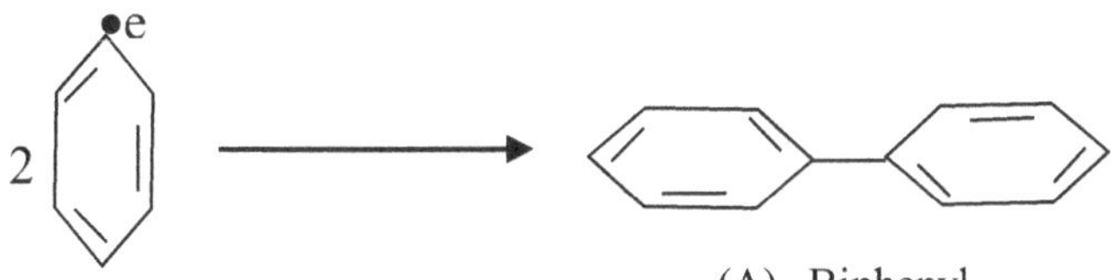

2 [radical benzene •e] ⟶ (A) Biphenyl 3.84a

Overall equation: 2 K + 2 Benzene ⟶ 2KH + Biphenyl 3.84b

It is only just a one stage Equilibrium mechanism system in which KH is a second product not identified. Usually all components in many experiments are never completely identified. If complete identifications are made as some do, so many doors will be opened. On the other hand, analytical equipment wherein oxidizing oxygen can be clearly distinguished from molecular oxygen or wherein radicals can clearly be identified and so many other cases have not yet been built. Most of these equipments already seem to exist. One of the major problems apart from knowing exactly what these equipments are doing, is that we yet do not know how to interpret the data we see, something which one did in Size Exclusion Chromatography many years ago, where the universal equation for separation according to size in a porous media was developed[14]. In the interlayer space of graphite intercalation compounds, a fraction of the benzene was forced to be in the energized state. After the first stage above, the followings were next obtained.

Stage 2:

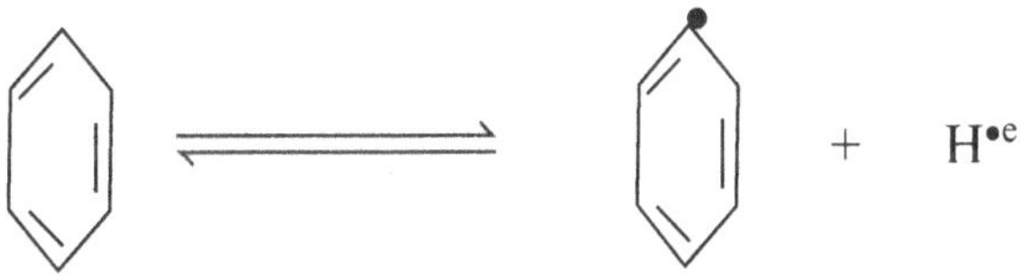

Benzene ⇌ (B) + $H^{\bullet e}$

$H^{\bullet e}$ + Biphenyl ⇌ (C) •e + H_2

(B) 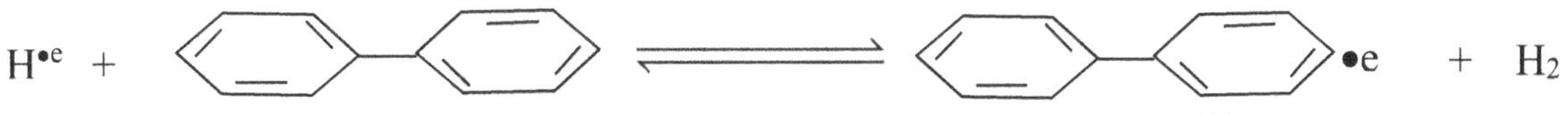 + (C) 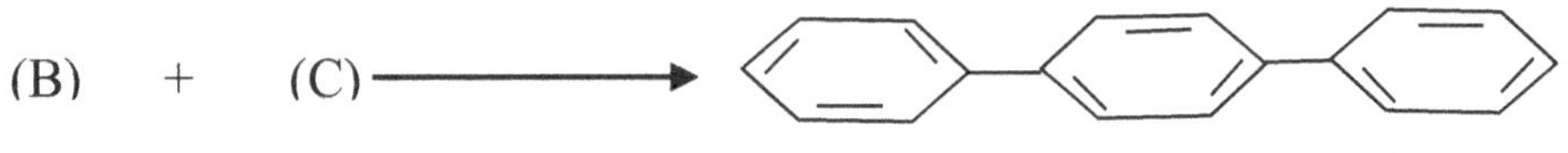 ⟶ (D) 3.85a

Overall equation: Benzene + Biphenyl ⟶ Terphenyl + H_2 3.85b

Stage 3:

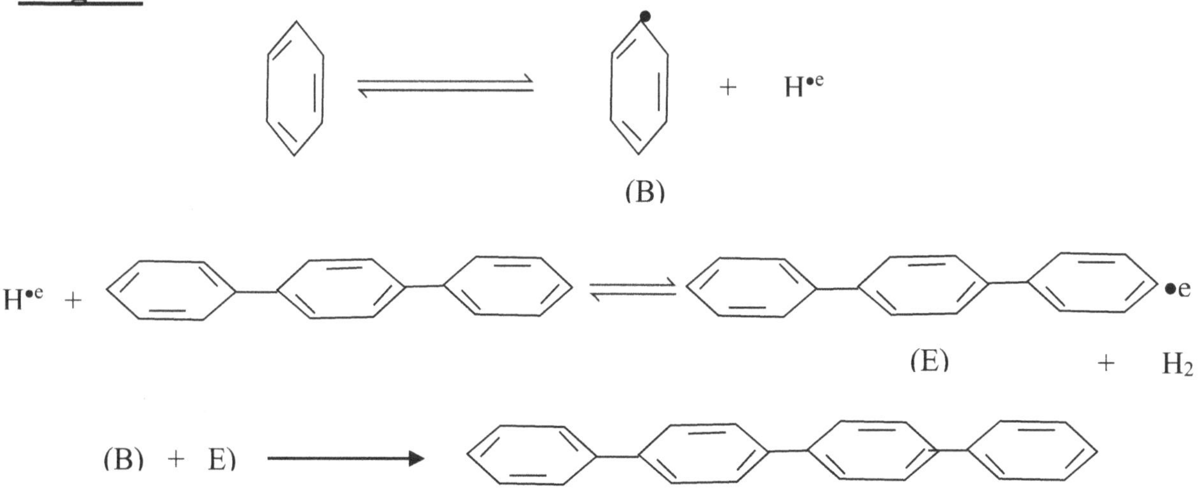

(B) + $H^{\bullet e}$

$H^{\bullet e}$ + ⇌ $\bullet e$

(E) + H_2

(B) + E) ⟶

3.86a

Overall equation: Benzene + Terphenyl ⟶ Quaterphenyl + H_2 3.86b

Overall overall equation: 4 Benzene + K ⟶ KH + $2H_2$ + Quaterphenyl 3.87

With presence of more benzene in the polymerization system in a good reactor and well controlled operating conditions, a long chain of p-polyphenyl will eventually be obtained along the chain. Like the case above, this is a STEP polymerization system. The carrier of the chain is H and the terminating agent is H. The statement given for the difference in the reactions above viz- "amphoteric nature of the graphite intercalation compounds, which are capable of both donating and accepting electrons" is a clear reflection of the development of Present-day Science, in which none of the statements makes sense. Electrons are inside the Nucleus of an atom. Only paired unbonded radicals can be donated to fill a vacant orbital. In fact, they are not donated but shared to form a dative bond making sure that the Boundary laws are not broken. What the graphite is doing can unquestionably be seen above. All these compounds just like humans have different "personalities" and responsibilities. ***These we have always known without knowing that we know them.*** The only difference is that we do not ask questions. Otherwise, what is the meaning of *an Oxidizing Agent* when there is plenty of oxygen around us, *Hydrogen catalysts* when we generate large quantities of H_2 every day in our refineries and many industries, *a nitrating agent* when there is plenty of N_2 around us and so on? We know these things, but yet do not know them!

In the electrochemical polymerization of benzene in the presence of aluminium chloride ($AlCl_3$)[15], $AlCl_3/CuCl_2/H_20$[16], ferric chloride ($FeCl_3$)[17-19],

Molybdenum pentachloride ($MoCl_5$)[20,21], p-polyphenyl was the polymeric product obtained. However unlike the case above, the polymer was said to contain a small amount of Chlorine, with the exception of the use of $AlCl_3$. In addition with respect to the others, small amounts of low molecular weight organic products containing 4,4′-dichlorobiphenyl, dichlorobenzene, and chlorobenzene, were also identified as products. Facile dehydrogenation of I, 4-cyclohexadiene to benzene using these "catalysts" was said to lend support to the proposal that the reaction proceeds by OXIDATIVE CATIONIC POLYMERIZATION. Like the case above also, the benzene ring could not be opened. Firstly, it is not oxidative, because there is no oxygen involved in the reactions. Secondly, no charges are involved in the system. Like Great Americans will say **"What the hell is going on here?** This in essence is a complete state of Confusion. All the reactions take place radically. The overall reactions provided for their polymerization reactions using some of these catalysts were as follows[17,20].

$$n\,C_6H_6 \quad + \quad nMoCl_5 \longrightarrow \left[\!-\!C_6H_4\!-\!\right]_n \quad + \; nMoCl_3 \; + \; 2n\,HCl \qquad 3.88a$$

$$n\,C_6H_6 \quad + \quad 2nFeCl_3 \longrightarrow \left[\!-\!C_6H_4\!-\!\right]_n \quad + \; 2nFeCl_2 \; + \; 2n\,HCl \qquad 3.88b$$

Without any doubt, this again is a STEP POLYMERIZATION SYSTEM, since we have a situation where there are small molecular by-products. For the cases above, while Al can form hydrides with H, Mo and Fe cannot form hydrides with H. While Mo is a Transition metal, Fe is a Transition transition metal. On the other hand, nothing was said about the influence of the water added with the non-ionic metallic chlorides as was highlighted in the water co-catalysis in the polymerization of benzene by ferric chloride[19]. As has already been shown herein, when water is present with metal alkyls such as ZnR_2, Electrostatically or Covalently charged-paired initiators are formed. However, with $FeCl_3$, apart from hydration resulting from the vacant orbitals in $FeCl_3$, there is no reaction at the operating conditions of below 100^0C. The equations above do not show the presence of Cl in the polymeric chain, if a small fraction was detected. The pyrolysis of the polymer in vacuo at 750^0-800^0C were said to give a sublimate in addition to residual material resembling carbon black. The sublimed product were said to contain biphenyl, low molecular weight p-polyphenyls including terphenyl, quarterphenyl, and quinquephenyl, in addition to uncharacterized higher molecular weight substances[17]. These are the products which have already been identified above (Equations 3.84 to 3.87a) via the use of K/Graphite. If Chlorine was identified with the polymers, this can only be found at the terminals of the chain or along the ortho-positions of some of the benzene rings along the chain. If carbon black was present during pyrolysis, then the benzene ring must have been opened as will shortly be shown

With the use of $FeCl_3$ wherein its molar ratio with benzene is 2 to 1 as shown in Equation 3.88b above[17], the followings are obtained.

Stage 1:

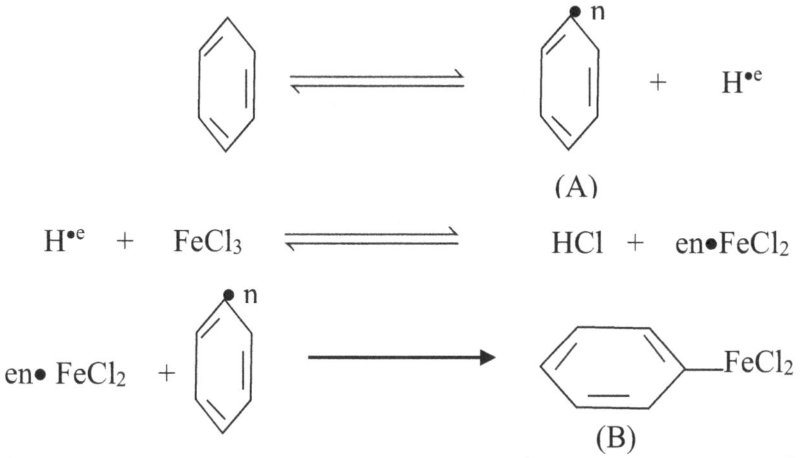

Benzene ⇌ (A) [n•] + $H^{\bullet e}$

(A)

$H^{\bullet e}$ + $FeCl_3$ ⇌ HCl + en•$FeCl_2$

en• $FeCl_2$ + (A) ⟶ (B) [benzene ring—$FeCl_2$]

(B) 3.89a

Overall equation: Benzene + $FeCl_3$ ⟶ HCl + (B) 3.89b

Stage 2: HCl ⇌ H •e + nn• Cl

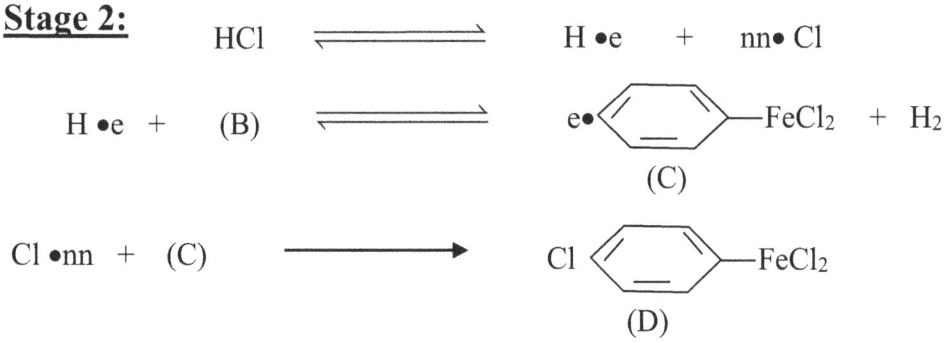

H •e + (B) ⇌ e•[benzene ring]—$FeCl_2$ + H_2

(C)

Cl •nn + (C) ⟶ Cl[benzene ring]—$FeCl_2$

(D) 3.90a

Overall equation: HCl + (B) ⟶ H_2 + (D) 3.90b

Overall overall equation: 100 Benzene + 100 $FeCl_3$ ⟶ 100H_2 + 100(D) 3.90c

With no benzene any more in the system, the polymerization route proceeds differently as shown below. The two stages above are indeed 100 stages in parallel when 100 moles of benzene are used. The (D) above now looks like a Step monomer with two functional groups.

Stage 3:

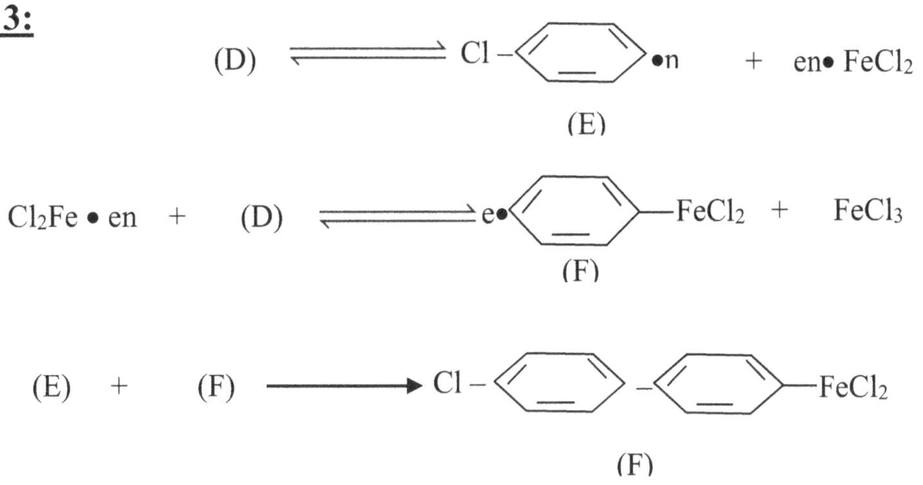

$$(D) \rightleftharpoons Cl-C_6H_4\bullet n \quad + \quad en\bullet FeCl_2$$

(E)

$$Cl_2Fe\bullet en \quad + \quad (D) \rightleftharpoons e\bullet C_6H_4-FeCl_2 \quad + \quad FeCl_3$$

(F)

$$(E) \quad + \quad (F) \longrightarrow Cl-C_6H_4-C_6H_4-FeCl_2$$

(F) 3.91

Stage 4:

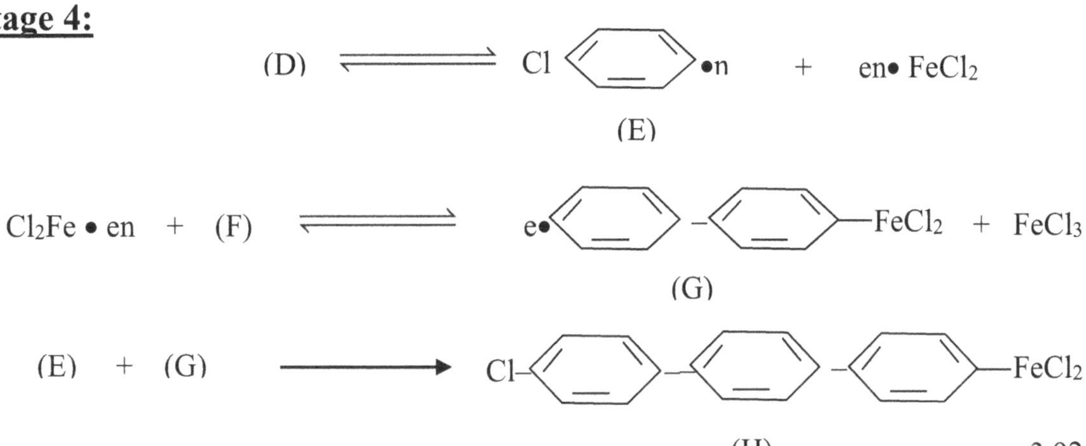

$$(D) \rightleftharpoons Cl\ C_6H_4\bullet n \quad + \quad en\bullet FeCl_2$$

(E)

$$Cl_2Fe\bullet en \quad + \quad (F) \rightleftharpoons e\bullet C_6H_4-C_6H_4-FeCl_2 \quad + \quad FeCl_3$$

(G)

$$(E) \quad + \quad (G) \longrightarrow Cl-C_6H_4-C_6H_4-C_6H_4-FeCl_2$$

(H) 3.92

Addition continues in a stepwise fashion until all (D)s are consumed. This will involve 102 stages to give the following polymeric chain and overall equation.

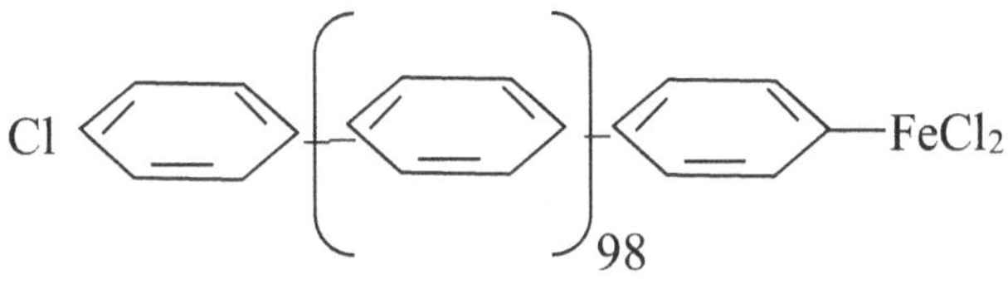

$$Cl\ C_6H_4-\left[C_6H_4\right]_{98}-C_6H_4-FeCl_2$$

(I)

Overall overall equation: $100\ Benzene + 100\ FeCl_3 \longrightarrow 100\ H_2 + (I) + 99\ FeCl_3$

3.93

Since $FeCl_3$ is a hydrogenation catalyst, H_2 formed can readily be kept in Equilibrium state of existence. For this reason, the reactions continue as follows-

Stage 103:

$$2H_2 \rightleftharpoons 2\,H\bullet e + 2\,n\bullet H$$

$$2H\bullet e + 2FeCl_3 \rightleftharpoons 2HCl + 2\,en\bullet FeCl_2$$

$$2en\bullet FeCl_2 \rightleftharpoons 2\,FeCl_2$$

$$2n\bullet H \longrightarrow H_2 \qquad 3.94a$$

Overall equation: $100\,H_2 + 100\,FeCl_3 \longrightarrow 100\,HCl + 100\,FeCl_2 + 50H_2$ 3.94b

Worthy of note above, is the last step wherein two hydrides instead of two atoms are adding together to form H_2. It is possible because radicals do not repel and a single element is involved.

Stage 104: This is the same as Stage 103 above to give the overall equation below.

Overall equation: $50\,H_2 + 50\,FeCl_3 \longrightarrow 50\,HCl + 50\,FeCl_2 + 25\,H_2$ 3.95

Stage 105: This is also the same as Stage 103 to give the overall equation below.

Overall equation: $24\,H_2 + 24\,FeCl_3 \longrightarrow 24\,HCl + 24\,FeCl_2 + 12H_2$ 3.96

Stage 106: This is also the same as Stage 103 above to give the overall equation below.

Overall equation: $12H_2 + 12\,FeCl_3 \longrightarrow 12\,HCl + 12\,FeCl_2 + 6\,H_2$ 3.97

Stage 107: This again is the same as Stage 103 above with the following overall equation.

Overall equation: $6H_2 + 6\,FeCl_3 \longrightarrow 6\,HCl + 6\,FeCl_2 + 3H_2$ 3.98

Recalling that we left one H_2 molecule from Stage 105, we now have $4H_2$ molecules left.

Stage 108: This is the same as the last stage.

Overall equation: $4H_2 + 4FeCl_3 \longrightarrow 4HCl + 4FeCl_2 + 2H_2$ 3.99

Stage 109: This is the same as Stage 103.

Overall equation: $2H_2 + 2FeCl_3 \longrightarrow 2HCl + 2FeCl_2 + H_2$ 3.100a

Overall overall equation: $100\ H_2 + 199\ FeCl_3 \longrightarrow 198\ HCl + 198\ FeCl_2 + H_2 + FeCl_3$ 3.100b

Overall overall equation: $100\ C_6H_6 + 200\ FeCl_3 \longrightarrow (I) + 198HCl + 198\ FeCl_2 + H_2 + FeCl_3$ 3.101

Stage 110: $H_2 \rightleftharpoons H\bullet e + n\bullet H$

$H\bullet e$ + Cl–[C₆H₄]–[C₆H₄]$_{98}$–[C₆H₄]–$FeCl_2$ $\rightleftharpoons$ HCl +

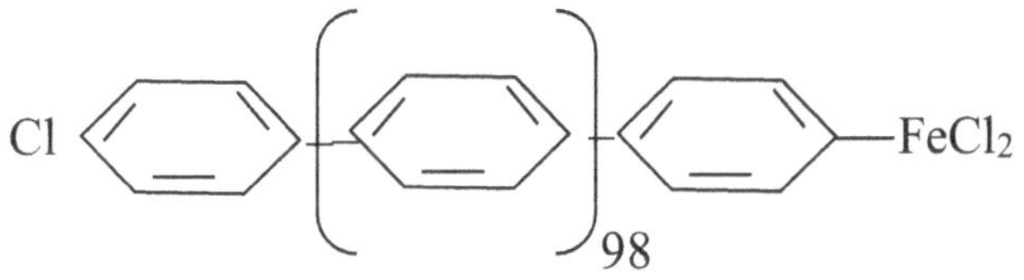

(I)

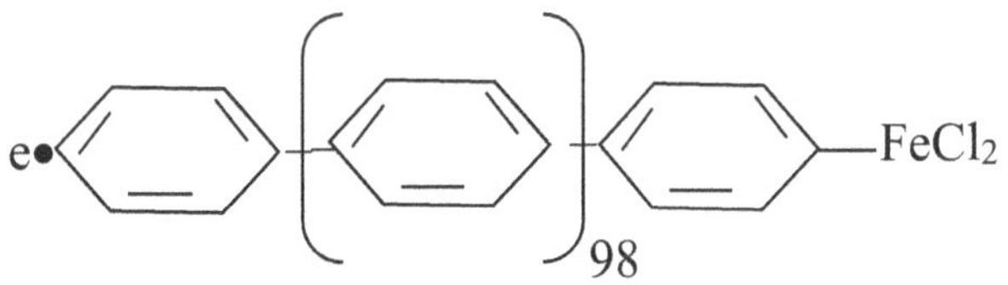

e•–[C₆H₄]–[C₆H₄]$_{98}$–[C₆H₄]–$FeCl_2$

(J)

(J) + $n\bullet H$ $\longrightarrow$ H–[C₆H₄]–[C₆H₄]$_{98}$–[C₆H₄]–$FeCl_2$

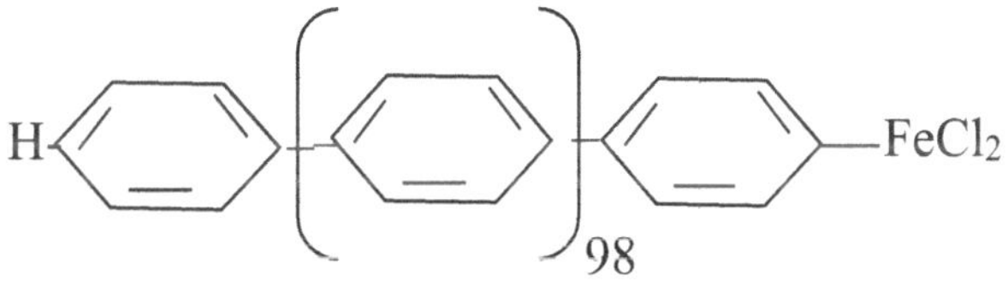

(K) 3.102

Stage 111:

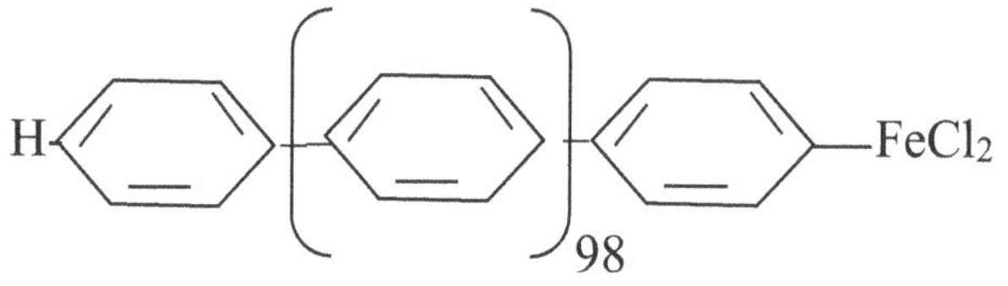

$$H-C_6H_4-[C_6H_4]_{98}-C_6H_4-FeCl_2 \rightleftharpoons$$

(K)

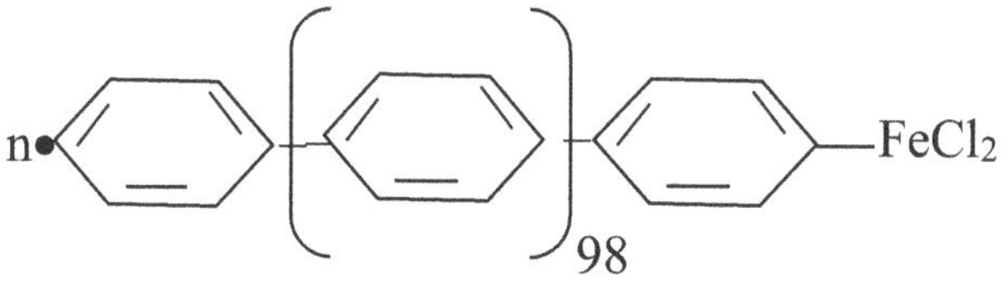

$$n\bullet C_6H_4-[C_6H_4]_{98}-C_6H_4-FeCl_2 \quad + \quad H\bullet e$$

(L)

$$H\bullet e \quad + \quad FeCl_3 \rightleftharpoons HCl \quad + \quad en\bullet FeCl_2$$

$$Cl_2Fe\bullet en \; + \; (L) \longrightarrow Cl_2Fe-C_6H_4-[C_6H_4]_{98}-C_6H_4-FeCl_2$$

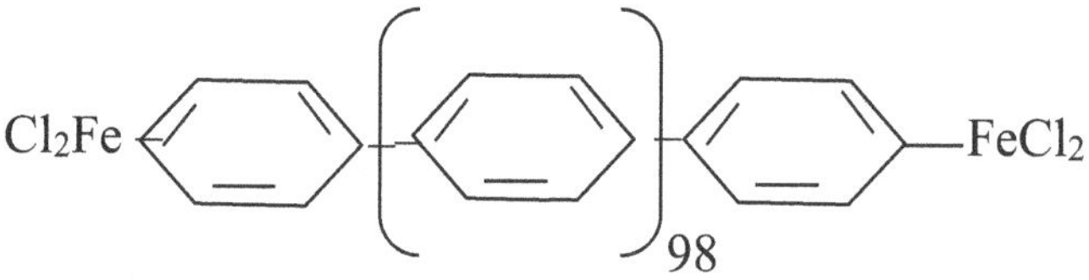

(M) 3.103a

Overall equation: $H_2 \; + \; (I) \; + \; FeCl_3 \longrightarrow 2HCl \; + \; (M)$ 3.103b

Final overall equation: $100\,C_6H_6 \; + \; 200\,FeCl_3 \longrightarrow 200HCl \; + \; (M) \; + \; 198FeCl_2$ 3.103c

Indeed, the products identified in Equation 3.88b are the product which have been obtained above under the following operating condition-

i) Temperature at 50^0C - 80^0C and I atmospheric pressure. If the temperature is raised to the hundreds, the situation will change drastically as was reportedly observed during pyrolysis of the products.

ii) Perfect mixing, whereby there is 100% contact between components, not as exists in many largely used reactors where there is a mixer or stirrer. For example, the bottom layer of stomachs of living systems is a mixing tank with no stirrers.

iii) Use of exact molar ratios. If there was more benzene than the $FeCl_3$, the situation will change drastically.

Equation 3.88b is almost well balanced because there is no chlorine in the polymer. The polymeric products obtained which have in the past been identified as "rust colored"[17] and "black solid"[18], is very well reflected by what the chain (M) is carrying at the terminals; the carrier of the chain being $FeCl_2$ and the terminating agent being $FeCl_2$. For the chain above the C/(H + Cl) atomic ratio is 1.485 for 100 monomer units. For 200 monomer units, it is 1.49 and for 500 monomer units, it is 1.497. The limiting value of 1.5 is that wherein there is no Cl. Where the polymer is the only product, the limit 1.5 cannot be attained if there must be Cl in the chain.

On the whole, there are one hundred and eleven series stages. In Stage 1, the $FeCl_3$ was suppressed by the energized state of existence of benzene to form (B) and hydrochloric acid, two products which can still react with themselves to form (D) and H_2 in Stage 2. If there had been no transition metal in the system, the H_2 formed would have remained in Stable state of existence throughout the course of the reaction. However, it can readily be kept in Equilibrium state of existence with the presence of the transition metal. The same will apply if $MoCl_5$ is used in place of $FeCl_3$. Though the H_2 is in equilibrium state of existence, it is not strong enough to suppress that of (D) which is now to be used as a STEP MONOMER. Polymerization begins from Stage 3 and continued until all the (D)s are consumed. If not all are consumed, because of the type of reactor, other small products such chlorobenzene begin to appear. For the number of moles of benzene and $FeCl_3$ used, polymerization stopped in Stage 102. The same will apply with the case of $MoCl_5$ with its (D) given as shown below.

Cl – (benzene ring) – $MoCl_4$

(D) –For $MoCl_5$ 3.104

With the full consumption of (D) and formation of the dead gigantic Step monomer (I), the H_2 molecules which have been sitting, now begins to work not yet with the polymer but with the excess $FeCl_3$. Because Fe cannot form hydrides with H, in Stage 103, ferrous chloride ($FeCl_2$) and H_2 were formed from H_2 and $FeCl_3$. One can see herein the wonders of NATURE, just as downstream, from some chemical reactions one was able to see that infinity (∞) does not exist as a number for any variable, but only for time and variables that are functions of time. The same Stage 103 was repeated until only one mole of H_2 and $FeCl_3$ were left behind, because both (odd moles) cannot react together and be productive. These were next used in the last two stages to stabilized the (I) formed from Stage 102. It is from Stage 103, that the use of $FeCl_3$ begins to differ from the use of $MoCl_5$ as shown below.

Overall equation: 100 Benzene + $100MoCl_5 \longrightarrow 100\ H_2$ + (I) + $99MoCl_5$

3.105

Stage 103:

$$2H_2 \rightleftharpoons 2\ H\bullet e + 2\ n\bullet\ H$$

$$2H\bullet e + 2MoCl_5 \rightleftharpoons 2HCl + 2\ en\bullet\ MoCl_4$$

$$2en\bullet\ MoCl_4 \rightleftharpoons MoCl_3 + MoCl_5$$

$$2n\bullet H \longrightarrow H_2 \qquad 3.106a$$

Overall equation: 100 Benzene + 100 $MoCl_5$ $\longrightarrow$ 51H_2 + 50 $MoCl_5$ + 49$MoCl_3$ + (I) + 98HCl 3.106b

Stage 104: Same as above.

Overall equation: 100 Benzene + 100 $MoCl_5$ $\longrightarrow$ 26 H_2 + 25 $MoCl_5$ + 74 $MoCl_3$ + (I) + 148 HCl 3.107

Stage 105: Same as above.

Overall equation: 100 Benzene + 100 $MoCl_5$ $\longrightarrow$ 14 H_2 + 13 $MoCl_5$ + 86 $MoCl_3$ + (I) + 172 HCl 3.108

Stage 106: Same as above.

Overall equation: 100 Benzene + 100 $MoCl_5$ $\longrightarrow$ 8H_2 + 7$MoCl_5$ + 92 $MoCl_3$ + (I) + 184 HCl 3.109

Stage 107: Same as above.

Overall equation: 100 Benzene + 100 $MoCl_5$ $\longrightarrow$ 5H_2 + 4$MoCl_5$ + 95 $MoCl_3$ + (I) + 190 HCl 3.110

Stage 108: Same as above.

Overall equation: 100 Benzene + 100 $MoCl_5$ $\longrightarrow$ 3 H_2 + 2$MoCl_5$ + 97 $MoCl_3$ + (I) + 194 HCl 3.111

Stage 109: Same as above.

Overall equation: 100 Benzene + 100 $MoCl_5$ $\longrightarrow$ 2H_2 + $MoCl_5$

$+ \quad 98\ MoCl_3 \quad + \quad (I) \quad + \quad 196HCl$ 3.112

Like the case of $FeCl_3$, the followings are obtained.

Stage 110:

$H_2 \rightleftharpoons H\bullet e + n\bullet H$

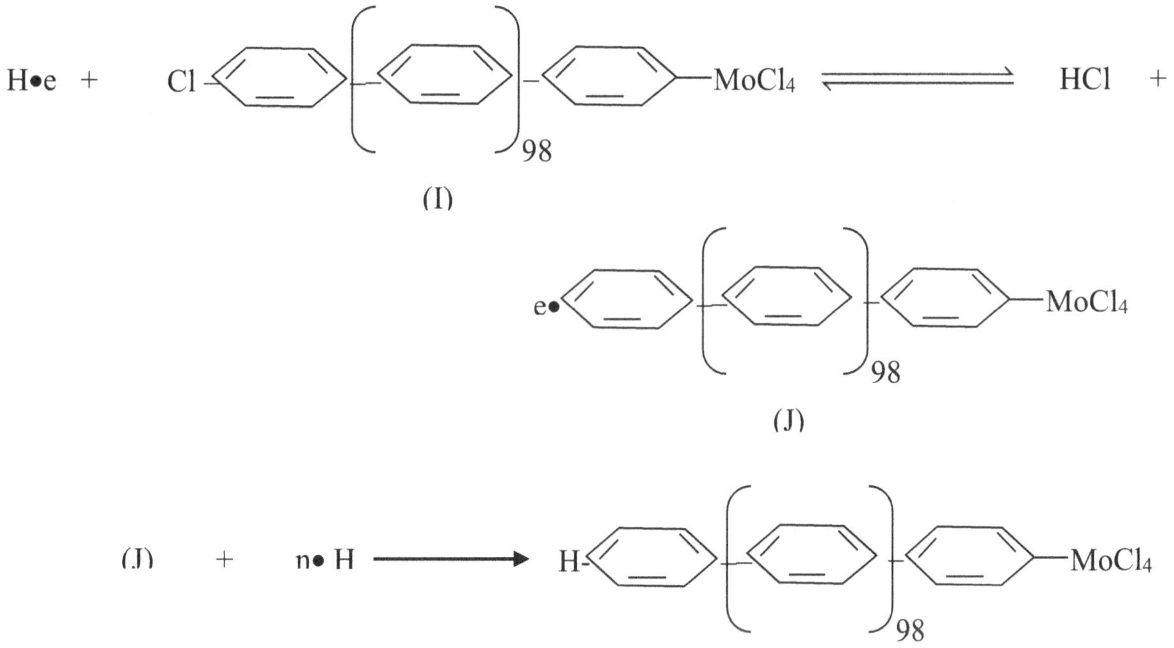

$H\bullet e$ + Cl–C_6H_4–[C_6H_4]$_{98}$–C_6H_4–$MoCl_4$ $\rightleftharpoons$ HCl +

(I)

e•–C_6H_4–[C_6H_4]$_{98}$–C_6H_4–$MoCl_4$

(J)

(J) + $n\bullet H$ ⟶ H–C_6H_4–[C_6H_4]$_{98}$–C_6H_4–$MoCl_4$

(K) 3.113

Stage 111:

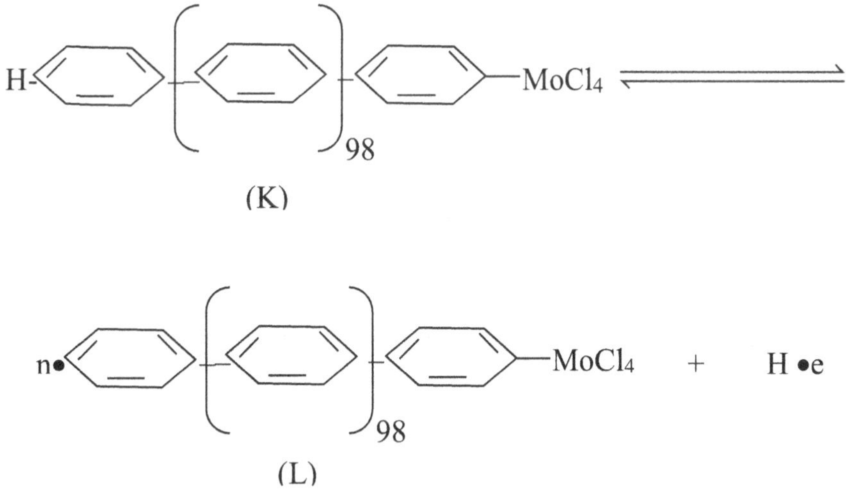

H–C_6H_4–[C_6H_4]$_{98}$–C_6H_4–$MoCl_4$ $\rightleftharpoons$

(K)

n•–C_6H_4–[C_6H_4]$_{98}$–C_6H_4–$MoCl_4$ + $H\bullet e$

(L)

$H\bullet e + MoCl_5 \rightleftharpoons HCl + en\bullet MoCl_4$

$Cl_4Mo\bullet en$ + (L) ⟶ Cl_4Mo–C_6H_4–[C_6H_4]$_{98}$–C_6H_4–$MoCl_4$

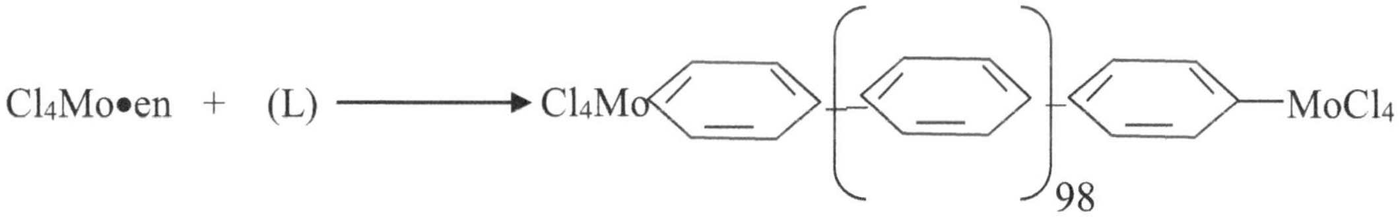

(M) 3.114a

Overall equation: $H_2 + MOCl_5 + (I) \longrightarrow 2HCl + (M)$ 3.114b

Overall overall equation: $100\ C_6H_6 + 100\ MoCl_5 \longrightarrow H_2 + 198\ HCl + 98\ MoCl_3 + (M)$ 3.115

Because Fe and Mo are transition metals, it is not surprising to see that the same numbers of stages are involved; yet $FeCl_3$ and $MoCl_5$ are uniquely different as reflected in Stage 103 of Equations 3.94a and 3.106a respectively. At the operating conditions of around 60^0C, $MoCl_4$ should not disproportionate, unless under pyrolytic conditions. But it did here because of the presence of even numbered moles in that stage and under equilibrium conditions inside the stage. It cannot disproportionate when there is only one mole. That was why half number of moles of what was used for $FeCl_3$ was used with $MoCl_3$. The mechanism provided for all these systems are worthy of note, because they open new doors to greater understanding of how Nature operates.

When $AlCl_3$ is used[17,18], the situation is completely different, because Al can form hydrides with H and therefore cannot be used as a hydrogen catalyst, i.e., cannot keep H_2 in Equilibrium state of existence. Hence, Stages such as Stages 103 to 109 will not take place. $AlCl_3$ cannot react with H_2 in its Stable state of existence and become productive as shown below.

Stage 1: $AlCl_3 \rightleftharpoons Cl_2Al\ \bullet e + nn\bullet\ Cl$

$Cl_2Al\ \bullet e + H_2 \rightleftharpoons Cl_2AlH + H\ \bullet e$

$H\ \bullet e + nn\bullet\ Cl \rightleftharpoons HCl$

[Reactive, stable and soluble] 3.116a

Overall equation: $AlCl_3 + H_2 \rightleftharpoons HCl + AlHCl_2$ 3.116b

With $AlCl_3$, after the production of the same type of (D) for Al in the first two Stages, Step polymerization commences in Stage 3 in the same fashion until all (D)s are consumed. This will involve 102 stages to give the following polymeric chain and overall equation.

$$Cl-C_6H_4-[C_6H_4]_{98}-C_6H_4-AlCl_2$$

(I)

Overall overall equation: $100\ \text{Benzene} + 100\ AlCl_3 \longrightarrow 100\ H_2 + (I) + 99\ AlCl_3$ 3.117

The chain cannot react with H_2 or $AlCl_3$ or both and be productive. No HCl can be formed in the real world. In the absence of a passive hydrogen catalyst, nothing can be done at the operating conditions. Here, the chain can be observed to be terminally chlorinated. Until now, universally, these reactions are believed to proceed by "Oxidative Cationic Polymerization of the aromatic nuclei". Only THE ALMIGTHY INFINITE GOD and HIS Messengers know what this means. And for THEM, it is meaningless and a display of IGNORANCE in our world which does not still know what AN ATOM is. Yet, we in our world think we have advanced when we have not started. Even from what we have seen so far, the ATOM has not yet been fully defined, because just as we have RADICAL configuration for the outer shell of ATOM, so also we have "ELECTRONIC" configuration for the Nucleus of the ATOM, because "electrons" only reside inside the Nucleus of the ATOM. It is from the ELECTRONIC configuration, we begin to see what Radiation is.

Thus, one has shown the second kind of STEP monomer which is Non-Polar/Non-ionic, but still di-functional, the functional group being H, and that monomer is Benzene.

Under vibratory milling conditions, benzene and pyridine were said to lead to solid products consisting of a mixture of fractions "soluble" in methanol and DMF (Dimethyl formamide) and an "insoluble" fraction. Chromatographic and spectroscopic analyses indicated that the soluble fraction is a mixture of compounds with linear structures obtained by opening of the aromatic ring. Interaction of these compounds with O_2, light, or high temperatures, leads to cross-linked structures[22]. [Notice the way the words "soluble" and "insoluble" have been used above.] This was said to be a ***mechanochemical polymerization.*** It is mechano- in the sense that a mechanical means has been used in place of an initiator to open the ring. How can such rings with no point of scission in its Stable state be opened? No matter what visible force is used, such rings can never be opened. To open such rings, one must create a point of scission. This can be done either by saturating the ring, that which we don't want to do here, or keep the ring in Equilibrium state of existence or activate the ring and do something immediately. With the case above, in the absence of any initiator (Chemical means), the benzene under such conditions was kept in Equilibrium state of existence and opened as shown below.

Stage 1:

$$C_6H_6 \rightleftharpoons C_6H_5^{\cdot n} + H^{\cdot e}$$

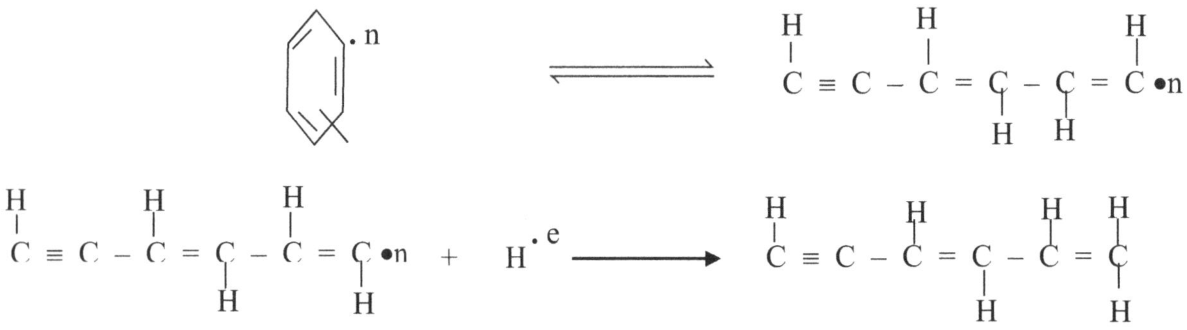

[Benzene radical (•n)] ⇌ $H-C \equiv C-C(H)=C(H)-C(H)=C(H)\bullet n$

$H-C \equiv C-C(H)=C(H)-C(H)=C(H)\bullet n \;+\; H^{\bullet e} \longrightarrow H-C \equiv C-C(H)=C(H)-C(H)=CH_2$

(A) Butadienyl acetylene 3.118a

Overall equation: 20 Benzene ⟶ 20 Butadienyl acetylene 3.118b

Stage 2:

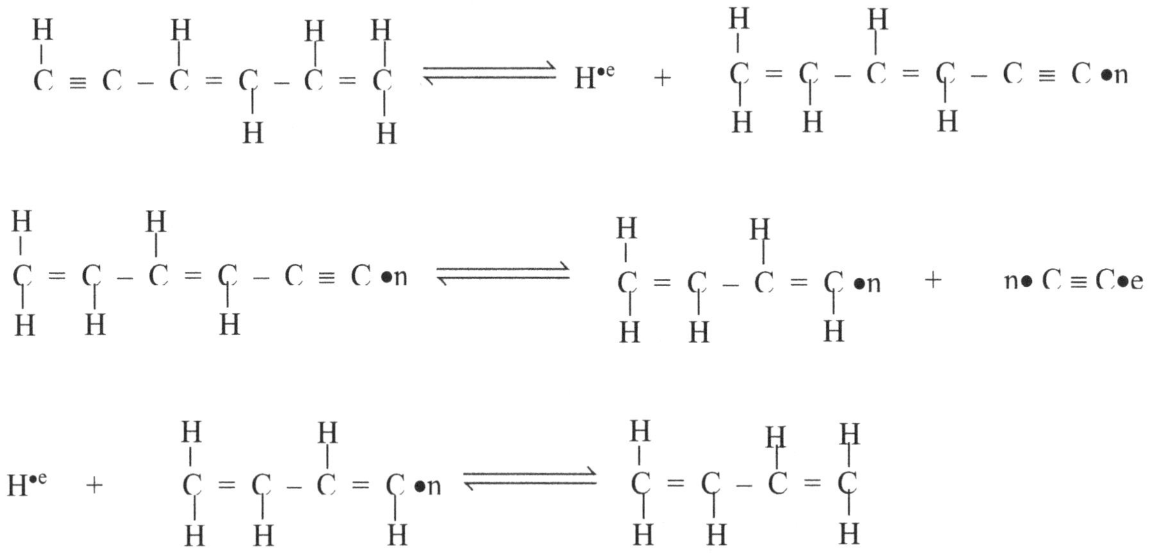

$H-C \equiv C-C(H)=C(H)-C(H)=CH_2 \rightleftharpoons H^{\bullet e} + CH_2=C(H)-C(H)=C(H)-C \equiv C\bullet n$

$CH_2=C(H)-C(H)=C(H)-C \equiv C\bullet n \rightleftharpoons CH_2=C(H)-C(H)=C(H)\bullet n \;+\; n\bullet C \equiv C\bullet e$

$H^{\bullet e} + CH_2=C(H)-C(H)=C(H)\bullet n \rightleftharpoons CH_2=C(H)-C(H)=CH_2$

(B) Butadiene

$n\bullet C \equiv C\bullet e \longrightarrow$ 2C (Carbon black) 3.119a

Overall equation: 10 Butadienyl acetylene ⟶ 20C + 10 Butadiene 3.119b

Stage 3:

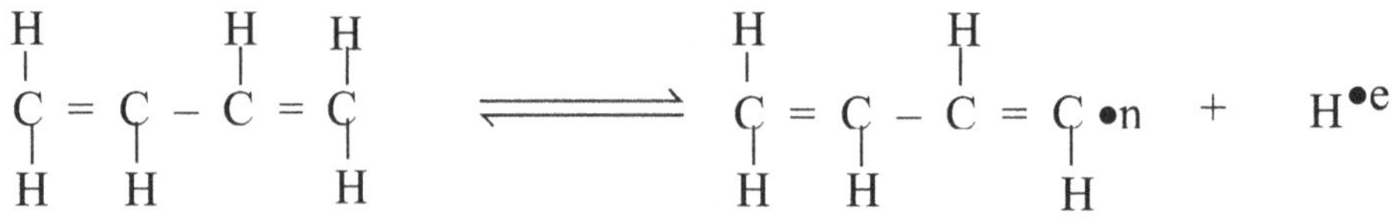

$CH_2=C(H)-C(H)=CH_2 \rightleftharpoons CH_2=C(H)-C(H)=C(H)\bullet n \;+\; H^{\bullet e}$

(C)

$$H^{\bullet e} + \overset{H}{C} \equiv C - \overset{H}{C} = \underset{H}{C} - \overset{H}{C} = \overset{H}{\underset{H}{C}} \rightleftharpoons \overset{H}{C} \equiv C - \overset{H}{C} = \underset{H}{C} - \underset{H}{C} = \overset{H}{C}\bullet n + H_2$$

(D)

$$(C) + (D) \longrightarrow \overset{H}{C} \equiv C - \overset{H}{C} = \underset{H}{C} - \overset{H}{C} = \underset{H}{C} - \overset{H}{C} = \underset{H}{C} - \overset{H}{C} = \overset{H}{\underset{H}{C}}$$

(E) Octatetraenyl acetylene 3.120a

Overall equation: **10 Butadiene + 10 Butadiene acetylene** $\longrightarrow$

10 Octatetraenyl acetylene + 10 H_2

3.120b

Stage 4:

$$\overset{H}{C} \equiv C - \overset{H}{C} = \underset{H}{C} - \overset{H}{C} = \underset{H}{C} - \overset{H}{C} = \underset{H}{C} - \overset{H}{C} = \overset{H}{\underset{H}{C}} \rightleftharpoons n\bullet C \equiv C -(\overset{H}{C} = \underset{H}{C})_3 - \overset{H}{C} = \overset{H}{\underset{H}{C}} + H^{\bullet e}$$

$$n\bullet C \equiv C -(\overset{H}{C} = \underset{H}{C})_3 - \overset{H}{C} = \overset{H}{\underset{H}{C}} \rightleftharpoons n\bullet \overset{H}{C} = \overset{H}{C} -(\overset{H}{C} = \underset{H}{C})_2 - \overset{H}{C} = \overset{H}{\underset{H}{C}} + n\bullet C \equiv C\bullet e$$

$$H^{\bullet e} + n\bullet \overset{H}{C} = \overset{H}{C} -(\overset{H}{C} = \underset{H}{C})_2 - \overset{H}{C} = \overset{H}{\underset{H}{C}} \rightleftharpoons \overset{H}{\underset{H}{C}} = \overset{H}{C} - (\overset{H}{C} = \underset{H}{C})_2 - \overset{H}{C} = \overset{H}{\underset{H}{C}}$$

(F)

$$n\bullet C \equiv C\bullet e \longrightarrow 2C \text{ (Carbon black)}$$ 3.121a

Overall equation: 5 Octatetraenyl acetylene $\longrightarrow$ 10 C + 5 Octatetraene

3.121b

Stage 5:

$$\overset{H}{\underset{H}{C}} = \overset{H}{C} -(\overset{H}{C} = \underset{H}{C})_2 - \overset{H}{C} = \overset{H}{\underset{H}{C}} \rightleftharpoons \overset{H}{\underset{H}{C}} = \overset{H}{C} -(\overset{H}{C} = \underset{H}{C})_2 - \underset{H}{C} = \overset{H}{C}\bullet n + H^{\bullet e}$$

(G)

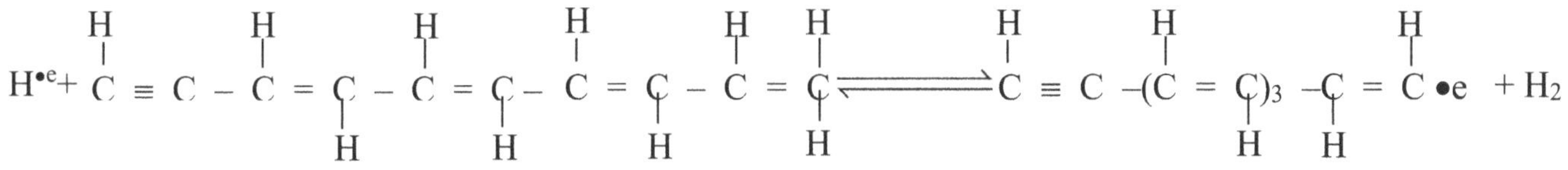

$$H^{\bullet e} + HC \equiv C - CH = CH - CH = CH - CH = CH - CH = CH_2 \rightleftharpoons HC \equiv C -(CH = CH)_3 - CH = CH^{\bullet e} + H_2$$

(H)

$$(G) \quad + \quad (H) \longrightarrow HC \equiv C -(CH = CH)_7 - CH = CH_2$$

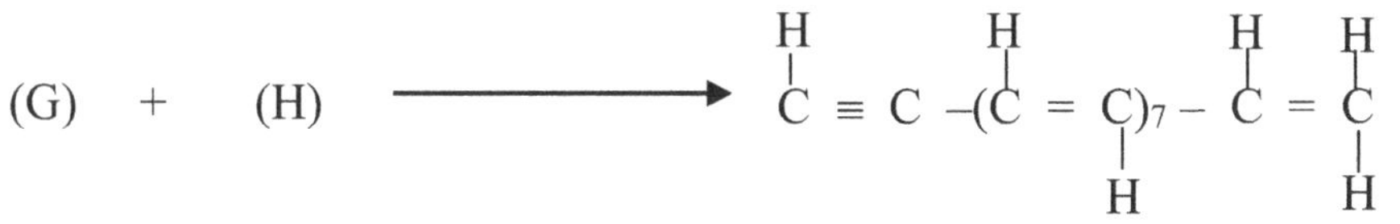

(I) 3.122a

Overall equation: 5Octatetraene + 5Octatetraenyl acetylene ⟶ $5H_2$ + 5Polyacetylene 3.122b

Overall overall equation: 20 Benzene ⟶ 30 C + $15H_2$ + 5Polyacetylene 3.123

One can see the number of stages involved in producing the Living polyacetylene from 20 moles of benzene. The living polyacetylene is said to be living, because the carrier of the chain is acetylene with H at its end. With this type of carrier, at least a good fraction of the polymer will always exist in Equilibrium state of existence, depending on the operating conditions, the fraction decreasing as the chain continues to grow. From the five stages above with the use of 20 moles of benzene, one mole of the polymer produced has eight acetylenic monomer units with acetylene being the carrier and H the terminating agent. With the 20 moles of benzene used, if polymerization was allowed to continue, only one mole of the type of (I) above will be left behind along with more C and H_2, bringing the total number of stages to nine. That type of (I) left alone will subsequently break down to a longer diene and C. Above, one has used an even number of moles of benzene. If the number of moles is odd, one knows what to expect.

A look at the five stages above, clearly indicates that this is not an Addition polymerization system as has been mistakenly thought to be the case in the past. It is not a Combination mechanism system as can be seen above. ***This is a Step polymerization system wherein all the stages take place via Equilibrium mechanism.*** In Stage 1, benzene was put in Equilibrium state of existence- the Energized state, and with force resulting from vibrations, the ring was opened to give (A) - a butadienyl acetylene. In Stage

2, a fraction of this was kept in Equilibrium state of existence. Above, one used fifty percent, in order to minimize the numbers of stages one will go through and for simplicity. That fraction broke down to give (B) butadiene and Carbon black (not Coke). In Stage 3, the remaining fraction in Stable state of existence was next attacked by the butadiene which was put in Equilibrium state of existence. In the process, H_2 was formed along with (E)- octatetraenyl acetylene. Just like (A), a fraction of (E) was kept in Equilibrium state of existence, and this decomposed to give an extended butadiene and carbon black in Stage 4. In Stage 5, just like Stage 3, (I) was produced. This continued until the process could not be repeated anymore. The polymer formed has very many cross-linking sites. It will not be surprising therefore that when the polymer is heated, cross-linked products will be obtained. But when it interacts with O_2 and light, the H on the acetylene will create some other problems, leading to formation of cross-links.

(ii) Use of Benzene as an ADDITION monomer

Having seen the use of Benzene as a STEP monomer, it will not be complete if its use as an ADDITION monomer is not started to be shown. Just as the ring of cyclooctatetraene has been reported to be opened[23] when the ring is made to carry a radical-pushing group, so also it can be done for benzene using special initiators. Note that why benzene is aromatic and cyclooctatetreane and others are not aromatic has been shown in the New Frontiers. Though how it is opened has been unknown till date, this has been explained above. The answer like all cases which humanity cannot provide answers to, is always so simple - something which we mistakenly classify as "COMMON SENSE". ***Things we ignore, is the origin of IGNORANCE.*** When radical pushing groups are placed on the ring, the ring becomes so strained to the point of attaining Max.RSE inside the ring. When that energy is made to exist in the ring, then the ring can no longer exist as a ring. It will open explosively.

H_3C– n• •e (I)

G •e .... n•Z (X, X, X, X, □, □)

(II) n• •e H_3C–

$$G - \overset{H}{\overset{|}{C}} = \underset{H}{\underset{|}{C}} - \overset{H}{\overset{|}{C}} = \underset{H}{\underset{|}{C}} - \overset{H}{\overset{|}{C}} = \underset{CH_3}{\underset{|}{C}}^{\bullet e} \text{———} n\bullet Z(\square)_2(X)_4 \longrightarrow$$

(II)

$$G - \overset{H}{\overset{|}{C}} = \underset{H}{\underset{|}{C}} - \overset{H}{\overset{|}{C}} = \underset{H}{\underset{|}{C}} - \overset{H}{\overset{|}{C}} = \underset{CH_3}{\underset{|}{C}} - \underset{H}{\underset{|}{C}} = \overset{H}{\overset{|}{C}} - \underset{H}{\underset{|}{C}} = \overset{H}{\overset{|}{C}} - \underset{H}{\underset{|}{C}} = \overset{CH_3}{\overset{|}{C}}^{\bullet e} \text{———} n\bullet Z(\square)_2(X)_4$$

(III)

Syndiotactic poly(Methyl-tri-acetylene) 3.124

The initiator used above is not ideal for the system, because it is an imaginary one, provided with two vacant orbitals. Initiators of the non-Electrostatic types for syndiotactic placement of Nucleophiles along the chain are not common. However, it has been used above to illustrate some basic fundamental principles. Note the point of scission on the ring and the center first activated. The point of scission is next to the point of attack. After opening of the ring, note the center carrying the electro-free-radical. It is the center carrying the radical-pushing group which made it possible to open the ring in this manner. Because of the presence of two reservoirs on the counter-center, hence syndiotactic placement of the group can be observed. The larger the radical-pushing capacity of the group placed on the ring, the easier it is for the ring to be opened. ***This is Free-radical polymerization via Combination mechanism.*** The only induction period will be the time for the initiator to be prepared in-situ.

For the first time, one has shown how benzene ring can be used as a monomer to produce unique polymers. Step by step, one is beginning to show how NATURE operates in the real and imaginary domains. Without going through each one of them, one will just be scratching the surface. ***All atoms, molecules, compounds, species both known and unknown, all have their different personalities, far greater than the issues of Chemical and Physical properties. The issues of Chemical and Physical properties, though very important, are still but a grain of sand in our abilities to understand what these things are.***

3.4.2 Definition of a Reversible Reaction

Herein, one is going to use the well-known reaction wrongly represented as follows-

$$2NH_3 \xrightleftharpoons[\text{Reversibility}]{\text{Wrong Representation of}} 3H_2 + N_2 \qquad 3.125$$

It is a wrong representation because it is important to know ***the operating conditions*** under which reversibility has taken place, since when the operating conditions are the same, it is an Equilibrium process either of Solubility or Insolubility as already shown. In the equation above, different operating conditions are required. Hence the equation above is properly represented as follows.

$$2NH_3 \underset{\text{Use of } H_2 \text{ catalyst or the like}}{\overset{\text{Heat}}{\rightleftarrows}} 3H_2 + N_2 \qquad 3.126$$

It is not an Equilibrium reaction of the one-sided type (Productive) or the half sided- doubled arrowed type (Solubilisation/Insolubilisation), but reversible reaction as indicated above and shown below.

a) <u>Decomposition of Ammonia</u>.

<u>Stage 1</u>:

$$NH_3 \underset{\text{Existence of Ammonia}}{\overset{\text{Equilibrium State of}}{\rightleftharpoons}} H\cdot e + nn\cdot NH_2$$

$$H\cdot e + \underset{(stable\ fraction)}{NH_3} \underset{\text{Due to Heat}}{\overset{\text{Abstraction}}{\rightleftharpoons}} H_2 + en\cdot NH_2$$

$$H_2N\cdot en + nn\cdot NH_2 \xrightarrow[\text{Existence of Hydrazine}]{\text{Combination State of}} \underset{(Hydrazine)}{H_2N-NH_2} \qquad 3.127a$$

$$\underline{Overall\ Equation}: 2NH_3 \longrightarrow H_2 + H_2N-NH_2 \qquad 3.127b$$

<u>Stage 2</u>:

$$H_2N-NH_2 \underset{\text{Existence of Hydrazine}}{\overset{\text{Equilibrium State of}}{\rightleftharpoons}} H\cdot e + \underset{(A)}{nn\cdot \overset{H}{\overset{|}{N}}-NH_2}$$

$$(A) \underset{\text{very harsh operating operating}}{\overset{\text{Release of } H\cdot n \text{ at}}{\rightleftharpoons}} H\cdot n + \underset{(B)}{nn\cdot \overset{H}{\overset{|}{N}}-\overset{en}{\overset{\cdot}{N}}-H}$$

$$H\cdot e + H\cdot n \underset{\text{Existence of } H_2}{\overset{\text{Equilibrium State of}}{\rightleftharpoons}} H_2$$

$$(B) \xrightarrow[\text{Release of Energy}]{\text{Deactivation/}} \underset{(C)}{H-N=N-H} + Heat \qquad 3.128a$$

$$\underline{Overall\ Equation}: H_2N-NH_2 \xrightarrow{Heat} H_2 + H-N=N-H + Energy \qquad 3.128b$$

Stage 3:

$$H-N=N-H \underset{State\ of\ Existence}{\overset{Its\ Equilibrium}{\rightleftharpoons}} H\cdot e + nn\cdot \ddot{N} = \ddot{N} - H$$

$$(D)$$

$$(D) \underset{harsh\ operating\ \leq nditions}{\overset{Release\ of\ H\cdot n\ at\ very}{\rightleftharpoons}} H\cdot n + nn\cdot \ddot{N} = \underset{\cdot\cdot}{N}\cdot en$$

$$(E)$$

$$H\cdot e + H\cdot n \underset{Existence\ of\ H_2}{\overset{Equilibrium\ State\ of}{\rightleftharpoons}} H_2$$

$$nn\cdot N = N\cdot en \xrightarrow[Release\ of\ Energy]{Deactivation} N \equiv N + Energy \qquad 3.129a$$

$$\underline{Overall\ Equation}: N_2H_2 \longrightarrow N_2 + H_2 + Energy \qquad 3.129b$$

Overall overall Equation: $2NH_3 \xrightarrow{Heat} N_2 + 3H_2 + \text{Energy} \qquad 3.129c$

At the operating conditions, notice that a fraction of around fifty percent of NH_3 is kept in Equilibrium state of existence while the remaining fraction is kept in a Stable state of existence. The operating conditions for many cases like this compound will determine the fractions to be kept in Equilibrium state of existence and therefore the optimum conversion level of the reaction ranging from 0 to 100 percent. NH_3 is unique in that only a small fraction is held in Equilibrium state of existence all the time depending on the operating conditions and the types of neighbours around its vicinity as already shown for ammonium hydroxide, ammonium chloride and more. In general, H is always released as an atom, i.e., as H•e and never as a hydride, i.e., as H•n. In the second and third stages, for the second or third time we are noticing the release of H as a hydride in the second step of the stages. With N carrying an electro-non-free-radical, heat must be released. This can only take place at operating conditions used for cracking of hydrocarbons in the Petroleum industries. These are temperatures of the ordcrs of a thousand and above. Notice the transfer species also removed from the ammonia. It is not the NH_2 group, the species held in its Equilibrium state of existence, but H as a hydride. If it was Na that was doing the abstraction, far less energy will be required, because H is abstracting itself differently and Na is far more electropositive than H. Hence, it is no surprise to find the harsh operating conditions used in the industries for decomposition of ammonia. In Nature, this can easily be done underground at even far below normal operating conditions as will be shown in the New Frontier, that which is called ***SURFACE or UNDERGROUND CHEMISTRY.***

b) Recovery of NH_3 from H_2 and N_2

Stage 1:

$$H_2 \underset{\substack{\text{needed here to keep } H_2 \\ \text{in Equilibrium State of Existence}}}{\overset{H_2 \text{ Catalyst is}}{\rightleftharpoons}} H\cdot e \; + \; H\cdot n$$

$$H\cdot e + N \equiv N \xrightleftharpoons{\text{Activation}} H - \underset{\cdot\cdot}{N} = \ddot{N}\cdot en$$

$$H - N = N\cdot en \; + H\cdot n \xrightarrow[\text{State}]{\text{Combination}} H - N = N - H \qquad 3.130a$$

Stage 2;

$$H_2 \underset{\substack{\text{Existence of } H_2 \\ (\text{Use of } H_2 \text{ Catalyst})}}{\overset{\text{Equilibrium State of}}{\rightleftharpoons}} H\cdot e \; + \; n\cdot H$$

$$H\cdot e + H - N = N - H \underset{H_2N_2}{\overset{\text{Activation of}}{\rightleftharpoons}} \underset{(A)}{H - \overset{\overset{H}{|}}{N} - \overset{\overset{H}{|}}{N}\cdot en}$$

$$(A) \; + \; H\cdot n \longrightarrow H_2N - NH_2 \qquad 3.130b$$

Stage 3:

$$H_2 \underset{H_2 \text{ Catalyst}}{\overset{\text{Use of}}{\rightleftharpoons}} H\cdot e \; + \; H\cdot n$$

$$H\cdot e + H_2N - NH_2 \xrightleftharpoons{\text{Abstraction}} NH_3 \; + \; en\cdot NH_2$$

$$H_2N\cdot en \; + \; H\cdot n \xrightarrow[\text{Existence of Ammonia}]{\text{Combination State of}} NH_3 \qquad 3.131a$$

$$\underline{\textit{Overall Equation}}: \; 3H_2 \; + \; N_2 \longrightarrow 2NH_3 \qquad 3.131b$$

Notice that different operating conditions have been used for the forward and backward reactions of Equation 3.125. In both cases, three stages are involved. This is not generally the case. For a system to be reversible, it must involve different operating conditions, and not same operating conditions. While the first takes place in the absence of any visible neighbour, the second takes place in the presence of a visible neighbour in this case H_2 catalysts which are Passive in character. There are lots to be observed in all the steps of the stages above. For example, in the second step of Stage 3 above, if Chlorine electro-non-free radical (Cl•en) had been used in place of H•e, hydrogen as opposed to the amine abstracted, would have been abstracted. In present-day modus operandi, one will say that if H was abstracted from NH_3 in Equation 3.127a, why was the same H not abstracted from H_2N-NH_2? That is not the way to do things-Theory of Semblance and so on. In the first case, when H_2N-NH_2 is held in Equilibrium state of existence, as shown in Equation 3.128a, NH_2 is not the one held. Secondly, N-N bond is far weaker than the N-H bond. Thirdly, NH_2 is just a group replacing one H atom in NH_3 and that group though far more radical-pushing than H, is electronegative in character. Energy is required to help the transition metal keep H_2 in Equilibrium state of existence and more energy is required to activate N≡N and N=N than to activate C=C bond.

3.4.3 Cracking of some aliphatic hydrocarbons.

Herein, the cracking of some hydrocarbons in their physical gaseous state are considered because cracking is a major step in the Petroleum industry.

(a) Cracking of Methane

Stage 1:

$$CH_4 \underset{\substack{\text{Existence of Methane}\\ \text{at Melting or boiling points}}}{\overset{\text{Equilibrium State of}}{\rightleftharpoons}} H\cdot e \ + \ n\cdot CH_3 \quad (\text{Above Boiling Point})$$

$$n\cdot CH_3 \underset{\text{Decomposition}}{\overset{\text{High Temperature}}{\rightleftharpoons}} H\cdot n \ + \ e\cdot \underset{\substack{|\\ H}}{\overset{\substack{H\\ |}}{C}}\cdot n \quad (\text{Higher Temperature})$$

$$e\cdot \underset{\substack{|\\ H}}{\overset{\substack{H\\ |}}{C}}\cdot n \underset{\text{Decomposition}}{\overset{\text{Higher Temperature}}{\rightleftharpoons}} H\cdot e + e\cdot \underset{\substack{\cdot\\ n}}{\overset{\substack{H\\ |}}{C}}\cdot n \ (\text{Moderate Temperature})$$

$$\underset{n}{H\cdot e + H\cdot n} \underset{\text{Existence of } H_2}{\overset{\text{Equilibrium State of}}{\rightleftharpoons}} \underset{n}{H_2}$$

$$e\cdot \underset{\substack{|\\ H}}{\overset{\substack{n\\ \cdot}}{C}}\cdot n \underset{\text{for Decomposition}}{\overset{\text{More Higher Temperature}}{\rightleftharpoons}} e\cdot \underset{\substack{\cdot\\ e}}{\overset{\substack{n\\ \cdot}}{C}}\cdot n \ + \ H\cdot n \ (\text{Very High Temperature})$$

$$\underline{(\text{Activated Carbon Black})}$$

$$H\cdot e + H\cdot n \underset{\text{Existence of } H_2}{\overset{\text{Equilibrium State of}}{\rightleftharpoons}} H_2$$

$$n\cdot \underset{\substack{\cdot\\ e}}{\overset{\substack{n\\ \cdot}}{c}}\cdot e \xrightarrow[\text{its centers one at a time}]{\text{Deactivation of one of}} :\overset{\substack{nn\\ \cdot}}{C}\cdot en \ + \ Heat \qquad 3.132a$$

$$\underline{(\text{Carbon black})}$$

$$\underline{\text{Overall Equation}}: CH_4 \longrightarrow \text{Carbon black} + 2H_2 + \text{Energy} \qquad 3.132b$$

Worthy of note again as seen so far is the structure of carbon black clearly different from the structures of Ground-state carbon and Excited-carbon as shown below. In the Stage above, there are seven steps. Based on the break-down steps above (2nd 3rd and 5th), the cracking of methane demands the use of very high temperatures of the order of 1200°C. ***The major reason for the use of high temperature is because of release of H as a hydride instead of its natural form. This can only be done by the use of stronger forces such as found in Physics already provided by NATURE.*** One can see the magnanimity of NATURE and the clear distinction between the Physical and Natural world. They both must co-exist for any to exist.

$$\underline{\text{Ground State Carbon}} - :\overset{nn}{\dot{C}} \cdot nn \;;\; \text{Excited Carbon} - n \cdot \underset{\underset{n}{\cdot}}{\overset{n}{\dot{C}}} \cdot n \;\; \underline{(\text{Charcaol})}$$

$$(A) \qquad\qquad\qquad\qquad (B) \qquad\qquad 3.133a$$

$$\text{Groud State Carbon} - :\overset{nn}{\dot{C}} \cdot nn \xrightleftharpoons{\text{Activation}} e \cdot \underset{\underset{n}{\cdot}}{\overset{n}{\dot{C}}} \cdot n$$

$$\underline{(C) - (\text{Activated Carbon})} \qquad 3.133b$$

$$\text{Carbon Black} - :\overset{nn}{\dot{C}} \cdot en \xrightleftharpoons{\text{Activation}} e \cdot \underset{\underset{n}{\cdot}}{\overset{n}{\dot{C}}} \cdot e$$

$$(D) \qquad\qquad \underline{(E) - (\text{Activated Carbon Black})} \qquad 3.133c$$

$$\underline{\text{Coal}} - e \cdot \underset{\underset{n}{\cdot}}{\overset{e}{\dot{C}}} \cdot e \, \underline{(\text{Semi} - \text{Metallic Carbon})} \;;\; \underline{\text{Coke}} - :\overset{en}{\dot{C}} \cdot en \, \underline{(\text{Metallic Carbon})}$$

$$(F) - \text{Activated Coke} \qquad\qquad (G) \qquad\qquad 3.133d$$

Activation can take place during Equilibrium or Combination or even Decomposition mechanism. The cases above were done under Equilibrium mechanism. The carbon center alone cannot carry Real charges; hence these structures carry radicals only. Ground-state carbon (A), Excited carbon or Charcoal (B), Activated carbon (C) and Carbon black (D) are some of the ones that can be used to produce aliphatic hydrocarbons. These are some or all of the different elements of carbon. All of them have different electro-negativity or positivity. While Ground state carbon can be activated to give activated carbon, Carbon black can be activated to give Activated carbon black (or Black carbon), coke can be activated to give Coal. Coal cannot be activated. From them all the allotropic forms of Carbon such as Amorphous, Graphite and Diamond carbons and even more are obtained as will be shown in the New Frontiers. Indeed, the real activated carbon black, which is polymeric in character, can be envisaged as follows from its element.

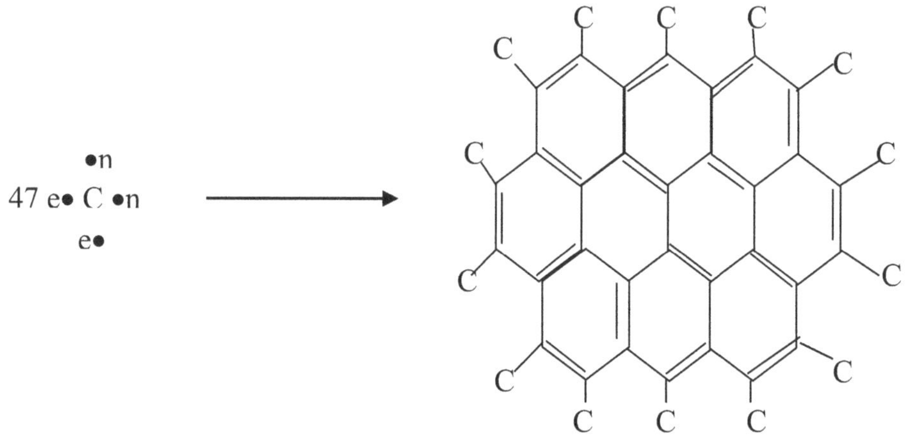

•n
47 e• C •n ⟶
e•

One type of Activated Carbon Black

3.134

There may be activated carbon blacks with different sizes of rings with stain energy below the Maximum Required Strain Energy, either double/single or triple/single or double/triple/single bonded. Coal from (F) which is a readily combustible rock, containing more than 50% by weight of carbonaceous material formed from compaction and indurations of variously altered plant remains similar to those in peat. Most coal are said to be fossil peats. Coke and coal can also form rings. However, the structure of the main carbon element of coal is different from that shown above, apart from the presence of other components which are impurities. The terminal C centers some of which are still carrying radicals are closed with very few elements of H which can rarely be identified, because their numbers are too small compared to C. From here, one can imagine what the structures of the different elements of Sulfur, Phosphorous, Iodine and more, will look like. So much works still have to be done. Some are black, some are white such as white phosphorus, some are red and so on. What is the origin of their colours that which is Physics of Light inside Chemistry? It is only inside Chemistry that one can see the origin. It is only inside Chemistry, that the Physics of Smell and Taste can be developed. How can we perceive the smell of a compound when the compound is not in Equilibrium state of existence?

Thus when methane is cracked, alkenes cannot be obtained. The only products are H_2 and Carbon black.

(b) Cracking of Ethane

Stage 1:

$$H-\underset{H}{\overset{H}{C}}-\underset{H}{\overset{H}{C}}-H \xrightleftharpoons[\text{Existence at Above Boiling point}]{\text{Equilibrium State of}} H\cdot e + n\cdot\underset{H}{\overset{H}{C}}-\underset{H}{\overset{H}{C}}-H \quad (A)$$

$$(A) \xrightleftharpoons[\text{Existence at Higher temperatures}]{\text{Equilibrium State of}} H\cdot n + n\cdot\underset{H}{\overset{H}{C}}-\underset{H}{\overset{H}{C}}\cdot e \quad [600^0C] \quad (B)$$

$$H\cdot e + H\cdot n \xrightleftharpoons[\text{Existence}]{\text{Equilibrium State of}} H_2$$

$$(B) \xrightarrow{\text{Deactivation}} H_2C=CH_2 + Heat \qquad 3.135a$$

$$\underline{Overall\ Equation}: H_6C_2 \longrightarrow H_2 + H_2C=CH_2 + Heat \quad (Ethylene\ or\ Ethene) \qquad 3.135b$$

Worthy of note are the followings-

(i) Hydrogen is one of the products.
(ii) No carbon black is produced unlike the case of methane.
(iii) Unlike methane, less degree of operating conditions (Temperature) are required here. This is of the order of 500°-600°C at one atmosphere.
(iv) In the last step of the stage, Energy is released via deactivation.
(v) An alkene (Ethene) is one of the products.

Like above, in the cracking of other higher normal hydrocarbons, only one **<u>alkene (Ethene) is produced along with hydrogen molecules and in some cases carbon black</u>** if the temperature is above 1000^0C. The cracking of aliphatic hydrocarbons is completely different from the cracking of their halogenated counterparts supported by universal experimental observations[24]. So also, is the cracking of Alkenes and Alkynes. For Alkenes and alkynes, higher operating conditions are required, for which for Alkynes (Starting from acetylene in particular) Carbon Black and Coke start appearing as carbon elements of the products along with hydrogen. The ethene obtained as product above can further be cracked at higher operating conditions to acetylene. When propane is cracked, ethene is obtained. All alkanes can be cracked. For example, tertiary butane can be cracked as shown below.

Stage 1:

$$\begin{matrix} & H & CH_3 & \\ & | & | & \\ H- & C & -C & -H \\ & | & | & \\ & H & CH_3 & \end{matrix} \underset{\text{Existence at Above Boiling point}}{\overset{\text{Equilibrium State of}}{\rightleftharpoons}} H\cdot e + n\cdot \begin{matrix} CH_3 & H & \\ | & | & \\ C & -C & -H \\ | & | & \\ CH_3 & H & \end{matrix}$$

$$ter-Butane \qquad\qquad\qquad\qquad (A)$$

$$(A) \rightleftharpoons H\cdot n + n\cdot \begin{matrix} CH_3 & H \\ | & | \\ C & -C\cdot e \\ | & | \\ CH_3 & H \end{matrix} \qquad e\cdot \begin{matrix} CH_3 & H \\ | & | \\ C & -C\cdot n \\ | & | \\ CH_3 & H \end{matrix}$$

$$(B)\ IMPOSSIBLE\ EXISTENCE \qquad (C)$$

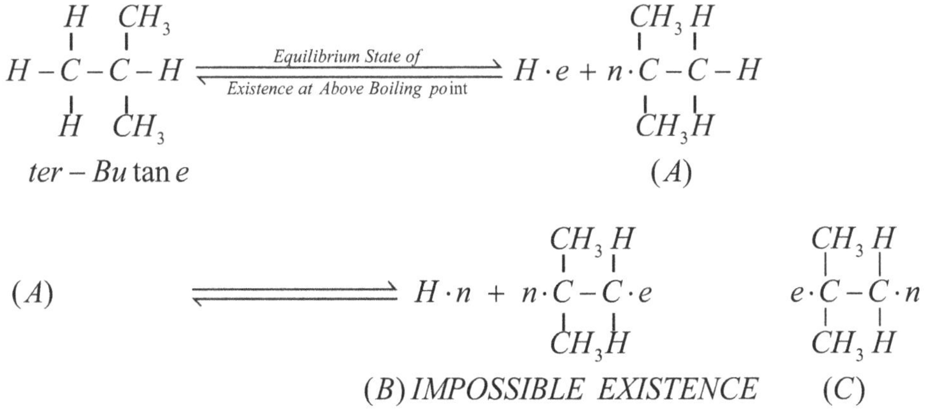

$$\underline{Overall\ equation}:\ ter-Butane \longrightarrow Cannot\ be\ cracked \qquad 3.136$$

Note that if CH_3 group was released, that to be expected, then cracking will be favoured, since a carbene which will rearrange will be formed to give propene and methane. For the stage above, the transfer species could not be rejected, because the C center that will release it cannot carry an electro-free-radical. (C) is the real activated state of isobutylene and not (B). As has been shown, monomers when activated, do not carry covalent charges or radicals indiscriminately on the Active C centers. How can one just put two dots (•) in front of for example two C centers of ethene, propene, butene and call them radicals? What the centers carry which are called RADICAL-PUSHING and RADICAL-PULLING in the New Frontiers, determine what types of radicals or charges the Active C centers carry. CH_3 groups above like H are radical-pushing groups of greater capacity than H. If the Equilibrium state of existence of the tertiary butane, had been such where the H held had been that from one of the CH_3 groups, then cracking will also be favoured as shown below.

Stage 1:

$$\begin{matrix} & H & CH_3 & \\ & | & | & \\ H- & C & -C & -H \\ & | & | & \\ & H & CH_3 & \end{matrix} \underset{\text{Existence at Above Boiling point}}{\overset{\text{Equilibrium State of}}{\rightleftharpoons}} H\cdot e + n\cdot \begin{matrix} H & CH_3 & \\ | & | & \\ C & -C & -H \\ | & | & \\ H & CH_3 & \end{matrix}$$

$$(A)$$

$$(A) \rightleftharpoons CH_3\cdot n + n\cdot \begin{matrix} H & CH_3 \\ | & | \\ C & -C\cdot e \\ | & | \\ H & H \end{matrix}$$

$$(B)$$

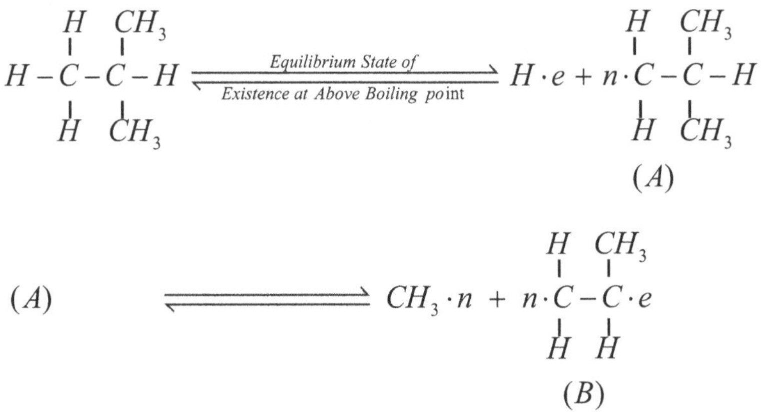

$$H\cdot e \; + \; CH_3\cdot n \rightleftharpoons CH_4$$

$$(B) \xrightarrow{Deactivation} H_2C = CH(CH_3) \; + \; Heat$$

POSSIBLE REACTION 3.137

The case above is the product, if it was iso-butane. In the stage above are the essences of EQUILIBRIUM STATES OF EXISTENCE and THE CENTRAL ATOM. Every compound that has the ability of showing its fingerprint, has one and only one fingerprint. It is not an issue of today one H is held, and tomorrow another H is held and so on. This makes no sense. The Central C atom in n-butane is different from the Central C atom in sec-butane and this in turn is different from that of ter-butane as shown below. It is the H atom directly connected to the Central C atom that is held in Equilibrium state of existence.

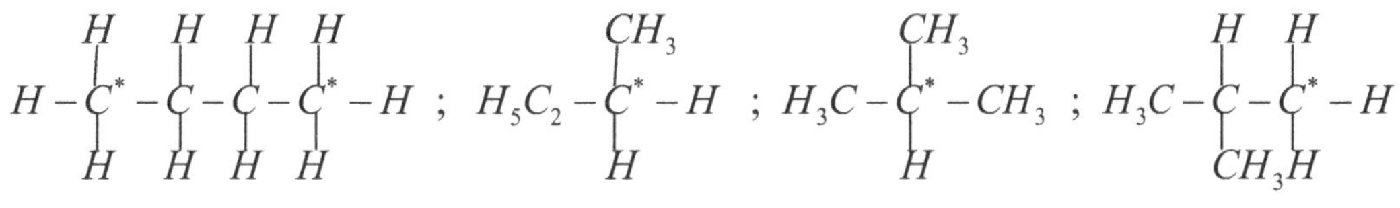

$$H-\overset{H}{\underset{H}{C^*}}-\overset{H}{\underset{H}{C}}-\overset{H}{\underset{H}{C}}-\overset{H}{\underset{H}{C^*}}-H \;;\; H_5C_2-\overset{CH_3}{\underset{H}{C^*}}-H \;;\; H_3C-\overset{CH_3}{\underset{H}{C^*}}-CH_3 \;;\; H_3C-\overset{H}{\underset{CH_3}{C}}-\overset{H}{\underset{H}{C^*}}-H$$

$(n-Butane)$ $(\text{sec}-Butane)$ $(ter-Butane)$ $(Iso-Butane)$

THE CENTRAL C – ATOM IN BUTANES 3.138a

In n-butane, any of the externally located C atoms can be the Central C atom. In sec-butane, there is only one, different also from the only one in t-butane. In iso-butane the Central C atom is carrying three H atoms and an iso-propyl group instead of n-propyl group as in n-butane. All this possibilities were provided by Nature, so that one can be able to get different kinds of products from C_4H_{10}. With n-butane, ethene is obtained along with ethane in the first stage. Sec-butane can also be cracked. With iso-butane, propene is obtained along with CH_4 when cracked. In that Stage above [Equation 3.137], worthy of note is that, instead of the H being released in the second step, it is CH_3 group that is released as nucleo-free-radicals, because based on the New Frontiers, the followings are important to note.

$$> \; H_{11}C_5\cdot e \; > \; H_9C_4\cdot e \; > \; H_7C_3\cdot e \; > \; H_5C_2\cdot e \; > \; H_3C\cdot e > H\cdot e$$

Order of radical – pushing capacity of Alkylane groups electro – free – radically

$$H\bullet n \; > \; n\cdot CH_3 \; > \; n\cdot C_2H_5 \; > \; n\cdot C_3H_7 \; > \; n\cdot C_4H_9 \; > \; n\cdot C_5H_{11} \; >$$

Order of radical – pushing capacity of Alkylane groups nucleo – free – radically

3.138b

When a group is to be released or rejected from a group of groups, it is the SMALLEST that is given out. You cannot give out the biggest from a group of all what you have. You give the smallest one at time in that order. It is H that is first released electro-free-radically amongst all the groups above being the smallest in capacity. Between H, CH_3 and C_2H_5, it is C_2H_5 which is the smallest that is first released nucleo-free-radically. One can be amazed at the WONDERS of NATURE. It looks like COMMON SENSE, but not at all.

Just as n-propane can be cracked, so also Iso-propane can be cracked. Isobutylene cannot be obtained from tertiary butane and cannot also be obtained from secondary butane. Sec-butane can be cracked as shown below.

Stage 1:

$$\underset{Sec-Butane)}{H-\underset{CH_3}{\overset{C_2H_5}{C}}-H} \underset{\text{Existence at Above Boiling point}}{\overset{\text{Equilibrium State of}}{\rightleftharpoons}} \underset{(A)}{H\cdot e + n\cdot\underset{CH_3}{\overset{C_2H_5}{C}}-H}$$

$$(A) \rightleftharpoons H_5C_2\cdot n \; + \; \underset{(B)}{n\cdot\underset{CH_3}{\overset{H}{C}}\cdot e}$$

$$H\bullet e \; + \; H_5C_2\bullet n \rightleftharpoons C_2H_6$$

$$(B) \xrightarrow[\text{with release of Heat}]{\text{Deactivation}} Ethene(C) \; + \; Heat$$

$$\underline{Overall\ equation}: Sec-Butane \longrightarrow C_2H_6 \; + \; H_2C=CH_2 \; + \; Heat \qquad 3.139$$

Note here again that the group that is rejected is not coming from C_2H_5. The reason is because, when done the equation will not exist. It is C_2H_5 group that is rejected nucleo-free-radically to form a carbene which rearranges and deactivates to form ethene and release heat.

Butenes cannot be obtained from ***all the*** butanes, just like all the ***normal*** members after ethane cannot give their corresponding 1-Alkenes. They can only be cracked to give ethene and lower hydrocarbons which will continue cracking at higher temperatures to finally give in some cases Carbon black and H_2. In general, it seems that all of the alkanes can be cracked to give selected alkenes.

(c) Cracking of Acetylene

Stage 1:

$$H-C\equiv C-H \underset{\text{Acetylene all the time}}{\overset{\text{Equilibrium State of}}{\rightleftharpoons}} H\cdot e + n\cdot C\equiv C-H \quad (A)$$

$$(A) \underset{\text{at very High Temperatures}}{\overset{\text{Equilibrim State of}}{\rightleftharpoons}} H\cdot n + n\cdot C\equiv C\cdot e \quad (B)$$

$$(B) \underset{\text{Existence at Higher temperatures}}{\overset{\text{Equilibrium State of}}{\rightleftharpoons}} 2n\cdot \overset{\overset{e}{\cdot}}{\underset{\underset{e}{\cdot}}{C}}\cdot n \quad (C)$$

$$H\cdot e + H\cdot n \underset{\text{Existence}}{\overset{\text{Equilibrium State of}}{\rightleftharpoons}} H_2$$

$$2(C) \xrightarrow{\text{Deactivation}} 2:\overset{\overset{en}{\cdot}}{C}\cdot nn + Energy \quad (Carbon\ Black) \qquad 1.140a$$

$$\underline{Overall\ Equation}: H-C\equiv C-H \longrightarrow Carbon\ Black + H_2 + Heat \qquad 1.140b$$

It is important to note how the stage or stages take(s) place in an orderly manner. In the last step above, energy is released due to deactivation. One has already shown where coke was obtained using for example, the high temperature reaction of Mg with CO. With hydrocarbons, the main carbon element obtained has largely been carbon black. No activated carbon can be obtained during cracking. Coal is most unique, because this is yet to be obtained synthetically. Sawdust from some woods may carry them. It can only be obtained synthetically if a molecule or mineral that carries it exist. Coal itself is a natural mineral.

Cracking of Hydrocarbons is a major subject area of Chemistry at the Advanced level. CHEMISTRY is too much a subject area to comprehend. Just above within the HYDROCABON FAMILY TREE, one covered just a grain of sand in Thermal decompositions of Hydrocarbons- the Alkanes, and a very small part of Alkenes and Alkynes. These introductions will not be complete if an additional kind is not considered. These also form a large subject area, because, there are different type of ringed hydrocarbons. One will consider just one of many contained in the New Frontiers.

3.4.4 Decomposition of Some Cyclic Hydrocarbons

(a) Decomposition of Cyclopentadiene at 600^0-950^0C

Consider the decomposition of cyclopentadiene a unique heavily unsaturated hydrocarbon which can be resonance stabilized discretely. It is nucleophilic (Female) in character. Its Equilibrium state of existence is as follows.

HC — CH ‖ ‖ HC CH (ring closed by CH_2) ⇌ H •e + n•C CH (HC — CH ring closed by CH_2) AND NOT H C CH (HC — CH ring closed by e•CH)

Not a cyclopentadienyl group Cyclopentadienyl group

3.141

When I_2 like some others are made to react with the cyclopentadiene when kept in Stable state of existence, the H removed is not that held in Equilibrium state of existence, but that in CH_2 and ***that is what makes it cyclopentadienyl in character.*** When heated, above 1000K, many kinds of products have been reportedly obtained[25-30]. The products include ***acetylene and H-absorption profiles*** at **1260 to 1600K** with pressures between 0.7 and 5.6 bar in a single pulse shock tube[25]. Another found ***benzene, acetylene and ethylene*** at **1300 to 1700K** in a small flow tube reactor[26]. In another case, in order of abundance, the products obtained were ***acetylene, ethylene, methane, allene, propyne, butadiene, propylene and benzene*** at **1080 to 1550K** and pressures behind the shock was between 1.7 to 9.6 atm.[27]. Thermal decomposition of cyclopentadiene to c-C_5H_5 (cyclopentadienyl radical) and H and the reverse, were studied quantum chemically at the G2M level of theory[28]. The ring opening of the cyclopentadienyl radical was found to be the crucial step in the mechanism[27, 28,30]. Even **Indene, benzene and naphthalene** were found to be the major products of decomposition in a laminar flow reactor operating in the temperature range of **600 to 950^0C**[30]. Obviously, the decomposition of the cyclopentadiene starting via Decomposition mechanism is impossible, since it has no point of scission to start with, and it cannot undergo any rearrangement to give another product. Its decomposition can only take place starting with Equilibrium mechanism. When held in Equilibrium state of existence, as shown in the last equation above, the group held is not a cyclopentadienyl group as mistakenly universally thought to be, but something else.

Shown below is the mechanism of decomposition of cyclopentadiene when all the fractions are held in Equilibrium state of existence.

Stage 1:

EQUILIBRIUM MECHANISM

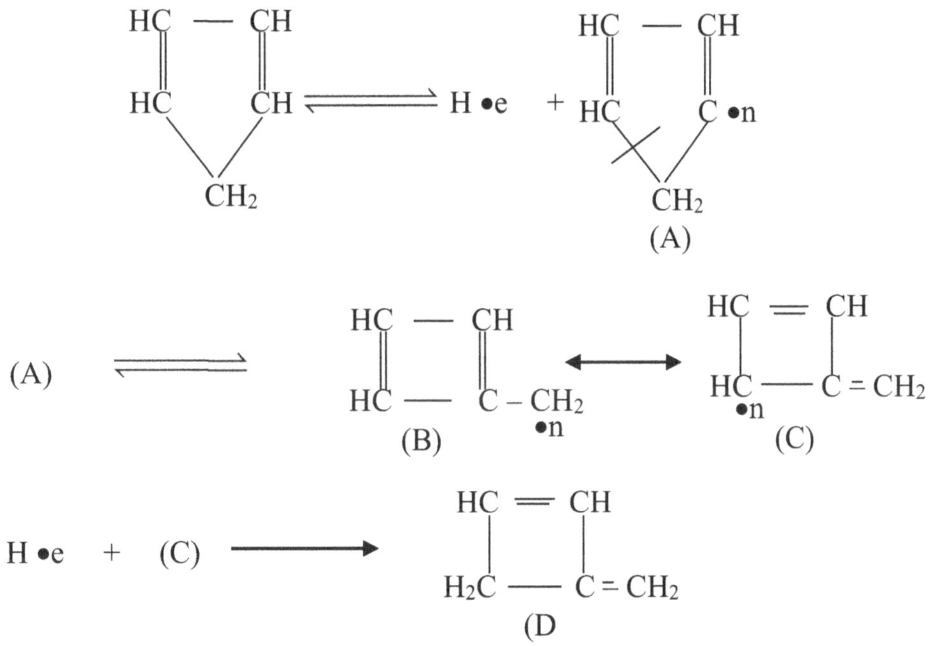

HC — CH ‖ ‖ HC CH (ring with CH_2) ⇌ H •e + HC — CH ‖ ‖ HC C •n (ring with CH_2, scission mark)

(A)

(A) ⇌ HC — CH ‖ ‖ HC — C – CH_2 •n (B) ⟷ HC = CH | | HC •n — C = CH_2 (C)

H •e + (C) ⟶ HC = CH | | H_2C — C = CH_2 (D) 3.142a

Overall equation: 2Cyclopentadiene ⟶ 2(D) 3.142b

The (D) formed is a four-membered ring too strained to exist as a ring, since it has close to the Max RSE in the ring. It has it because of the load the ring is carrying in addition to the visible π-bond. The load is the methylene group cumulatively placed on the ring. Because of the load, the ring is instantaneously opened in the next stage as shown below, since it has one point of scission. All these are new concepts which will become fully clear and simple to understand in the New Frontiers, since one is using only one example to show the essence of the New Frontiers.

DECOMPOSITION MECHANISM

Stage 2:

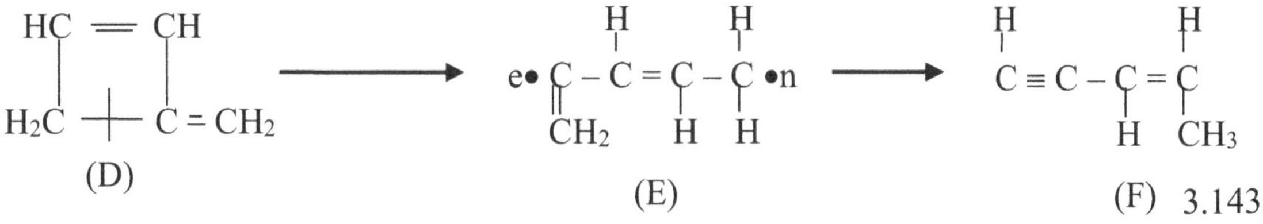

HC = CH | | H_2C + C = CH_2 (D) ⟶ e• C(=CH_2) – C = C(H)(H) – C(H)(H) •n (E) ⟶ C(H) ≡ C – C(H)(H) = C(H)(CH_3) (F) 3.143a

Overall equation: 2Cyclopentadiene ⟶ (E) + (F) 3.143b

A fraction of the (E) formed via instantaneous opening of the ring, is made to undergo what the New Frontiers calls ***Molecular rearrangement of the THIRD KIND of the first type***, which is that in which H•e (Atom) is moved from one C center to the next carbon center without breaking any of the laws of Nature as shown above to form a stable molecule (F). The remaining fraction of (E) which could not undergo the rearrangement above, are forced to decompose into small pieces in a systematic order as shown below via Decomposition mechanism.

DECOMPOSITION MECHANISM

Stage 3a:

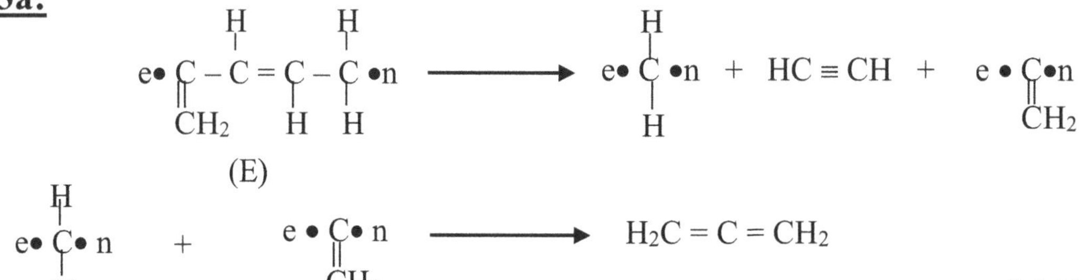

$$\underset{(E)}{e\bullet\overset{H}{\underset{CH_2}{\overset{|}{\underset{||}{C}}}}-C=\overset{}{\underset{H}{\underset{|}{C}}}-\overset{H}{\underset{H}{\overset{|}{\underset{|}{C}}}}\bullet n} \longrightarrow e\bullet\overset{H}{\underset{H}{\overset{|}{\underset{|}{C}}}}\bullet n + HC\equiv CH + e\bullet\underset{CH_2}{\underset{||}{C}}\bullet n$$

$$e\bullet\overset{H}{\underset{H}{\overset{|}{\underset{|}{C}}}}\bullet n + e\bullet\underset{CH_2}{\underset{||}{C}}\bullet n \longrightarrow H_2C=C=CH_2 \qquad 3.144a$$

Overall equation: 2Cyclopentadiene ⟶ Acetylene + Propyne + (F) 3.144b

The stable (F) formed is also made to exist in Equilibrium state of existence and decomposed as shown below.

EQUILIBRIUM MECHANISM

Stage 3b:

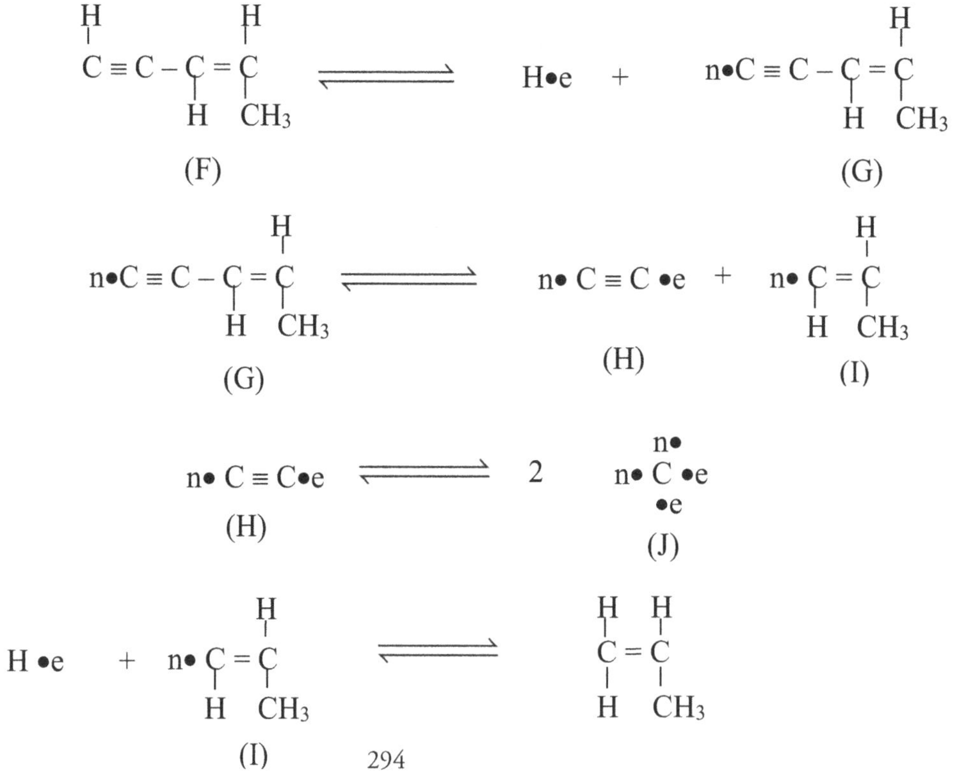

$$\underset{(F)}{\overset{H}{\overset{|}{C}}\equiv C-\underset{H}{\underset{|}{C}}=\overset{H}{\underset{CH_3}{\overset{|}{\underset{|}{C}}}}} \rightleftharpoons H\bullet e + \underset{(G)}{n\bullet C\equiv C-\underset{H}{\underset{|}{C}}=\overset{H}{\underset{CH_3}{\overset{|}{\underset{|}{C}}}}}$$

$$\underset{(G)}{n\bullet C\equiv C-\underset{H}{\underset{|}{C}}=\overset{H}{\underset{CH_3}{\overset{|}{\underset{|}{C}}}}} \rightleftharpoons \underset{(H)}{n\bullet C\equiv C\bullet e} + \underset{(I)}{n\bullet\underset{H}{\underset{|}{C}}=\overset{H}{\underset{CH_3}{\overset{|}{\underset{|}{C}}}}}$$

$$\underset{(H)}{n\bullet C\equiv C\bullet e} \rightleftharpoons 2\ \underset{(J)}{n\bullet\overset{n\bullet}{\underset{\bullet e}{C}}\bullet e}$$

$$H\bullet e + \underset{(I)}{n\bullet\underset{H}{\underset{|}{C}}=\overset{H}{\underset{CH_3}{\overset{|}{\underset{|}{C}}}}} \rightleftharpoons \overset{H}{\underset{H}{\overset{|}{\underset{|}{C}}}}=\overset{H}{\underset{CH_3}{\overset{|}{\underset{|}{C}}}}$$

$$2 \quad \overset{n\bullet}{\underset{\bullet e}{n\bullet C\bullet e}} \longrightarrow 2 \quad :\overset{\bullet en}{C}\bullet nn$$

(J) Carbon Black 3.145a

Overall equation: 2Cyclopentadiene ⟶ Acetylene + Propyne + Propene + 2 Carbon Black 3.145b

One had expected to see acetylene or cumulenic units when decomposed via Equilibrium mechanism. However, based on the operating conditions (600^0-950^0C) close to those used in the Petroleum industries for Cracking of hydrocarbons (above 1000^0C to 1750^0C), one is not surprised at the products obtained above based on the applications of the current developments in the NEW FRONTIERS. When the operating conditions are such that all the cyclopentadienes are held in Equilibrium state of existence, the products obtained above are essentially the main products. The system above is a series/parallel network system. In the first stage, methylene cyclobutene (D) was first obtained. In stage 2, the (D) decomposed into (E) via instantaneous opening of the ring as shown followed by what is called ***Molecular rearrangement of the Third Kind in the New Frontiers.*** All these will become very clear in the New Frontiers. The example above is an illustration of what are to be seen in the New Frontiers. Half of (E) rearranged to Methyl vinyl acetylene (F), leaving the remaining half or fraction which could not rearrange behind in its activated state to decompose. In stages 3a and 3b, both in parallel, i.e. at the same time, the fraction of (E) left and (F) formed decompose to give the products shown. ***The number of stages above is indeed four, because the rearrangement of the allene produced in Stage 3a to methyl acetylene (Propyne) was not shown.*** This will shortly be shown. Where some products such as benzene and butadiene for example were observed, these were formed based on their operating conditions-temperature, pressure, solvents used, presence of unwanted components, types of reactors, and so on. As will shortly be shown with the case of Toluene, the presence of benzene is a result of large number of moles of cyclopentadiene decomposed, for which three moles of acetylene were used to produce benzene as shown below.

Stage 4 or5:

$$HC \equiv CH \rightleftharpoons n\bullet \overset{H}{\overset{|}{C}} = \underset{\overset{|}{H}}{C} \bullet e$$

(A

$$(A) \quad + \quad HC \equiv CH \rightleftharpoons n\bullet \overset{H}{\overset{|}{C}} = \underset{\overset{|}{H}}{C} - \overset{H}{\overset{|}{C}} = \underset{\overset{|}{H}}{C} \bullet e$$

(B)

$$(B) \quad + \quad HC\equiv CH \rightleftharpoons n\bullet \overset{H}{\underset{}{C}} = \underset{H}{C} - \overset{H}{C} = \underset{H}{C} - \overset{H}{C} = \underset{H}{C}\bullet e$$

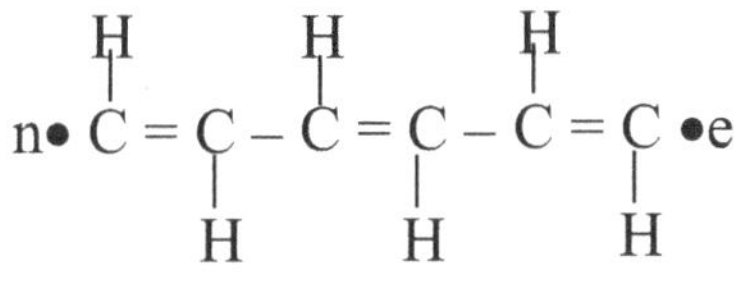

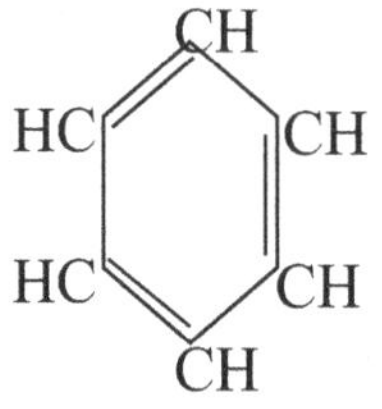

(C)

$$(C) \longrightarrow \text{Benzene (HC, CH ring)} \qquad 3.146a$$

Benzene

Overall equation: $3HC\equiv CH \longrightarrow$ Benzene 3.146b

Three moles of acetylenes in Activated state of existence based on the operating conditions (just a fraction since some are also in Equilibrium state of existence and probably none in Stable state of existence), combine together in one single stage to give benzene. Cyclobutadiene could not be formed because of the large SE in the ring. Reactions take place on molar basis molecularly. While one can observe that the ring has been "expanded" from five to six, as was observed during the reported ring expansion reactions in the thermal decomposition of tert-butyl-1,3-cyclopentadiene[31] and ring expansion in 1- or 2- or 3-methylcyclopentadiene radicals; quantum chemical and kinetic calculations[32], ***these have nothing to do with the mechanisms for ring expansion, wherein the original ring of five is taken as it is and worked on mechanically to expand it without opening the former ring. That is ring expansion.*** The reactions in Stage 4 or 5 above like the cases just referred to above have nothing to do with expansion of rings. One did not expand the cyclopentadiene ring to benzene. One clearly can see what is going on in Present –day Science, issues that should not arise if the mechanisms of how NATURE operates were known. One can clearly see the intrinsic use of the word "mechanical" above.

When a fraction of the cyclopentadiene is in Stable state of existence leaving the remaining fraction in Equilibrium state of existence, then the followings take place.

EQUILIBRIUM MECHANISM

Stage 1:

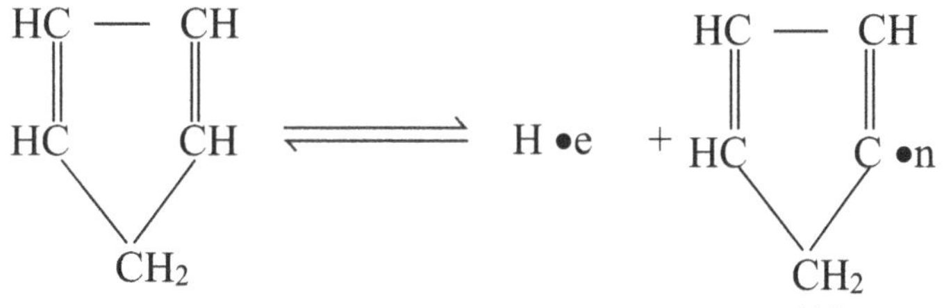

HC — CH
|| ||
HC CH ⇌ H •e + HC — CH (with C •n), CH_2
CH_2 (A)

H• e + Cyclopentadiene ⇌ (B) + H_2

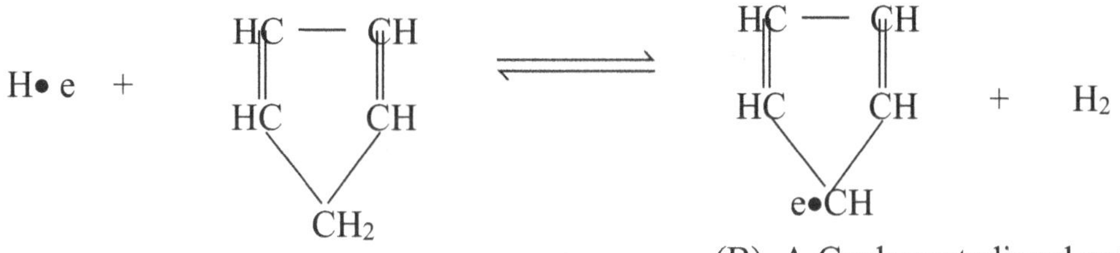

(B) A Cyclopentadienyl radical

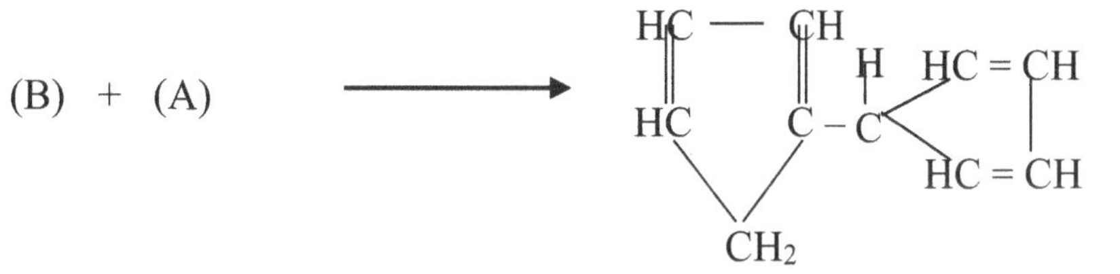

(B) + (A) ⟶ (C)

(C) 1-Cyclopentadienyl cyclopentadiene 3.147a

Overall equation: 2 Cyclopentadiene ⟶ H_2 + (C) 3.147b

The stage above is like the case of ammonia's decomposition.

Stage 2:

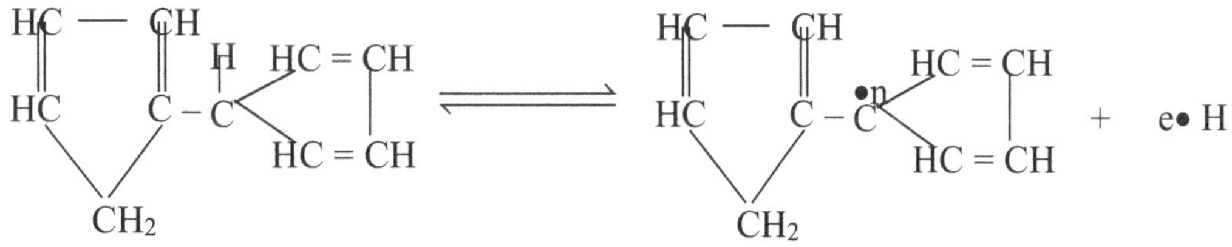

(C) 1-Cyclopentadienyl ⇌ (D) + e• H

(D) $\underset{\text{Stabilization}}{\overset{\text{Resonance}}{\rightleftharpoons}}$ (E)

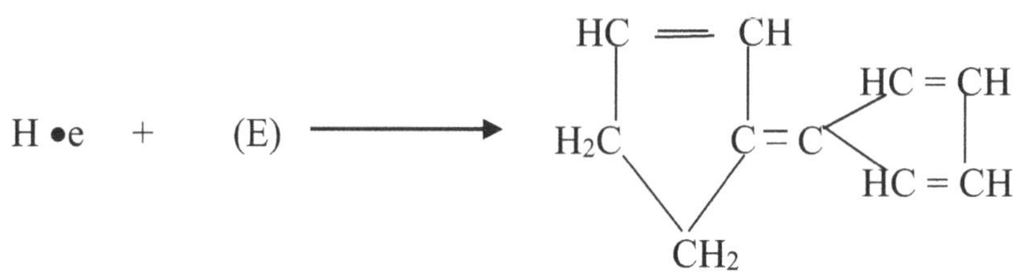

$$H \bullet e \;+\; (E) \longrightarrow$$ (F) Cyclopentadienylene Cyclopentene

3.148a

Overall equation: (C) ⟶ (F) 3.148b

Via resonance stabilization phenomenon, (C) molecularly rearranged to (F). (F) being so highly strained, because of the "methylene group" cumulatively placed to two rings, one of which has only one point of scission, decomposes as shown below. Only the ring with one internally located double bond has one point of scission. The second ring which is more strained has none.

DECOMPOSITION MECHANISM

Stage 3:

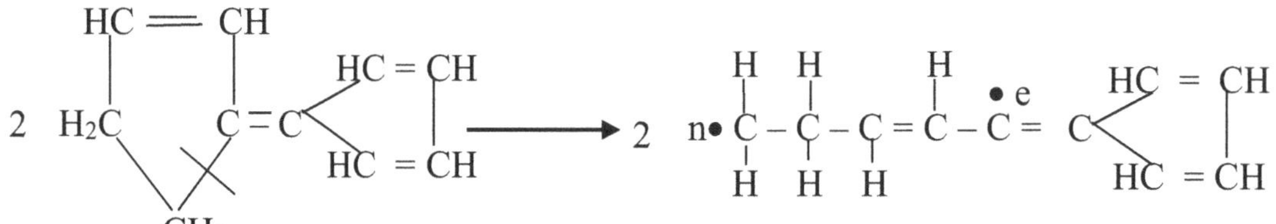

2 (F) Cyclopentadienylene Cyclopentene ⟶ 2 (G)

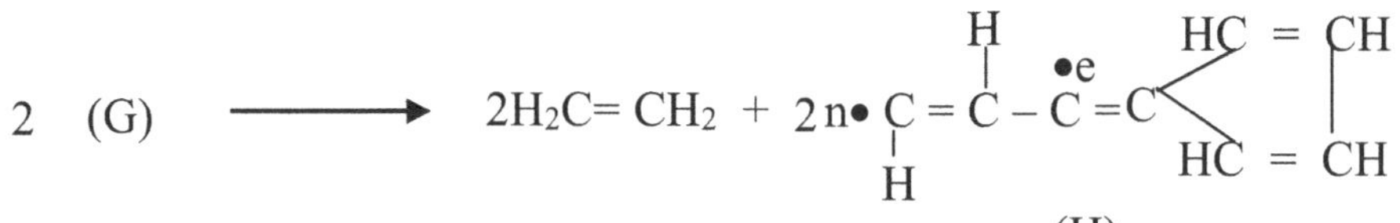

$$2\,(F) \longrightarrow 2\; n\bullet CH_2-CH_2-CH=C(\bullet e)-C=C\langle(HC=CH)(HC=CH)\rangle$$

(G)

$$2\;(G) \longrightarrow 2H_2C{=}CH_2 \;+\; 2n\bullet CH=CH-C(\bullet e)=C\langle(HC=CH)(HC=CH)\rangle$$

(H)

$$2\;(H) \longrightarrow 2\,HC{\equiv}CH \;+\; 2\,C(\bullet n)(\bullet e){=}C\langle(HC=CH)(HC=CH)\rangle$$

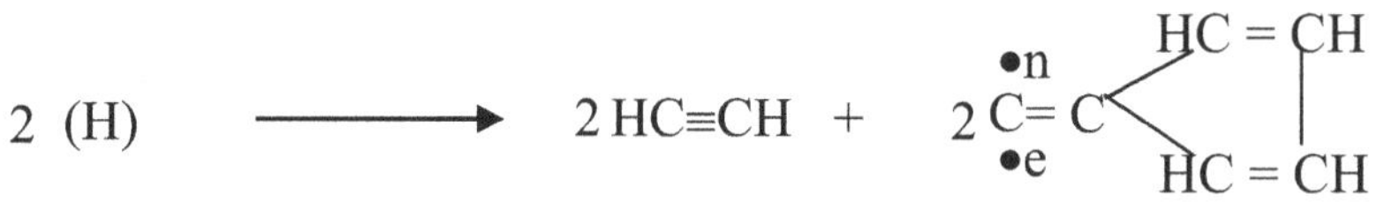

(I)

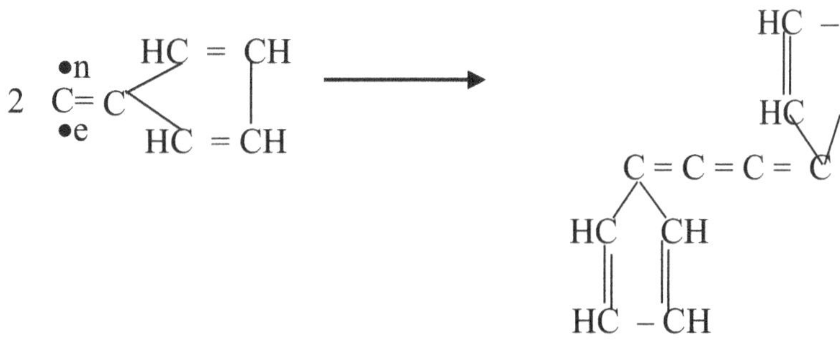

2 •n •e C=C< (HC = CH, HC = CH) ⟶ C = C = C = C (1,4-di-cyclopentadienyl)

(I*) 1,4-Di-cyclopentadienyl cumulene

3.149a

Overall equation: 2F ⟶ $2HC{\equiv}CH + 2H_2C{=}CH_2 + (I^*)$ 3.149b

Overall overall equation: 12Cyclopentadiene ⟶ $6H_2 + 6HC{\equiv}CH + 6H_2C{=}CH_2 + 3(I^*)$ 3.150

EQUILIBRIUM MECHANISM

Stage 4:

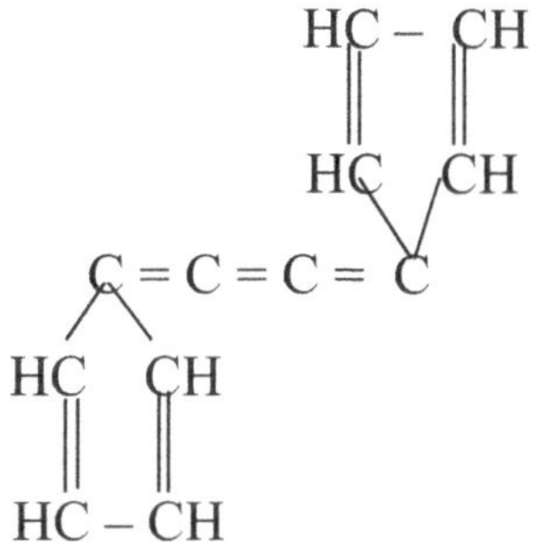

(I*) 1,4-Di-cyclopentadienyl cumulene $\underset{}{\overset{Activation}{\rightleftharpoons}}$ (J)

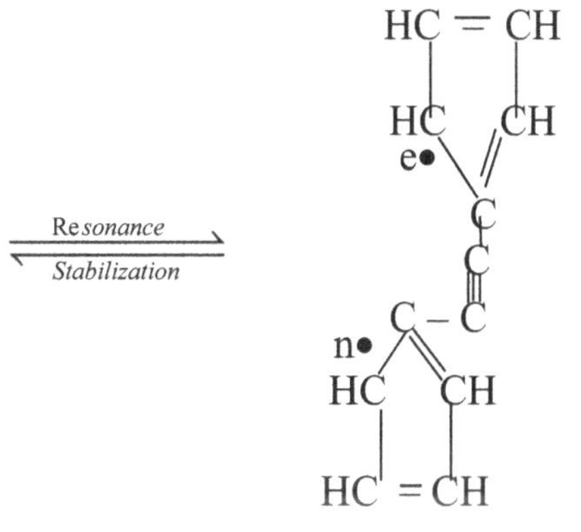

$\underset{Stabilization}{\overset{Resonance}{\rightleftharpoons}}$ (K)

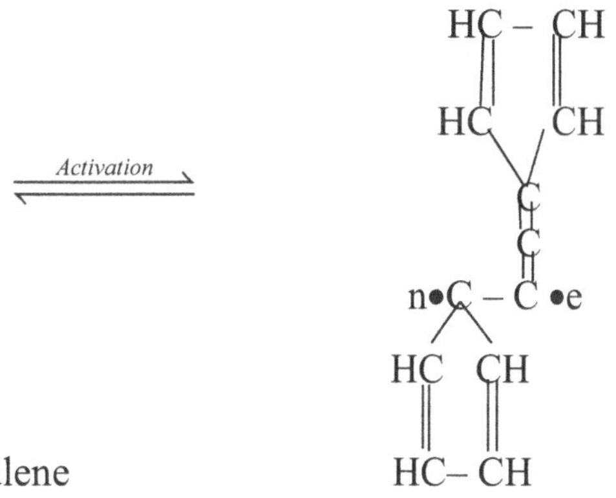

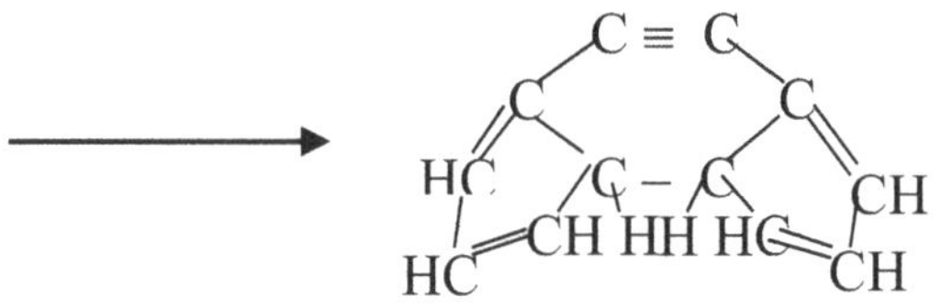

(L) 3.151a

Overall equation: (I) ⟶ (L) 3.151b

Stage 5:

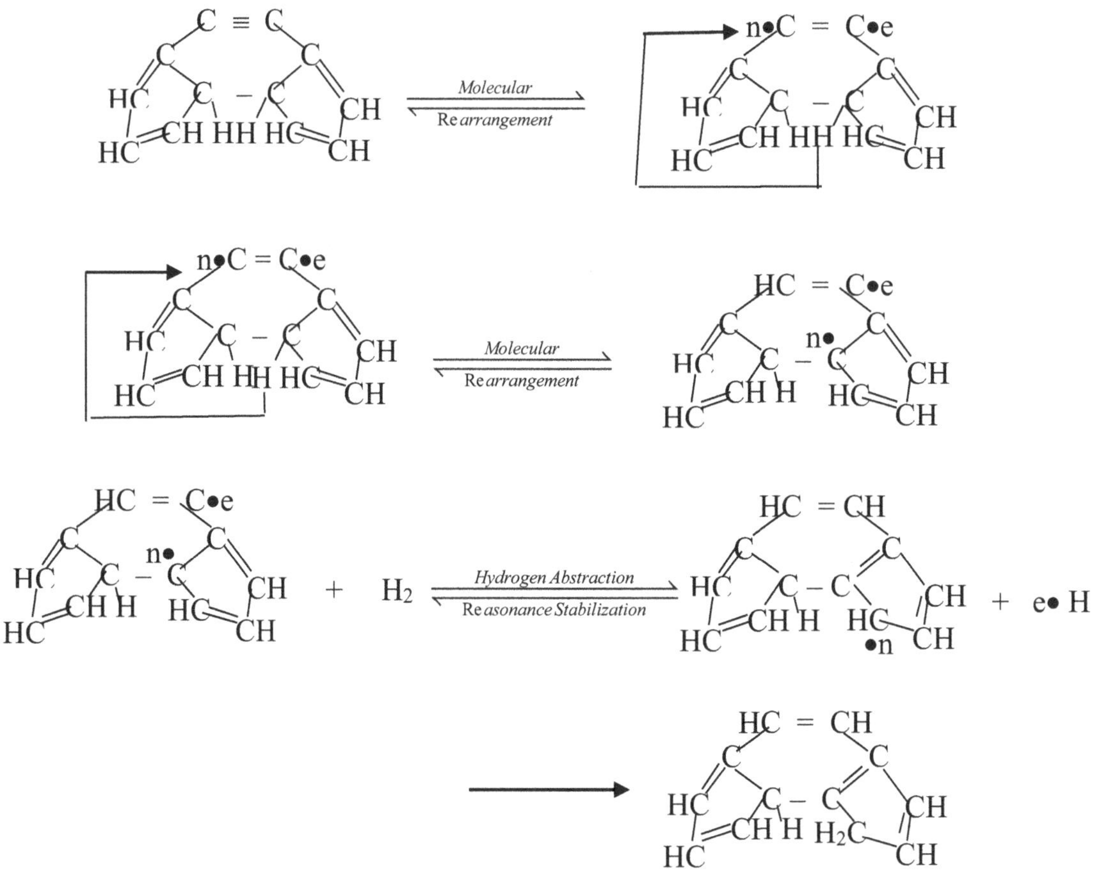

$\xrightleftharpoons[\text{Rearrangement}]{\text{Molecular}}$

$\xrightleftharpoons[\text{Rearrangement}]{\text{Molecular}}$

+ H_2 $\xrightleftharpoons[\text{Reasonance Stabilization}]{\text{Hydrogen Abstraction}}$ + e• H

(M) 3.152a

Overall overall equation: 12Cyclopentadiene ⟶ $6H_2C{=}CH_2$ + $6C_2H_2$ + 3(M) + $3H_2$ 3.152b

Stage 6:

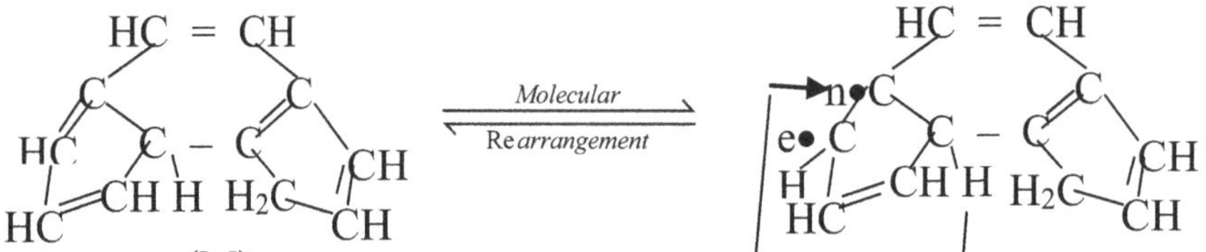

(M) $\xrightleftharpoons[\text{Rearrangement}]{\text{Molecular}}$

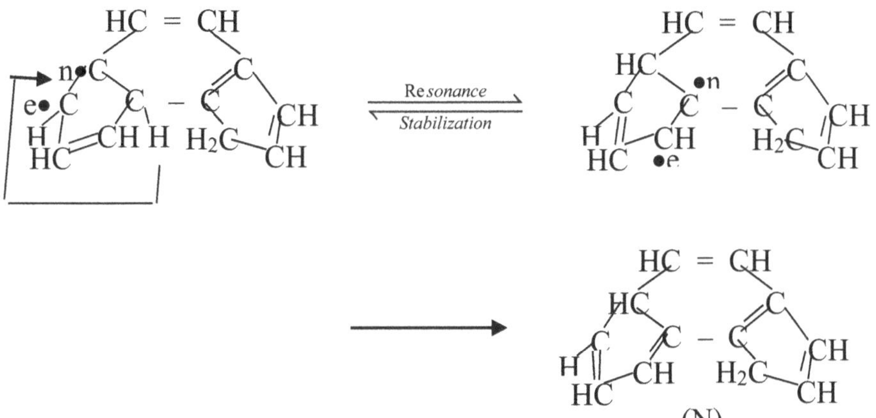

Resonance Stabilization

(N) 3.153

Stage 7:

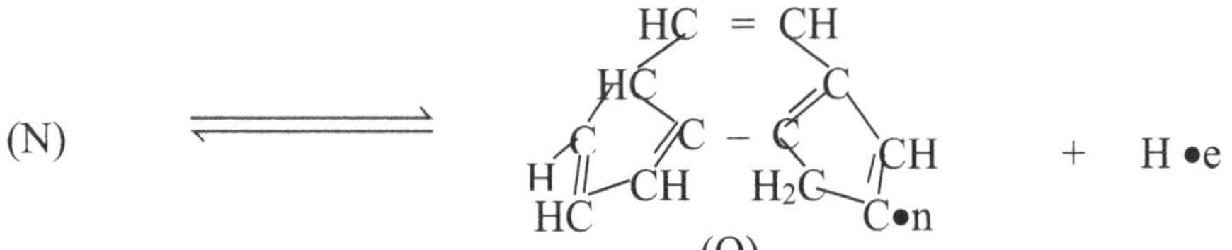

(N) ⇌ (O) + H •e

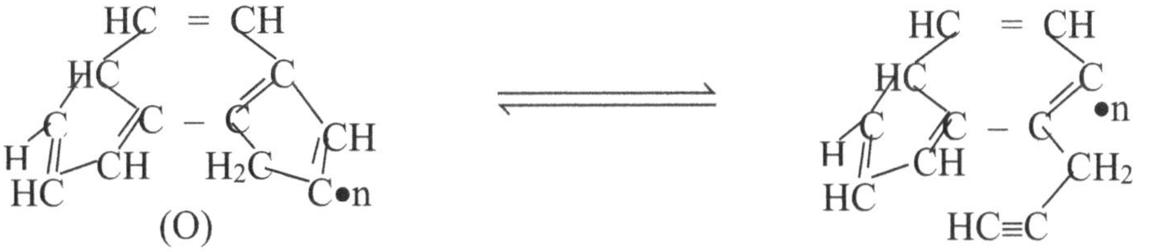

(O) ⇌ (P)

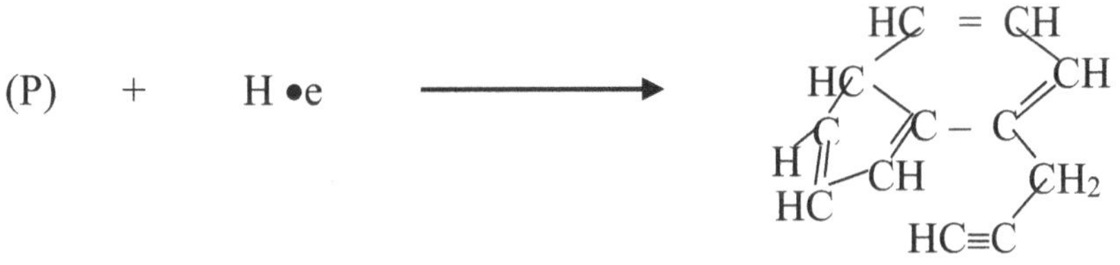

(P) + H •e ⟶ (Q) 3.154

Stage 8:

(Q) ⇌ (R)

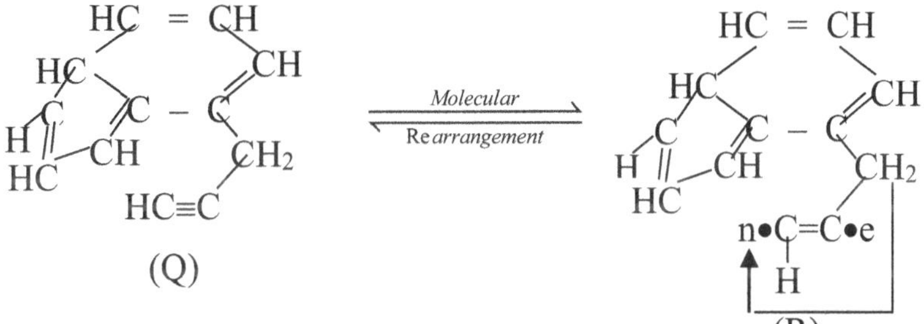

Molecular Rearrangement

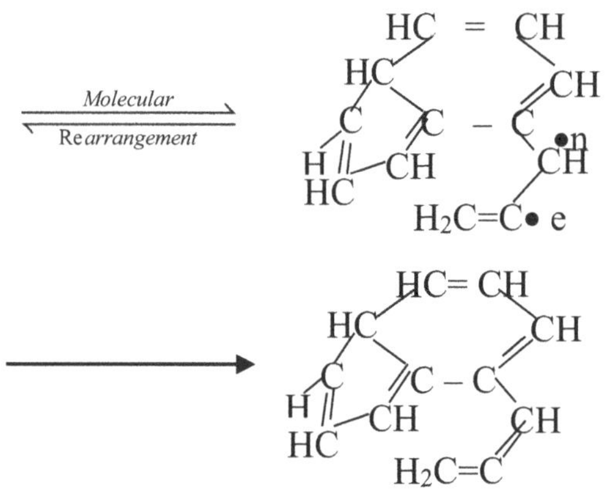

Molecular Rearrangement

HC = CH, HC, CH, C, C – C •n, H HC, CH, CH, $H_2C=C$• e

HC= CH, HC, CH, C, C – C, H HC, CH, CH, $H_2C=C$

(S) 3.155

While a simple cumulene such as allene will rearrange to a more stable acetylene, the acetylene above based on what it is carrying is less nucleophilic than the cumulene formed. In fact, the cumulene once formed cannot rearrange back to the former acetylene.

Stage 9:

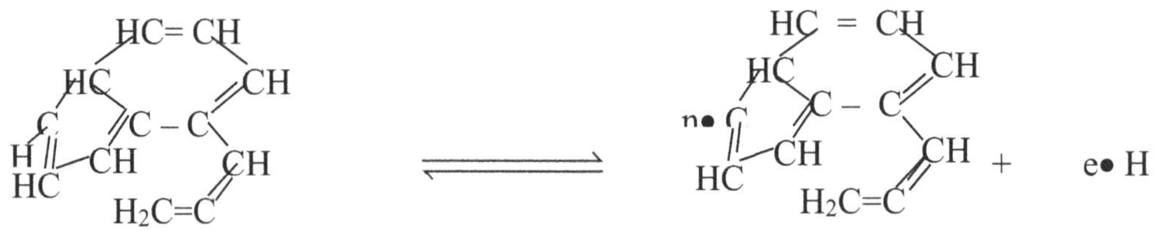

HC= CH, HC, CH, C, C – C, H HC, CH, CH, $H_2C=C$ ⇌ HC = CH, HC, CH, n• C, C – C, HC, CH, CH, $H_2C=C$ + e• H

(T)

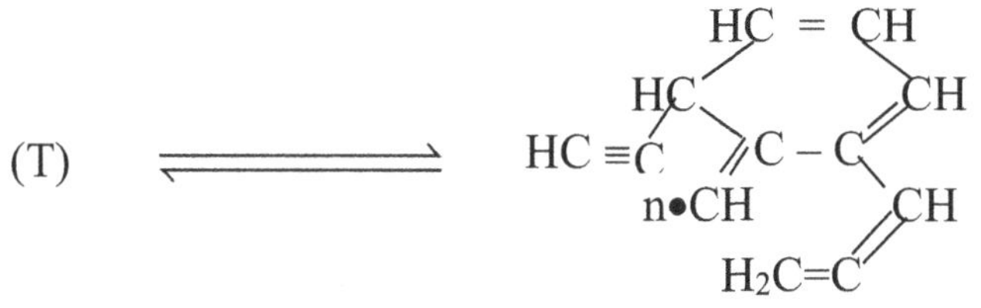

(T) ⇌ HC = CH, HC, CH, HC ≡C, C – C, n•CH, CH, $H_2C=C$

(U)

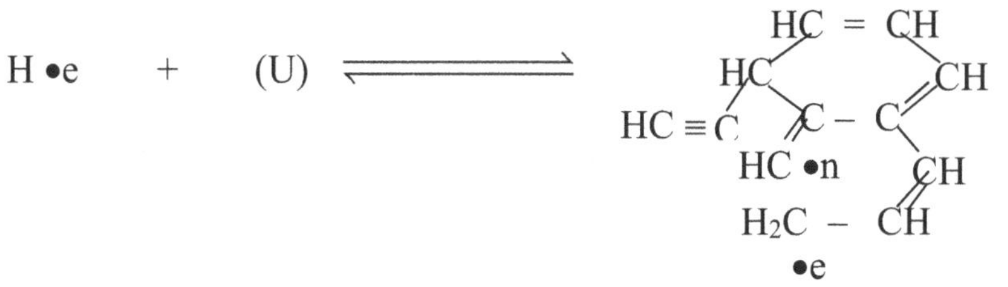

H •e + (U) ⇌ HC = CH, HC, CH, HC ≡ C, C – C, HC •n, CH, H_2C – CH, •e

(V)

⟶ HC = CH, HC, CH, HC ≡ C, C – C, HC, CH, H_2C – CH

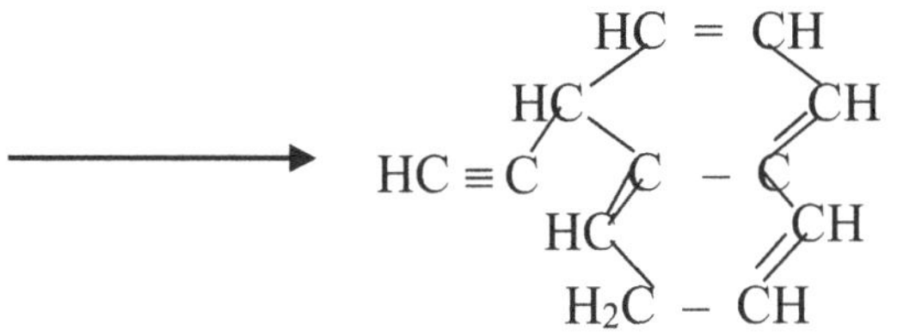

(W) 3.156

Stage 10:

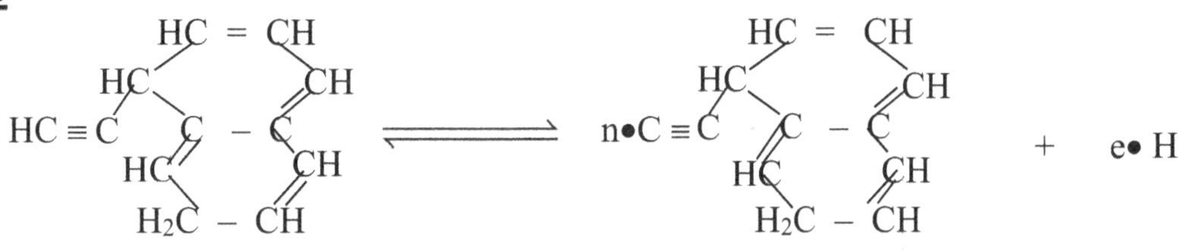

HC = CH, HC, HC ≡ C, C – C, CH, HC, CH, H_2C – CH (W) ⇌ n•C ≡ C (ring: HC = CH, HC, CH, C – C, HC, CH, H_2C – CH) (X) + e• H

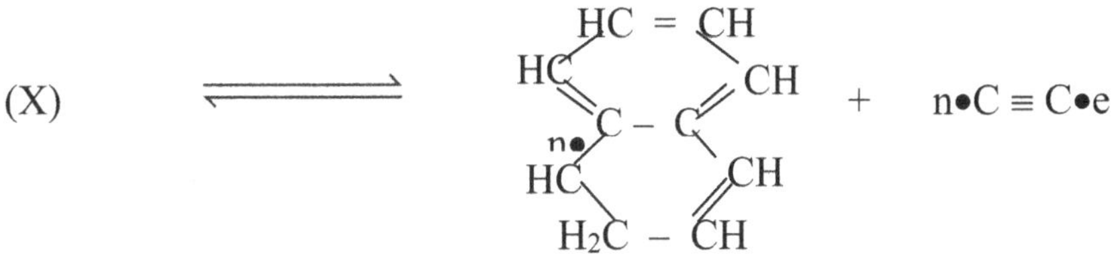

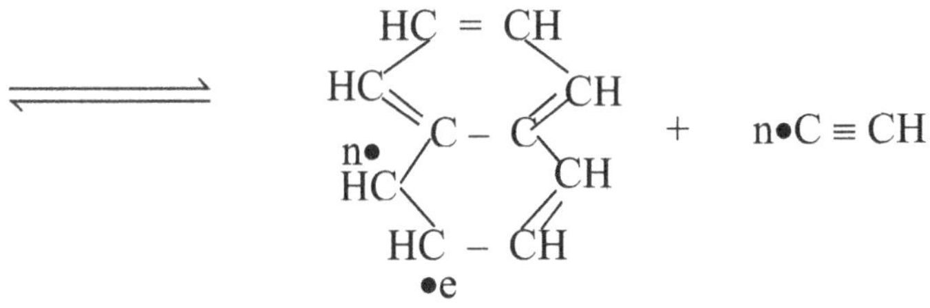

(X) ⇌ (HC = CH, HC, CH, n• C – C, HC, CH, H_2C – CH) (Y) + n•C ≡ C•e

⇌ (HC = CH, HC, CH, C – C, n• HC, CH, HC – CH •e) (Z) + n•C ≡ CH

H•e + n• C ≡ CH ⇌ H C ≡ CH

(HC = CH, HC, CH, C – C, n• HC, CH, HC – CH •e) ⟶ (HC = CH, HC, CH, C – C, HC, CH, HC – CH)

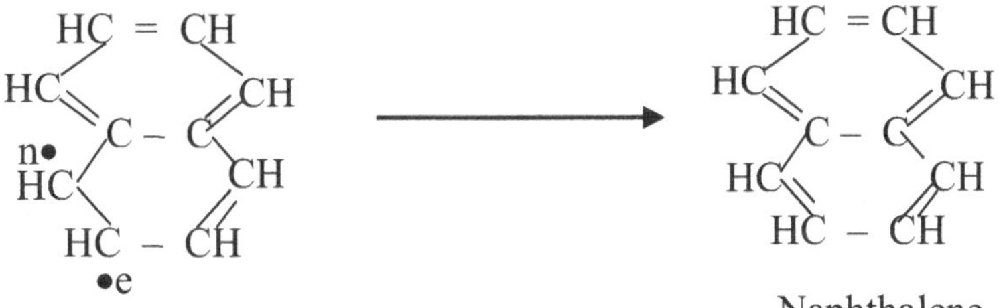

Naphthalene 3.157a

Overall overall equation: 12Cyclopentadiene ⟶ $6H_2C{=}CH_2$ + $9C_2H_2$ + $3H_2$ + $3C_{10}H_8$ 3.157b

Stage 11:

$$H_2C = CH_2 \rightleftharpoons H{\bullet}e + n{\bullet}CH = CH_2$$

$$n{\bullet}CH = CH_2 \rightleftharpoons n{\bullet}CH = HC{\bullet}e + H{\bullet}n$$

$$H{\bullet}e \quad + \quad H{\bullet}n \rightleftharpoons H_2$$

$$n{\bullet}CH = HC{\bullet}e \longrightarrow HC \equiv CH \qquad 3.158a$$

Overall overall equation: 12 Cyclopentadiene $\longrightarrow$ $15HC \equiv CH$ + $9H_2$ + $3C_{10}H_8$ 3.158b

Stage 12: Same as Stage 4 or 5 of Equation 3.146a.

Overall overall equation: 12Cyclopentadiene $\longrightarrow 9H_2 + 5C_6H_6 + 3C_{10}H_8$ 3.158b

Twelve stages coincidentally were involved above for the synthesis of naphthalene at such operating condition close to what obtains in the cracking of hydrocarbons in the Petroleum industry. Note that throughout the stages, no ring was expanded in size. Though 9,10-dihydrofulvalene exists as shown below, it cannot be obtained during decomposition of cyclopentadiene, based on the new foundations being laid in the New Frontier.

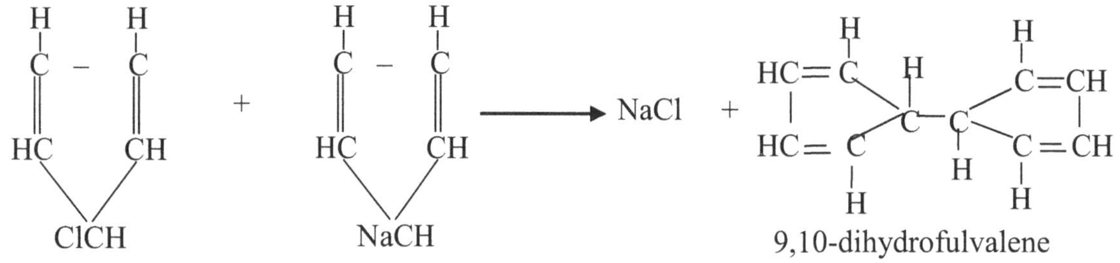

ClCH (cyclopentadienyl, H–C=CH / HC=CH ring) + NaCH (cyclopentadienyl, H–C=CH / HC=CH ring) $\longrightarrow$ NaCl + 9,10-dihydrofulvalene (two HC=CH–CH=CH rings joined at C(H)–C(H))

9,10-dihydrofulvalene

3.159

It is 1,10-dihydrofulvalene which was called 1-Cyclopentadienyl cyclopentadiene (C) in Stage 1 of Equation 3.147a that was first formed. In order to provide a point of scission for it, it was made to exist in Equilibrium state of existence to give cyclopendienylene cyclopentene in Stage 2, via resonance stabilization. Notice where the electro-free-radical came from to grab the nucleo-free-radical. It came from the second ring. In the third stage, decomposition began to start giving smaller products and most importantly (I^*), herein called 1,4-Di-cyclopentadienylene cumulene. It from this, naphthalene was formed. The first product was H_2 from Stage 1, followed by ethene and next acetylene and then the (I^*). In Stage 4, with (I^*) activated via one of the externally located activation centers which of course is the weakest nucleophilic center, rearranged the radicals via resonance stabilization to give well placed radicals which closed and formed a six-membered ring without yet opening the cyclopentadiene rings which are symmetrically placed on the ring (L). One can imagine how NATURE operates. None of the rings have been opened to form the six-membered ring carrying a triple bond well placed. At first one will wonder what the triple bond is doing there. In Stage 5, it was the triple bond the least nucleophilic of all the centers which are resonance stabilized that was activated to begin molecular rearrangement and in order to provide a productive stage and much more, H_2 formed from Stage 1 was partly abstracted to form (M). In the process, one of the double bonds was removed from one of the five-membered rings and placed on the six-membered ring. Until this point, no point of scission has yet been provided. In Stage 6, (M) underwent both molecular rearrangement and resonance stabilization to give (N) which still has no point of scission. However, via Equilibrium state of existence, the same five membered rings with only one double bond were finally opened in Stage 7. In Stage 9, the second ring was opened with commencement of formation of a six-membered ring. This continued to Stage 10 where the naphthalene was formed with release of acetylene. No soot was formed. In Stage 11, the ethene decomposed to acetylene and H_2. Finally, in Stage 11 all the acetylenes formed all along were now used to give benzene. The products obtained above will be the main products when the number of moles of cyclo-pentadiene is even. When odd, the situation is slightly different.

When the number of moles of cyclopentadiene involved in an "ideal reactor" is odd, then Indene and some other products begin to appear in the natural world. For the mechanism provided above, one used twelve moles of cyclopentadiene in a capillary reactor, the types that largely exist in NATURE. In the physical world of today, it does not matter whether it is odd or even, because most of the reactors are not close to what exist in Nature. For example, how many pumps are flexible physically? In fact, without going through the path of the route above, once benzene and the smaller components have been formed, naphthalene, anthracene and more complex rings can readily be obtained. For example, from acetylene alone, the followings can be obtained-

i) Benzene, using three moles
ii) Styrene, using four moles, from which Cyclooctatetraene is obtained, and this molecularly rearranges to styrene.
iii) And many more complex products when the acetylenes are combined with other components such as benzene.

It should be noted, that ethene is never a major product from methylene. The methylenes formed can be used with the benzene ring for example to give cycloheptatriene. This is where the real ring expansion takes place as shown below.

Stage 1:

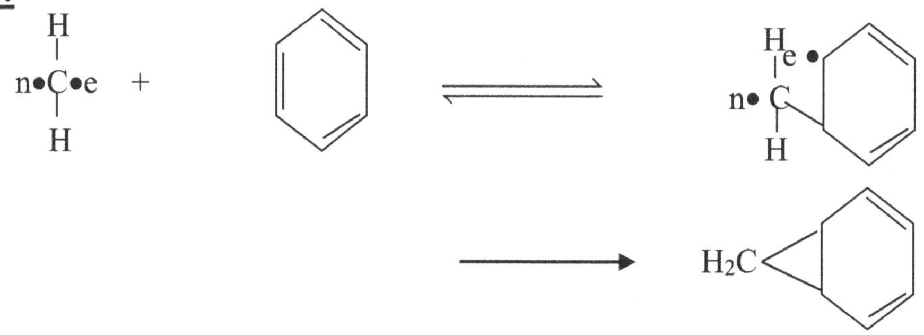

H
n•C•e + [benzene] ⇌ [H, e•, n•C, H adduct with ring]
H

⟶ H_2C< [bicyclic ring]

Stage 2: 3.160

H_2C< [bicyclic ring] ⇌ H_2C e• n [opened ring]

⇌ n HC e• [ring]

⟶ CH_2 [seven-membered ring]

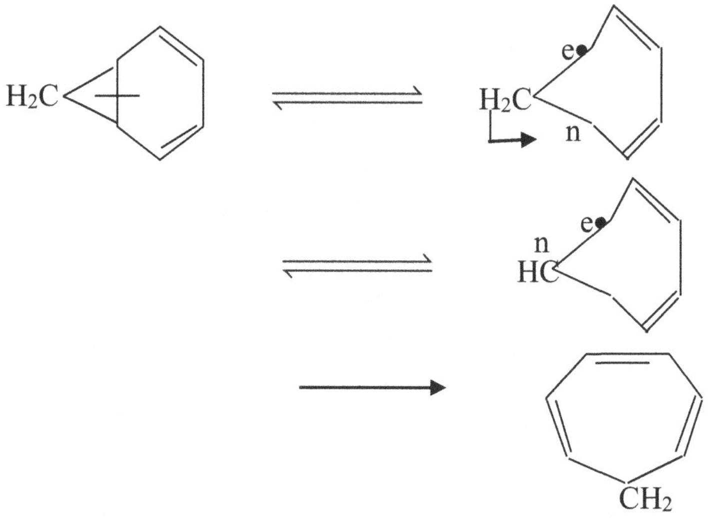

Cycloheptatrie 3.161a

Overall Equation: Methylene + Benzene ⟶ Cycloheptadiene

3.161b

The methylene can be used along with acetylene and benzene as follows.

Stage 1:

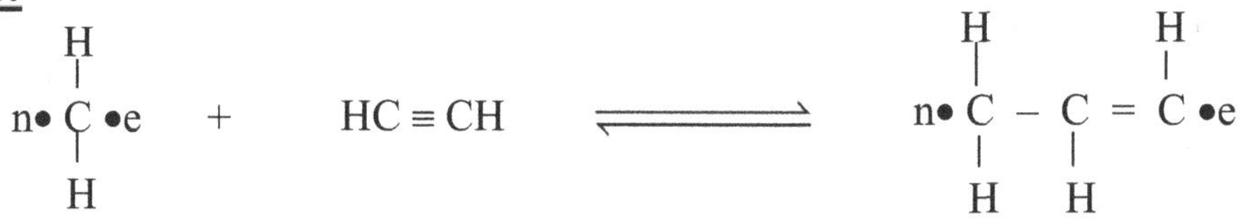

$$\text{n}\bullet\overset{\text{H}}{\underset{\text{H}}{\text{C}}}\bullet\text{e} \quad + \quad \text{HC}\equiv\text{CH} \quad \rightleftharpoons \quad \text{n}\bullet\overset{\text{H}}{\underset{\text{H}}{\text{C}}} - \underset{\text{H}}{\text{C}} = \overset{\text{H}}{\text{C}}\bullet\text{e}$$

(A)

(A) + [benzene] ⟶ HC, HC, H_2C, H, H [fused ring product]

(B) 3.162

The methods of addition of components in systems are very important. In some, all can be put into the system at the same time. In others, they can be added one at a time. All these depend on the level of understanding the mechanisms of the reactions.

Stage 2:

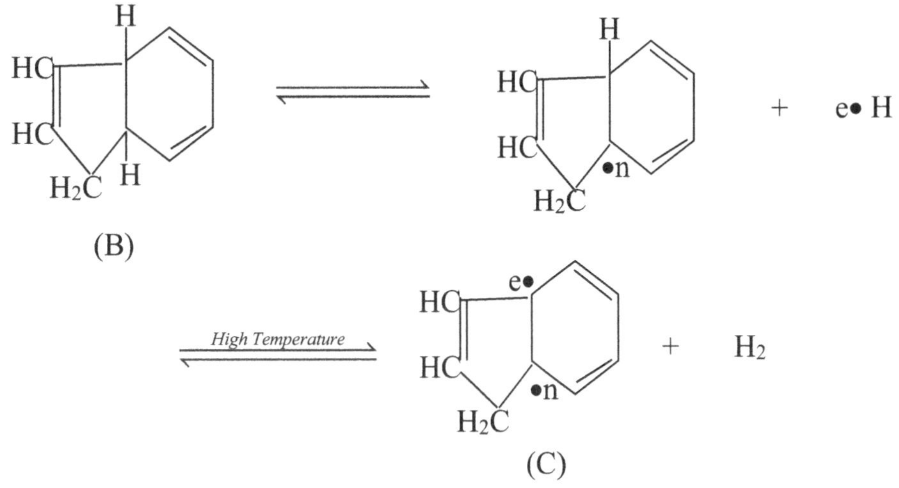

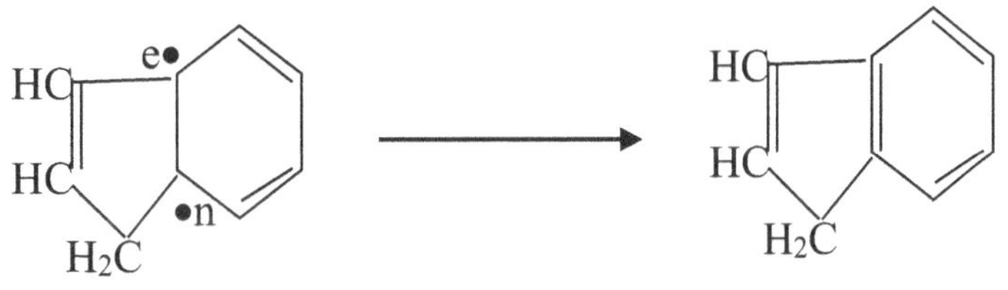

(B) ⇌ [structure with •n] + e• H

High Temperature ⇌ (C) + H_2

(C) ⟶ INDENE 3.163a

Overall equation: Methylene + Acetylene + Benzene ⟶ I*ndene* + H_2 3.163b

When benzene and two moles of acetylene are available in the system, the followings are obtained.

Stage1:

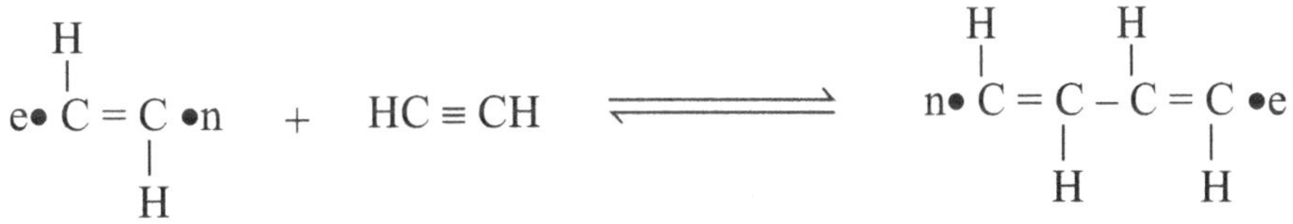

e• C(H) = C(H) •n + HC ≡ CH ⇌ n• C(H) = C(H) – C(H) = C(H) •e (A)

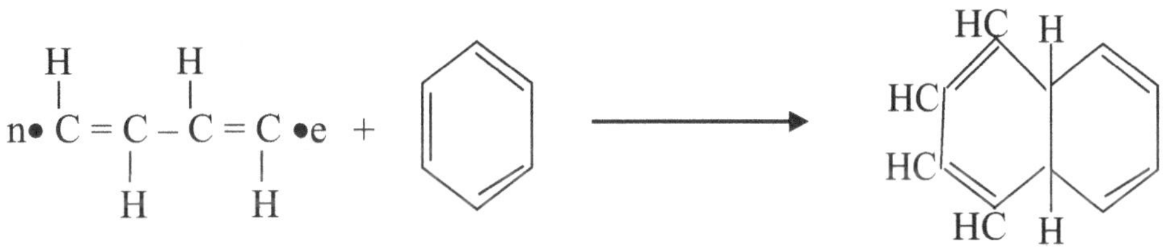

n• C(H) = C(H) – C(H) = C(H) •e (A) + [benzene] ⟶ (B) 3.164

Stage 2:

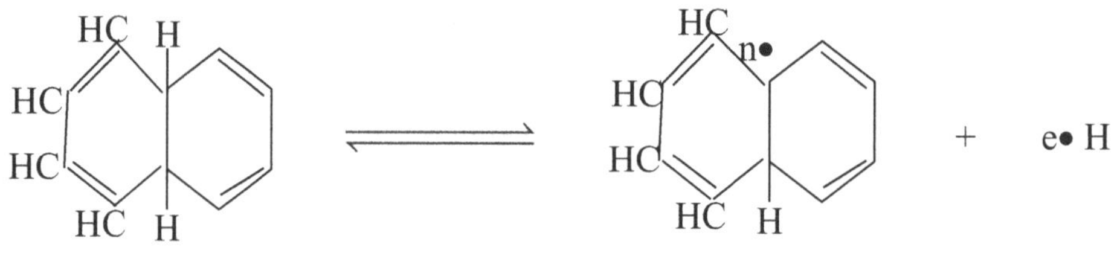

(B) ⇌ + e• H

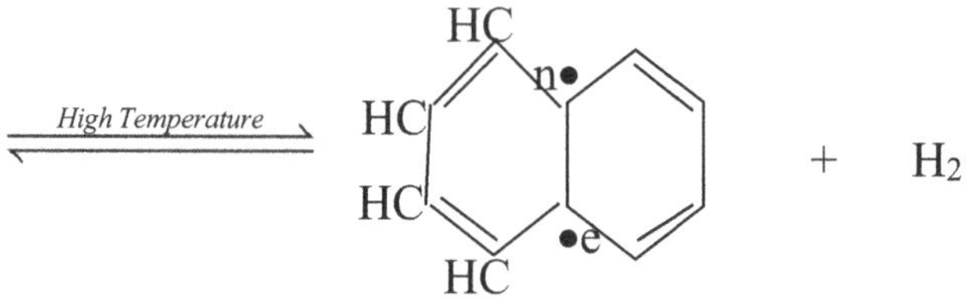

High Temperature ⇌ (C) + H_2

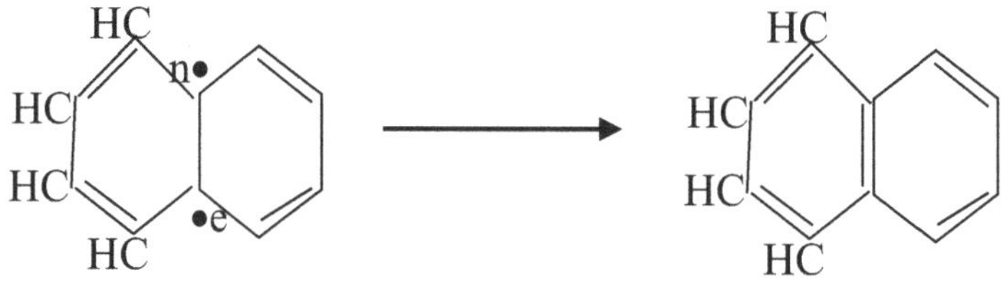

(C) ⟶ NAPHTHALENE 3.165a

Overall equation: 2Acetylene + Benzene ⟶ Naphthalene + H_2 3.165b

At the operating conditions, the monomers are more in self-activated states. It is the more nucleophilic that diffuses to the less nucleophilic monomer using the electro-free-radical ends.

(b) Molecular rearrangement of the First kind of Third type using Allene.

All types of molecular rearrangement of the first kind cannot all yet be shown. The first kind of the first type has already been shown when the need arose. In all of them, notice that within the hydrocarbon family, the transfer species were largely coming from the alkylane groups (In present-day Science they are called alkyl groups). They are called alkylane or alkanyl because their origin is from alkanes. Based on the New Frontiers, there are also those coming from the alkene family and these are called alkyl-ene or alkenyl groups. There are also those coming from the alkyne family and these are called alkyl-yne or alkynyl groups. The molecular rearrangement of the first kind of the third type is that in which the transfer species is coming from an alkyl-yne group as shown below and referred to in Equation 3.144a where allene rearranged to methyl acetylene (Propyne).

Stage1:

$$\underset{(Allene)}{\begin{matrix} H & & H \\ | & & | \\ C & = C = & C \\ | & & | \\ H & & H \end{matrix}} \xrightleftharpoons[\text{Existence (Very Unstable)}]{\text{Its Activated State of}} \underset{(A)}{\begin{matrix} & & H \\ & & | \\ e\cdot C & - & C\cdot n \\ \| & & | \\ H-C & & H \\ | & & \\ H & & \end{matrix}}$$

$$(A) \xrightleftharpoons[\text{hydrogen atom}]{\text{Movement of}} \underset{(B)}{\begin{matrix} CH_3 & & \\ | & & \\ e\cdot C & = & C\cdot n \\ & & | \\ & & H \end{matrix}}$$

$$(B) \xrightarrow{\text{Deactivation}} \underset{(Methyl\ Acetylene)}{H_3C - C \equiv C - H} + Heat \qquad 3.166a$$

$$\underline{Overall\ Equation}: \underset{(LESS\ Stable)}{H_2C = C = CH_2} \xrightarrow[\text{First Kind of Third Type}]{\text{Molecular Rearrangement of the}} \underset{(MORE\ Stable)}{H_3C - C \equiv C - H} + Heat \qquad 3.166b$$

Unlike the first type of molecular rearrangement of the first kind already shown, here the molecular rearrangement involves transfer of H a transfer species from an alkyl-yne group ($=CH_2$) as opposed to an alkyl-ane group (such as $-C_2H_5$). Hence, the classifications of third and first types respectively. The allene, being very unstable like acetylene (but unlike acetylene which cannot rearrange) rearranges instantaneously to give methyl acetylene at higher temperatures. The corresponding source of the second type is a resonance stabilization group- ($-CH=CH_2$) the alkyl-ene group (Not the alkylene family where methylene is the first member followed by ethylene followed by propylene and so on- hence the change in name to alkenyl above). The group when it is not used to provide resonance stabilization for a compound can rarely provide transfer species, because of the strong capacity of $=CH_2$ group. However, there are similar groups like it when the monomer or compound carries hetero atom(s).

3.4.5 Carbon Dioxide Absorption by Aqueous Amine

Carbon dioxide as has been said in the first chapter is not to be feared whether we decide to grow more vegetation or not, although growing more vegetation is very important. What should be feared in the same family is carbon monoxide which only methane does not emit when combusted. However, in view of the major concern which the world has placed on it, there is need to show the amount of millions which have been wasted positively or negatively with respect to its classification as one of the Green House gases, just as human beings are wrongly classified as higher animals. Plants and animals were long created before homo-sapience were created.

There is need to look at this, because a glimpse of the literature shows lots of confusions where many of the equations used for modelling do not exist,[33,34] a very great challenge to the discipline of Chemical engineering. Herein, one will use Mono ethanol amine (MEA) for the absorption. The absorption involves a chemical but not physical reaction. This is unlike what takes place in many Unit Separation systems where physical reactions are involved. MEA can readily be obtained from ethylene oxide and ammonia via Equilibrium mechanism. The ring is opened via the functional center of ethylene oxide by H.e coming from ammonia in its second step. In the third step of the stage, MEA is obtained.

Stage 1:

$$NH_3 \rightleftharpoons H\cdot e + nn\cdot NH_2$$

$$H\cdot e + \underset{O}{H_2C\!-\!CH_2} \xrightleftharpoons[\text{functional center}]{\text{Ring opening via}} H-O-\underset{H}{\overset{H}{C}}-\underset{H}{\overset{H}{C}}\cdot e \quad (A)$$

$$(A) + nn\cdot NH_2 \longrightarrow \underset{(MEA)}{H-O-CH_2-CH_2-NH_2} \qquad 3.167a$$

$$\underline{Overall\ equation}: NH_3 + H_2COCH_2 \xrightarrow{(Heat)} HOCH_2CH_2NH_2 \qquad 3.167b$$

Stage1:

$$HOCH_2CH_2NH_2 \xrightleftharpoons[\text{Existence of the Amine}]{\text{Equilibrium State of}} H\cdot e + \underset{(A)}{HOCH_2CH_2\underset{H}{N}\cdot nn}$$

$$H\cdot e + O=C=O \xrightleftharpoons[CO_2]{\text{Activation of}} \underset{(B)}{H-O-\underset{O}{\overset{\|}{C}}\cdot e}$$

$$(A) + (B) \xrightarrow{\text{Combination State}} \underset{(C)}{HOCH_2CH_2\overset{H}{N}-\overset{O}{\overset{\|}{C}}OH} \qquad 3.168a$$

$$\underline{Overall\ Equation}: CO_2 + MEA \longrightarrow (C) \qquad 3.168b$$

CO_2 is removed as (C) by the amine. (C) is both an amine and an acid. With the presence of only one H atom on the N center, one will be forced to think that Equilibrium state of existence can no longer be favoured from that center in the presence of H on the acidic side. Equilibrium state of existence cannot be favoured from the alcoholic side because of the presence of the amine group. However, the CO_2 can be recovered, since Equilibrium state of existence of (C) is as shown below.

Stage 1:

$$(C) \xrightleftharpoons[\text{Existence Of (C)}]{\text{Equilibrium State Of}} H\cdot e + \underset{(D)}{HOCH_2CH_2NHCOO\cdot nn}$$

$$(D) \xrightleftharpoons{\text{Release Of } CO_2} \underset{(E)}{HOCH_2CH_2\overset{H}{\overset{|}{N}}\cdot nn} + \underset{(F)}{e\cdot \overset{O}{\overset{||}{C}} - O\cdot nn}$$

$$(E) + H\cdot e \xrightleftharpoons[\text{Existence of the Amine}]{\text{Equilibrium State Of}} Amine$$

$$(F) \xrightarrow{\text{Deactivation}} CO_2 + Energy \qquad 3.169a$$

$$\underline{Overall\ Equation}: (C) \longrightarrow CO_2 + Amine + Energy \qquad 3.169b$$

The stage is said to be favoured, because that is the finger print of (C). If indeed it is not, then to recover the CO_2, may have to require replacing the H on N and the H on OH group with an R group. Unlike CO_2, H_2S is removed by Electrostatic addition of H_2S to the MEA in stable state of existence in a manner similar to the case of formation of ammonium hydroxide of Equations 3.49a and 3.49b.

As can be seen, this is not a Unit Separation system, since chemical reaction is involved. This is a reactor for removal or capture of CO_2. In capturing it, millions of material and human resources are wasted even to the oblivious point of setting up a center of excellence for its capture somewhere in our world. This is what we humans in particular emit every nano of a second universally! We could as well wear special gloves on our noses to capture the CO_2 we emit. The major problem in this family of compounds is CO. It is easier to activate CO than it is to activate CO_2. CO is one of the poisons both to humans and the environment; yet it can be very advantageously used if and only if all systems are CLOSED and not OPENED.

It is still important to establish the correctness of the Equilibrium state of existence of (C) of the last equation above. For this purpose, one will consider the synthesis of urea from Carbon dioxide and ammonia.

(a) Synthesis of Urea from CO_2 and NH_3

Stage 1:

$$NH_3 \rightleftharpoons H\cdot e + nn\cdot NH_2$$

$$H\cdot e + O{=}\overset{\overset{\displaystyle O}{||}}{C} \rightleftharpoons H-O-\overset{\overset{\displaystyle O}{|}}{C}\cdot e$$

$$(A)$$

$$(A) + nn\cdot NH_2 \longrightarrow HO-\overset{\overset{\displaystyle O}{||}}{C}-NH_2$$

$$\textit{Carbamic acid} \qquad 3.170a$$

$$\underline{\textit{Overall equation}}: CO_2 + NH_3 \longrightarrow HO(CO)NH_2 \qquad 3.170b$$

Stage 2:

$$HO-\overset{\overset{\displaystyle O}{||}}{C}-NH_2 \rightleftharpoons H\cdot e + nn\cdot O-\overset{\overset{\displaystyle O}{||}}{C}-NH_2$$

$$(B)$$

$$H\cdot e + NH_3 \rightleftharpoons e\cdot NH_4$$

$$(A) + e\cdot NH_4 \longrightarrow H_2N-\overset{\overset{\displaystyle O}{||}}{C}-O^{\ominus}.......{}^{\oplus}NH_4$$

$$(\textit{Ammonium carbamate}) \qquad 3.171a$$

$$\underline{\textit{Overall equation}}: HO(CO)NH_2 + NH_3 \longrightarrow \textit{Ammonium carbamate} \qquad 3.171b$$

If the ammonia had been allowed to exist in Equilibrium state of existence in the presence of carbamic acid which is similar to (C) of Equation 3.168a, then the need for Stage 2 above would not have arisen as shown below.

(Stage 2a:)

$$NH_3 \rightleftharpoons H\cdot e + nn\cdot NH_2$$

$$H\cdot e + HO-\overset{\overset{O}{\|}}{C}-NH_2 \rightleftharpoons \underset{(B)^*}{H_2N-\overset{\overset{O}{\|}}{C}\cdot e} + H_2O$$

$$(B)^* + nn\cdot NH_2 \longrightarrow \underset{(Urea)}{H_2N-\overset{\overset{O}{\|}}{C}-NH_2} \qquad 3.172$$

The stage above could not take place, because the Equilibrium state of existence of carbamic acid is far stronger than that of ammonia. The C atom on the carbonyl center is still the CENTRAL ATOM. (Note that there is no $(CH_2)_n$ between the carbonyl group and amine). Both groups are too close for comfort in (C) of Equation 3.168a. It is the H atom on the acidic side in carbamic acid that is held, despite the presence of two H atoms on the N center, otherwise the formation of ammonium carbamate would have been impossible. The ammonia which was suppressed in Stage 2, is now made to exist in Equilibrium state of existence as shown below in the next stage.

Stage 3:

$$NH_3 \rightleftharpoons H\cdot e + nn\cdot NH_2$$

$$H\cdot e + H_2N-\overset{\overset{O}{\|}}{C}-O^{\ominus}........{}^{\oplus}NH_4 \rightleftharpoons \underset{(C)}{H_2N-\overset{\overset{O}{\|}}{C}\cdot e} + H-O^{\ominus}......{}^{\oplus}NH_4$$

$$(C) + nn\cdot NH_2 \longrightarrow \underset{(Urea)}{H_2N-(CO)-NH_2} \qquad 3.173a$$

$$\underline{Overall\ equation}: Ammonium\ carbamate + NH_3 \longrightarrow Urea + NH_4OH \qquad 3.173b$$

$$\underline{Overall\ overall\ equation}: 3NH_3 + CO_2 \longrightarrow H_4NOH + H_2N(CO)NH_2 \qquad 3.173c$$

One can notice why pressure as close to 2500 psi was required to push the aqueous or oil-water slurry of CO_2 and NH_3 through an autoclave. Ammonia had to be kept in both Stable and Equilibrium states of existence. Instead of two moles of ammonia being thought to be required for the exact synthesis of urea, one can observe that it cannot be used. At least, three moles of ammonia are required. Worthy of note is that the product above contains no water, though it is in the ammonium hydroxide shown above as the second product. However, with the reaction carried out in an aqueous media at 170-180^{0}C, in a slurry of ammonia and carbon dioxide in the autoclave via circulation, the aqueous solution after leaving the reactor, is said to be distilled to remove ammonium carbonate and decomposed ammonium carbamate, which dissociate to CO_2 and ammonia, dissolved in water and then recirculated[35]. The ammonium carbamate decomposes in one single Equilibrium mechanism stage to first release ammonia followed by the formation of another ammonia in Equilibrium state, followed by formation of CO_2 by deactivation along with heat. If ammonium carbonate is one of the products instead of ammonium hydroxide, then this must have been obtained as follows.

Stage 4:

$$HO^{\odot}.......^{\oplus}NH_4 \rightleftharpoons H\cdot e + \underset{(C)}{HO^{\odot}.......^{\oplus}\underset{\bullet n}{N}H_3}$$

$$H\cdot e + O=\overset{O}{\overset{\|}{C}} \rightleftharpoons \underset{(D)}{H-O-\overset{O}{\overset{\|}{C}}\cdot e}$$

$$(C) \rightleftharpoons HO\cdot nn + NH_3$$

$$(D) + nn\cdot OH \longrightarrow \underset{(Carbonic\ acid)}{HO-\overset{O}{\overset{\|}{C}}-OH} \qquad 3.174a$$

$$\underline{Overall\ equation}: HONH_4 + CO_2 \longrightarrow HO(CO)OH + NH_3 \qquad 3.174b$$

Stage 5:

$$HO-\overset{O}{\overset{\|}{C}}-OH \rightleftharpoons H\cdot e + \underset{(E)}{HO-\overset{O}{\overset{\|}{C}}-O\cdot nn}$$

$$H\cdot e + NH_3 \rightleftharpoons e\cdot NH_4$$

$$H_4N\cdot e + (E) \longrightarrow \underset{(Ammonium\ bicarbonate)}{H_4N^{\oplus}...........^{\odot}O-\overset{O}{\overset{\|}{C}}-OH} \qquad 3.175a$$

Stage 6:

$$HO-\overset{O}{\overset{\|}{C}}-O^{\ominus}.......^{\oplus}NH_4 \rightleftharpoons H\cdot e \; + \; nn\cdot O-\overset{O}{\overset{\|}{C}}-O^{\ominus}....^{\oplus}NH_4 \quad (F)$$

$$H\cdot e \; + \; NH_3 \rightleftharpoons e\cdot NH_4$$

$$H_4N\cdot e \; + \; (F) \longrightarrow H_4N^{\oplus}......^{\ominus}O-\overset{O}{\overset{\|}{C}}-O^{\ominus}.....^{\oplus}NH_4 \quad (Ammonium\ carbonate) \qquad 3.175b$$

$$\underline{Overall\ equation}: \; HONH_4 + CO_2 + NH_3 \longrightarrow (H_4N)_2CO_3 \qquad 3.176a$$

$$\underline{Overall\ overall\ equation}: 4NH_3 + 2CO_2 \longrightarrow H_2N(CO)NH_2 + (H_4N)_2CO_3 \quad 3.176b$$

Observe that the molar ratio of ammonia to carbon dioxide is now 2 to 1, but however involving the use of more moles of the components. It was at the beginning that ammonia was forced to exist in Equilibrium state of existence by the use of harsh operating conditions. With the immediate presence of these types of acids (Carbonic and Carbamic acids), it no longer could exist in Equilibrium state of existence anymore. Hence, it was possible to produce the ammoniums. Even the ammonium bicarbonate could not exist in Equilibrium state of existence from the ammonium side, because of the presence of the carboxylic end. Their placements as already said are too close for comfort. That is why urea is a very unique compound.

One can see the wonderful benefit obtained in providing the mechanisms of all reactions. No reaction should be taken as something simple. Go through the steps and see the beauty of NATURE. One can observe the billions of money and human resources wasted by all these industries when the mechanisms of reactions are not known. Find the cheapest way to produce the urea so that any consumer can afford to buy it.

3.4.6 Dilute, Concentrated and Fuming Acids

We are all aware of different types of acids in terms of concentration- ***dilute, concentrated, and fuming acids***. This applies more to inorganic acids than organic acid carrying C, because of the non-polar character of C element than N, P, S elements and unique character and placement of C in the Periodic Table. ***C is the only atom which can both be activated and excited to give so many elements.*** Carbonic, Carbamic acids are no organic or inorganic acids, but hybrids of them. All of them contain Oxygen (O) and for this reason, O is never the CENTRAL ATOM in any of them. The presence of (CH_2) groups in carboxylic acids does a great deal to weaken the strength of the acids. That group plays very major roles in hydrocarbons and many other compounds-ringed and non-ringed. Though the existences of all kinds of acids are known, ***the structures of the dilute and fuming ones are still unknown.*** The same almost apply to bases. For example, shown below are the structures of dilute, concentrated and fuming sulfuric acids.

(a) H_2SO_4

Stage 1:

$$H_2SO_4 \rightleftharpoons H{\bullet}e + nn{\bullet}OSO_3H \quad \underline{[Concentrated]}$$
$$(A)$$

$$H{\bullet}e\ H_2O \rightleftharpoons en{\bullet}OH_3$$
$$(B)$$

$$(A) + (B) \longrightarrow H_3O^{\oplus} ------ {}^{\ominus}OSO_3H \quad \underline{[Fifty\ percent\ Diluted]} \qquad 3.177a$$

Stage 2:

$$H_3O^{\oplus} ----- {}^{\ominus}OSO_3H \rightleftharpoons HO_3SO^{\ominus} ---- {}^{\oplus}\underset{\cdot nn}{O}H_2 \quad + \quad H\cdot e$$
$$(C)$$

$$H\cdot e \quad + \quad H_2O \rightleftharpoons en\cdot OH_3$$
$$(B)$$

$$(B) \quad + \quad (C) \longrightarrow HO_3SO^{\ominus} ---- {}^{\oplus}\overset{\displaystyle H \atop |}{O_{\ominus}} - H$$
$$\vdots$$
$$H - {}^{\oplus}\underset{\displaystyle | \atop H}{O} - H$$

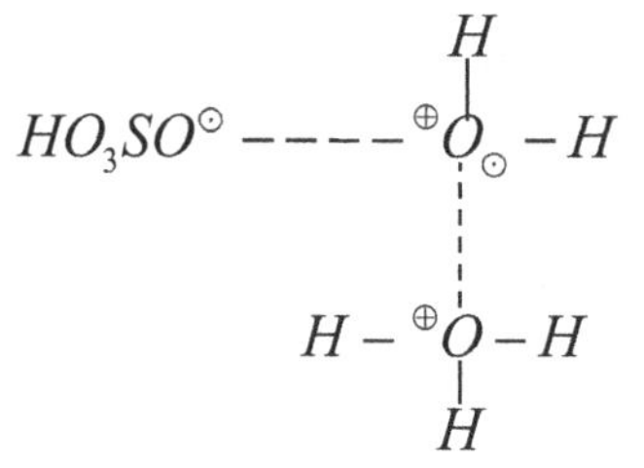

$$(One\ third\ diluted) \qquad 3.177b$$

$$\underline{Fuming\ Sulphuric\ acid:} \quad 2SO_3 \quad + \quad 2H_2O$$
$$(Fumes) \qquad 3.177c$$

Stage 1:

$$HO-\overset{\overset{\displaystyle O^{\odot}}{|}}{\underset{\underset{\displaystyle O^{\odot}}{|}}{S^{2\oplus}}}-OH \xrightleftharpoons[\text{of existence}]{\text{Equilibrium State}} H\cdot e \;+\; nn\cdot O-\overset{\overset{\displaystyle O^{\odot}}{|}}{\underset{\underset{\displaystyle O^{\odot}}{|}}{S^{2\oplus}}}-OH$$

(A)

$$H\cdot e \;+\; H_2SO_4 \rightleftharpoons H_2O \;+\; HO-\overset{\overset{\displaystyle O^{\odot}}{|}}{\underset{\underset{\displaystyle O^{\odot}}{|}}{S^{2\oplus}}}\cdot e$$

(B)

$$(B) \rightleftharpoons SO_3 \;+\; e\cdot H \;+\; Energy$$

$$(A) \rightleftharpoons HO\cdot nn \;+\; nn\cdot O-\overset{\overset{\displaystyle O^{\odot}}{|}}{\underset{\underset{\displaystyle O^{\odot}}{|}}{S^{2\oplus}}}\cdot e$$

(C)

$$H\cdot e \;+\; nn\cdot OH \rightleftharpoons H_2O$$

$$(C) \xrightarrow[\text{Release of heat}]{\text{Deactivation}} SO_3 \;+\; Energy \qquad 3.178a$$

$$\underline{Overall\ equation}: 2H_2SO_4 \longrightarrow 2H_2O \;+\; 2SO_3 \;+\; Energy \qquad 3.178b$$

Two moles of concentrated sulfuric acid are required to give what is called fuming sulfuric acid a mixture of two moles of water and two moles of sulfur trioxide with heat and fumes from the SO_3. The SO_3 can provide oxidizing oxygen only under Equilibrium conditions as shown below. When activated, it gives a non-productive stage. The oxidizing oxygen is held in equilibrium with SO_2. The oxidizing oxygen can only be used in situ for specific compounds as shown below to give a productive stage.

Stage 1:

$$O^{\ominus} - \overset{\overset{O}{\|}}{S}{}^{2\oplus} - {}^{\ominus}O \xrightleftharpoons{Activation} nn\cdot\ddot{\underset{\cdot\cdot}{O}}\cdot en \quad + \quad :\overset{\overset{O}{\|}}{S}{}^{\oplus} - {}^{\ominus}O \qquad 3.178c$$

[*Stable, reactive and soluble*]

$$nn\cdot O\cdot en \; + \; :CO \rightleftharpoons nn\cdot O - \overset{\overset{O}{\|}}{C}\cdot e$$

$$\longrightarrow \quad CO_2 \quad + \quad Heat \qquad 3.178d$$

$$\underline{Overall\ equation}:\; SO_3 \; + \; CO \longrightarrow SO_2 \; + \; CO_2 \; + \; Heat \qquad 3.178e$$

(b) HCl and $HClO_4$

The cases for dilute and concentrated hydrochloric acid have already been shown and recalled below.

HCl

Stage 1:

$$\underset{}{HCl \rightleftharpoons H\bullet e + nn\bullet Cl} \; [Concentrated] \qquad \underset{(A)}{} \qquad 3.179a$$

$$H\bullet e \; + \; H_2O \rightleftharpoons \underset{(B)}{en\bullet OH_3}$$

$$(A) \; + \; (B) \longrightarrow H_3O^{\oplus} ------ {}^{\ominus}Cl \; [Half\ Diluted] \qquad 3.179b$$

The dotted bonds above are electrostatic bonds of the negative type. There is no fuming HCl, because of the absence of polar bond(s). Perchloric acid ($HClO_4$) one of the strongest acids known (See Figure 2.1), is a colourless mobile acid[36]. Unlike HCl, it has three parts-dilute, concentrated and fuming which is very reactive and highly explosive when heated at 1 atmosphere[36]. The fuming acid here is a very strong oxidizing agent.

$HClO_4$

Stage 1:

$$HClO_4 \rightleftharpoons H\cdot e \; + \; \underset{(A)}{nn\cdot OClO_3} \; [Concentrated\ HClO_4]$$

$$H \cdot e + H_2O \rightleftharpoons en \cdot OH_3 \quad (B)$$

$$(A) + (B) \longrightarrow O_3ClO^{\ominus} .... {}^{\oplus}OH_3 \quad (Fifty\ percent\ diluted) \qquad 3.180a$$

$$\underline{Overall\ equation}: HClO_4 + H_2O \longrightarrow O_3ClO^{\ominus} ...... {}^{\oplus}OH_3 \qquad 3.180b$$

Fuming $HClO_4$ $4HClO_4 \longrightarrow 2Cl_2 + 2H_2O + 7\underline{O}_2 + \text{Energy}$ 3.181

Stage 1:

$$HClO_4 \rightleftharpoons H \cdot e + nn \cdot O - \overset{O^{\ominus}}{\underset{O^{\ominus}}{Cl^{3\oplus}}} - {}^{\ominus}O \quad (A)$$

$$H \cdot e + HOClO_3 \rightleftharpoons H_2O + e \cdot \overset{O^{\ominus}}{\underset{O^{\ominus}}{Cl^{3\oplus}}} - {}^{\ominus}O \quad (B)$$

$$(A) + (B) \longrightarrow O_3Cl - O - ClO_3 \quad (Di-Chlorine\ Heptoxide) \qquad 3.182a$$

$$\underline{Overall\ equation}: 2HClO_4 \longrightarrow O_3Cl - O - ClO_3 + H_2O \qquad 3.182b$$

What present-day-Science calls chlorine heptoxide is not the real name. It does not exist. The real name based on its structure is di-chlorine heptoxide.

Stage 2:

$$2O^{\odot}-\overset{\overset{O}{|}\odot}{\underset{\underset{O^{\odot}}{|}}{Cl^{3\oplus}}}-O-{}^{3\oplus}\overset{\odot\overset{O}{|}}{\underset{\underset{{}^{\odot}O}{|}}{Cl}}-{}^{\odot}O \rightleftharpoons 2O^{\odot}-\overset{\overset{O}{|}\odot}{\underset{\underset{O^{\odot}}{|}}{Cl^{3\oplus}}}-O\cdot en \; + \; 2n\cdot\overset{\overset{O}{|}\odot}{\underset{\underset{O^{\odot}}{|}}{Cl^{3\oplus}}}-{}^{\odot}O$$

$$(A) \qquad\qquad (B)$$

$$(A) \rightleftharpoons \underline{\underline{O_2}} + 2O_3Cl\cdot e + Energy$$

$$2O_3Cl\cdot e + 2n\cdot ClO_3 \longrightarrow 2O_3Cl-ClO_3$$

$$[Di-Chlorine\ Hexoxide] \qquad 3.183a$$

$$\underline{Overall\ equation}: 4HClO_4 \longrightarrow \underline{\underline{O_2}} + 2H_2O + 2O_3Cl-ClO_3 + Energy \qquad 3.183b$$

The di-chlorine hexoxide a dimer of chlorine trioxide, is the least explosive oxide of chlorine, but like the other oxides, it will explode if brought into contact with organic compounds[36]. As it seems, it is very stable when alone, but unstable when activated.

Stage 3:

$$2O^{\odot}-\overset{\overset{O}{|}\odot}{\underset{\underset{O^{\odot}}{|}}{Cl^{3\oplus}}}-{}^{3\oplus}\overset{\odot\overset{O}{|}}{\underset{\underset{{}^{\odot}O}{|}}{Cl}}-{}^{\odot}O \underset{by\ Heat}{\overset{Activated}{\rightleftharpoons}} 2O^{\odot}-\overset{\overset{O}{|}\odot}{\underset{\underset{O^{\odot}}{|}}{Cl^{3\oplus}}}\cdot e \; + \; 2:\overset{\overset{O}{|}\odot}{\underset{\underset{O^{\odot}}{|}}{Cl^{2\oplus}}}-O\cdot nn$$

$$(A) \qquad\qquad (B)$$

$$(A) \rightleftharpoons \underline{\underline{O_2}} + 2:\overset{\overset{O}{|}\odot}{\underset{\underset{O^{\odot}}{|}}{Cl^{2\oplus}}}\cdot en + Energy$$

$$(C)$$

$$(B) + (C) \longrightarrow 2O_2Cl-O-ClO_2$$

$$(D)-Di-Chlorine\ pentoxide \qquad 3.184a$$

$$\underline{Overall\ equation}: 4HClO_4 \longrightarrow 2H_2O + 2\underline{\underline{O_2}} + 2O_2Cl-O-ClO_2 + Energy \quad 3.184b$$

Di-chlorine pentoxide has never been known to exist. But above, it is one of the intermediate products when the number of moles is four or more.

Stage 4:

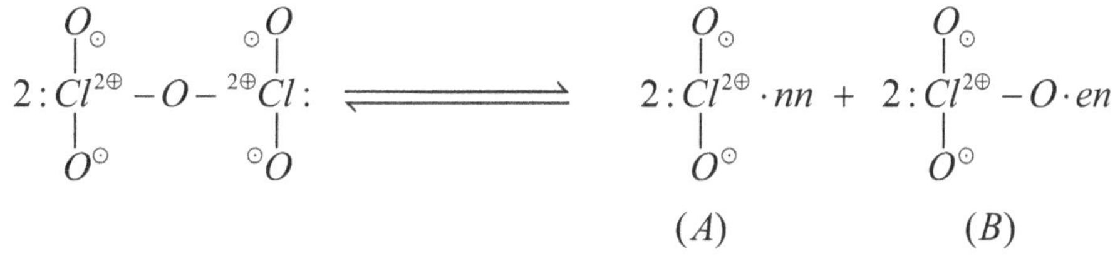

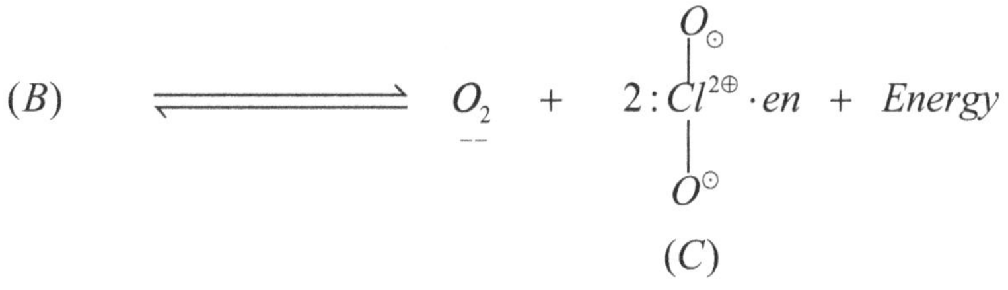

$$2:\overset{\overset{O^{\odot}}{|}}{\underset{\underset{O^{\odot}}{|}}{Cl^{2\oplus}}}-O-{}^{2\oplus}\overset{\overset{{}^{\odot}O}{|}}{\underset{\underset{{}^{\odot}O}{|}}{Cl}}: \rightleftharpoons 2:\overset{\overset{O^{\odot}}{|}}{\underset{\underset{O^{\odot}}{|}}{Cl^{2\oplus}}}\cdot nn + 2:\overset{\overset{O^{\odot}}{|}}{\underset{\underset{O^{\odot}}{|}}{Cl^{2\oplus}}}-O\cdot en$$

$$(A) \qquad (B)$$

$$(B) \rightleftharpoons \underline{O_2} + 2:\overset{\overset{O^{\odot}}{|}}{\underset{\underset{O^{\odot}}{|}}{Cl^{2\oplus}}}\cdot en + Energy$$

$$(C)$$

$$(A) + (C) \longrightarrow 2O_2Cl-ClO_2$$

$$(D)-Di-Chlorine\ tetroxide \qquad 3.185a$$

$\underline{Overall\ equation}$: $4HClO_4 \longrightarrow 2H_2O + 3\underline{O_2} + 2O_2Cl-ClO_2 + Energy$ 3.185b

The di-chlorine tetroxide (Cl_2O_4) a dimer of chlorine dioxide (ClO_2) is like di-chlorine hexoxide. It is stable when alone, but unstable when activated or a compound is around its neighbourhood.

Stage 5:

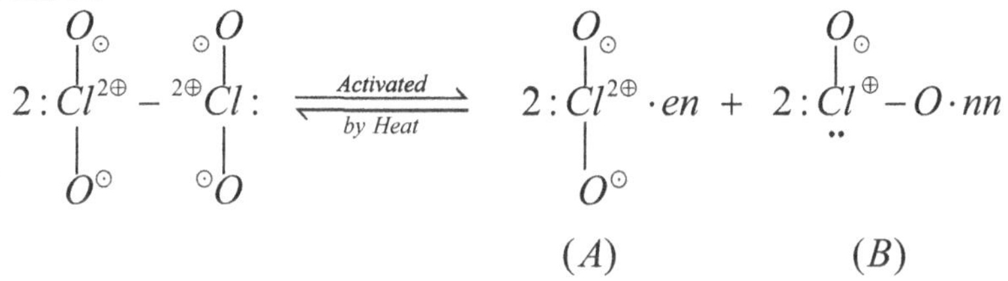

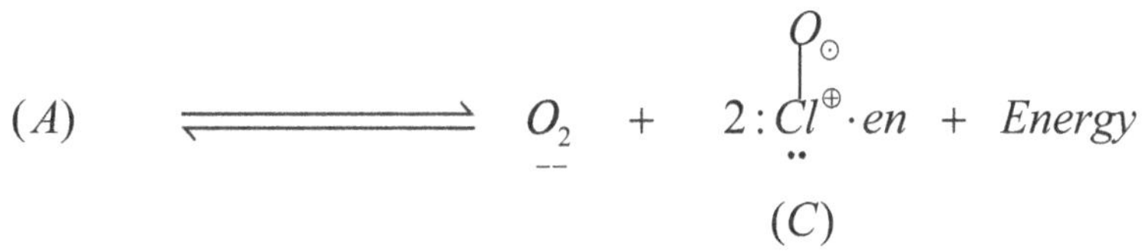

$$2:\overset{\overset{O^{\odot}}{|}}{\underset{\underset{O^{\odot}}{|}}{Cl^{2\oplus}}}-{}^{2\oplus}\overset{\overset{{}^{\odot}O}{|}}{\underset{\underset{{}^{\odot}O}{|}}{Cl}}: \underset{by\ Heat}{\overset{Activated}{\rightleftharpoons}} 2:\overset{\overset{O^{\odot}}{|}}{\underset{\underset{O^{\odot}}{|}}{Cl^{2\oplus}}}\cdot en + 2:\overset{\overset{O^{\odot}}{|}}{\underset{\cdot\cdot}{Cl^{\oplus}}}-O\cdot nn$$

$$(A) \qquad (B)$$

$$(A) \rightleftharpoons \underline{O_2} + 2:\overset{\overset{O^{\odot}}{|}}{\underset{\cdot\cdot}{Cl^{\oplus}}}\cdot en + Energy$$

$$(C)$$

$$(B) + (C) \longrightarrow OCl-O-ClO$$

$$(D)-Di-Chlorine\ trioxide \qquad 3.186a$$

$\underline{Overall\ equation}$: $4HClO_4 \longrightarrow 2H_2O + 4\underline{O_2} + 2OCl-O-ClO + Energy$ 3.186b

Di-chlorine trioxide here has also not been known to exist. But here, it is one of the intermediates in the fuming acids in-situ preparation. By itself, it is very unstable. Hence the next stage follows.

Stage 6:

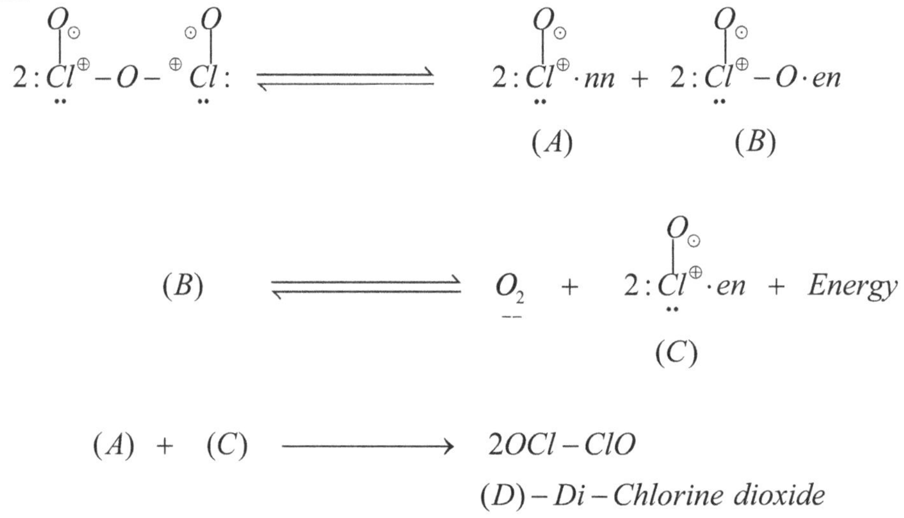

$$2:\underset{\cdot\cdot}{\overset{O\odot}{\overset{|}{Cl}}}{}^{\oplus}-O-{}^{\oplus}\underset{\cdot\cdot}{\overset{\odot O}{\overset{|}{Cl}}}: \rightleftharpoons 2:\underset{\cdot\cdot}{\overset{O\odot}{\overset{|}{Cl}}}{}^{\oplus}\cdot nn + 2:\underset{\cdot\cdot}{\overset{O\odot}{\overset{|}{Cl}}}{}^{\oplus}-O\cdot en$$

$$(A) \qquad (B)$$

$$(B) \rightleftharpoons \underline{O_2} + 2:\underset{\cdot\cdot}{\overset{O\odot}{\overset{|}{Cl}}}{}^{\oplus}\cdot en + Energy$$

$$(C)$$

$$(A) + (C) \longrightarrow 2OCl-ClO$$

$$(D) - Di - Chlorine\ dioxide \qquad 3.187a$$

$$\underline{Overall\ equation}: 4HClO_4 \longrightarrow 2H_2O + 5\underline{O_2} + 2O_2Cl-ClO_2 + Energy \qquad 3.187b$$

The di-chlorine dioxide above which is one of the intermediate products, is not known to exist in present-day Science. Indeed, all the intermediates which have been identified so far have both transient and full existence, depending on the operating conditions. The (D) above is again very unstable and explosive in character.

Stage 7:

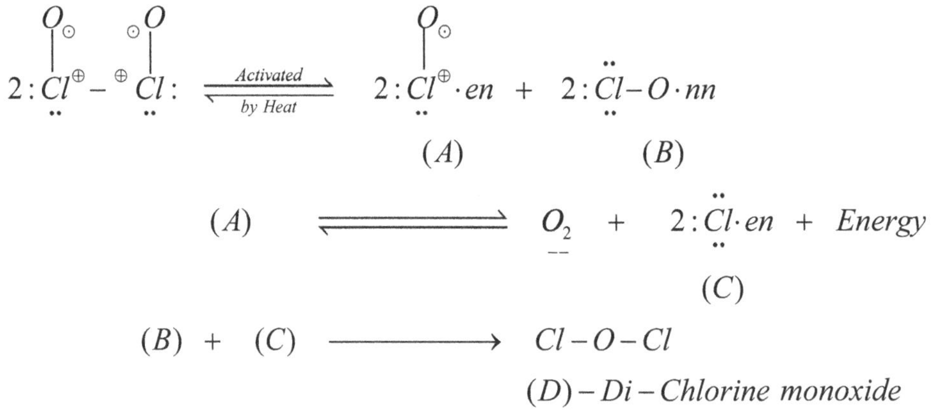

$$2:\underset{\cdot\cdot}{\overset{O\odot}{\overset{|}{Cl}}}{}^{\oplus}-{}^{\oplus}\underset{\cdot\cdot}{\overset{\odot O}{\overset{|}{Cl}}}: \underset{by\ Heat}{\overset{Activated}{\rightleftharpoons}} 2:\underset{\cdot\cdot}{\overset{O\odot}{\overset{|}{Cl}}}{}^{\oplus}\cdot en + 2:\underset{\cdot\cdot}{\overset{\cdot\cdot}{Cl}}-O\cdot nn$$

$$(A) \qquad (B)$$

$$(A) \rightleftharpoons \underline{O_2} + 2:\underset{\cdot\cdot}{\overset{\cdot\cdot}{Cl}}\cdot en + Energy$$

$$(C)$$

$$(B) + (C) \longrightarrow Cl-O-Cl$$

$$(D) - Di - Chlorine\ monoxide \qquad 3.188a$$

$$\underline{Overall\ equation}: 4HClO_4 \longrightarrow 2H_2O + 6\underline{O_2} + 2Cl-O-Cl + Energy \qquad 3.188b$$

The di-chlorine monoxide (Present-day Science calls it chlorine hemioxide or monoxide, both of which are wrong) produced above, is known and said "to be extremely soluble in water, with which it forms hypochlorous acid (HOCl) and is even more soluble in carbon tetrachloride"[36]. How can something be soluble and at the same time productive? This is the state of present-day-Science- a state of confusion. The di-chlorine monoxide is known to be explosive when alone and explodes violently on contact with easily oxidized substances such as hydrogen, sulfur, ammonia, nitric oxide, etc. We already met this compound when the mechanisms for the use of NCl_3 as a bleaching agent in three stages were provided (See Equation 3.62c). Therein, it was a third stage for that system. It is again recalled below as another last stage.

Stage 8:

$$2:\ddot{\underset{..}{Cl}}-\ddot{\underset{..}{O}}-\ddot{\underset{..}{Cl}}: \rightleftharpoons 2:\ddot{\underset{..}{Cl}}\cdot nn + 2:\ddot{\underset{..}{Cl}}-O\cdot en$$

$$2:\ddot{\underset{..}{Cl}}-O\bullet en \rightleftharpoons 2:\ddot{\underset{..}{Cl}}\bullet en + \underline{\underline{O_2}} + Energy$$

$$2:\ddot{\underset{..}{Cl}}\bullet en + 2:\ddot{\underset{..}{Cl}}\bullet nn \xrightarrow[\text{Oxidizing oxygen}]{\text{Suppressed by}} 2Cl_2 \qquad 3.189a$$

$$\underline{\text{Overall equation}}: 4HClO_4 \longrightarrow 2Cl_2 + 2H_2O + 7\underline{\underline{O_2}} + Energy \qquad 3.189b$$

Under another operating condition (mild), the last stage above is reversible. One can shockingly observe that fuming perchloric acid is indeed a very strong acid, a very strong oxidizing agent, and a very strong source of energy since all the seven stages after the first stage explode very violently. One can imagine the great precautions that will be needed and taken in using this type of acid. At the end of eight stages, seven oxidizing oxygen molecules were obtained. One can imagine what will take place after all the explosions with the subsequent use of the oxidizing oxygen molecules. The use of all these examples has displayed the great wonders which humanity is about to see in the New Frontiers.

(c) Nitric Acid

For nitric acid, different structures also exist for dilute, concentrated, and fuming types. Unlike fuming sulfuric acid, fuming nitric acid is a very strong oxidizing agent. SO_3 in fuming sulfuric acid as has been shown is weak source of oxidizing oxygen. Its oxidizing oxygen unlike all the cases seen so far, can only be used inside a stage only for some compounds.

Stage 1:

$$H-O-\overset{\overset{O^{\odot}}{|}}{N^{\oplus}}=O \rightleftharpoons H\cdot e + nn\cdot O-NO_2 \text{ [Concentrated Nitric acid]}$$
$$(A) \qquad 3.182a$$

$$H\cdot e + H_2O \rightleftharpoons en\cdot OH_3$$
$$(B)$$

$$(A) + (B) \longrightarrow O_2NO^{\odot}........{}^{\oplus}OH_3 \text{ [Fifty percent diluted]} \qquad 3.182b$$

$$\underline{Fuming\ nitric\ acid}: 2O=N-N=O + 2H_2O + 3\underline{O_2}(Oxidizing\ oxygen) \qquad 3.183$$

Stage 1:

$$HNO_3 \rightleftharpoons H\cdot e + nn\cdot O-NO_2$$
$$(A)$$

$$H\cdot e + HNO_3 \rightleftharpoons H_2O + e\cdot NO_2$$
$$(B)$$

$$(A) + (B) \longrightarrow O_2N-O-NO_2 \qquad 3.184a$$

$$\underline{overall\ equation}: 2HNO_3 \longrightarrow O_2N-O-NO_2 + H_2O \qquad 3.184b$$

Stage 2:

$$2O=\overset{\overset{O^{\odot}}{|}}{N^{\oplus}}-O-{}^{\oplus}\overset{\overset{{}^{\odot}O}{|}}{N}=O \rightleftharpoons 2O=\overset{\overset{O^{\odot}}{|}}{N^{\oplus}}\cdot n + 2en\cdot O-{}^{\oplus}\overset{\overset{{}^{\odot}O}{|}}{N}=O$$
$$(C)$$

$$(C) \rightleftharpoons \underline{O_2} + 2e\cdot NO_2 + Energy$$

$$2O_2N\cdot e + 2n\cdot NO_2 \longrightarrow 2O_2N-NO_2 \qquad 3.185a$$

$$\underline{Overall\ equation}: 4HNO_3 \longrightarrow 2O_2N-NO_2 + \underline{O_2} + 2H_2O + Energy \qquad 3.185b$$

Stage 3:

$$2O=\overset{\overset{O^{\odot}}{|}}{N^{\oplus}}-{}^{\oplus}\overset{\overset{{}^{\odot}O}{|}}{N}=O \rightleftharpoons 2O=\overset{\overset{O^{\odot}}{|}}{N^{\oplus}}\cdot e + 2n\cdot{}^{\oplus}\overset{\overset{{}^{\odot}O}{|}}{N}=O$$
$$(D)$$

$$(D) \rightleftharpoons \underline{O_2} + 2en\cdot\ddot{N}O + Energy$$

$$2n\cdot NO_2 \rightleftharpoons 2O=N-O\cdot nn$$

$$2O\ddot{N}\cdot en \ + \ 2nn\cdot O-N=O \longrightarrow 2O=N-O-N=O \qquad 3.186a$$

$$\underline{Overall\ equation}: 4HNO_3 \longrightarrow 2O=NON=O \ + \ 2\underline{\underline{O_2}} \ + \ 2H_2O \ + \ Energy \qquad 3.186b$$

Stage 4:

$$2O=N-O-N=O \rightleftharpoons 2O=N-O\cdot en \ + \ 2nn\cdot N=O$$

$$(E)$$

$$(E) \rightleftharpoons \underline{\underline{O_2}} \ + \ 2en\cdot NO \ + \ Energy$$

$$2ON\cdot en \ + \ 2nn\cdot NO \longrightarrow 2ON-NO \qquad 3.187a$$

$$\underline{Overall\ equation}: 4HNO_3 \longrightarrow 2ON-NO \ + \ 3\underline{\underline{O_2}} \ + \ 2H_2O \ + \ Energy \qquad 3.187b$$

One can clearly see why fuming nitric acid is a very strong oxidizing agent. One can observe the great significance and effect of number of moles of compounds on some compounds, Mathematics being their natural language of communication. One can observe the great significance of many things we have always ignored. One can observe the origins of the fumes, an area which we have not yet begun to pay attention to. This is a subject area of Physics in Chemistry. One can see why these fuming acids must be stored in CLOSED SYSTEM. They would have been very hazardous to the environment if not in CLOSED SYSTEMS. Hence, as is beginning to become clear, most industries in particular Chemical and related industries universally have to be dismantled and rebuilt again to keep all of them as CLOSED SYSTEMS, apart from other reasons related to the New Frontiers. We have just only begun laying new foundations for humanity, because nothing has ever been known, something very difficult to say. So many in particular the academia think they know when indeed they know nothing, because the more you know, the more the realization that you know nothing. In fact, the situation universally is worse than this, because all what they think they know are all nothing, but illusions. One is yet to see a Black that is as black as Carbon Black or Black carbon and a White that is as white as White phosphorus! The white light we see is not white, but a combination of all colours excluding black.

3.5 Conclusions

With the new classifications for Compounds, introduction to molecular rearrangements, flow of electric currents in solutions or fluids, explanation of the phenomena of dissolution/miscibilization, solubilisation/insolubilisation, the mechanisms of cracking of some hydrocarbons, the concept of reversibility, and so much too countless to list, this brings one to the end of the introduction of new concepts in CHEMISTRY and from CHEMISTRY as contained in the New Frontiers.

There are however still more new concepts some of which depend on the ones shown so far. Without the new concepts, one cannot provide the mechanisms for all reactions-chemical or non-chemical in general or indeed provide the mechanisms of how NATURE operates. We leave in a NATURAL and PHYSICAL world which is also COMPLEX. How can NATURE provide everything such as fruits on a mango tree and expect NATURE to be the one to pluck the fruit to put into your mouth? Something must be done physically to at least pluck the fruit and ....do something. Notice that so far, one has provided the real structures of so many compounds too countless to list; simple compounds such as IF_7, CO, NH_4OH, SO_2, inorganic acids, carbon elements such as carbon black, activated carbon, coal, coke, and so on.

Based on the applications of past and current universal data, this marks the beginning of very new developments for not only Chemistry and Chemical engineering, but for all disciplines, since Chemistry deals with the study of the LAWS OF NATURE in the REAL and IMAGINARY DOMAINS. It is indeed the MOTHER of all disciplines. This indeed is a new Chemistry for application to all disciplines. All these require the use of eye of the needle, unlike the types the world up to the present moment has been using almost since antiquity. To the world SEEING IS BELEIVING, when in NATURE the reverse is the case- BELEIVING IS SEEING. Universally as has been said herein and truthfully repeated again, all things and disciplines whether Sciences, Engineering, Medical Sciences, Social Sciences, History, Geography, etc., have all been ***ninety five percent ART*** and ***five percent SCIENCE***, analogous to our ability of the use of our brains where only five percent of our brain is what we have been able to use so far with ninety five percent left unused. This indeed is "The beginning of a NEW DAWN for humanity" and not "The dawn of a NEW BEGINNING for humanity", because the world since antiquity has been in slumber.

3.6 Foods for thought

Without asking questions after a chapter, one can never ascertain the orderliness and correctness of the materials presented. The greatest asset in any human is ability to ask questions and never accept that no solution exists. For every problem, there is one and only one solution applying the same fundamental principles based on the Laws of Nature, while for every solution there are countless numbers of problems from outside wherein principles all against the Laws of Nature are applied, that which is absolute nonsense. The final questions in this last chapter complete the triangle in one's introduction into the world of a New Science.

PB3.1 What is the new classification for compounds, comparing it with present-day classifications? Present-day text books for high schools are nowhere close to the new concepts, because of missing links. What are these missing links?

PB3.2 Based on the new classifications for compounds, there are three members in the family. Why three instead of more or less? What is the significance of the number "Three"? What are the driving forces favouring the existence of ionic characters in compounds? Give five examples each of Polar/Non-ionic, Polar/Ionic and Non-polar/Non-ionic compounds, showing the CENTRAL ATOM in all of them.

PB3.3 In Mathematics, we are very much aware of the Real and Imaginary domains-viz Real and Complex numbers. In Physics, we are yet to see the real and imaginary domains, though it is there and begun to emerge in recent years. In Chemistry, we have begun herein to see the real and imaginary domains. Discuss extensively on the concepts above.

PB3.4 Provide the real structures of the following compounds and elements comparing with those of present-day Science:

i) Concentrated Sulfuric acid.
ii) Concentrated nitric acid.
iii) Ammonium hydroxide.
iv) Carbon monoxide.
v) Ozone.
vi) Sulfur dioxide.
vii) Zinc dust.
xi) Dilute sulfuric acid.
xii) Dilute nitric acid.
xiii) Dilute hydrochloric acid.
xiv) Grignard reagent.
xv) Iodine hepta-fluoride (IF_7).
xvi) Carbon black.
xvii) Coal.

viii) Lithium tetra aluminium hydride
ix) Charcoal.
x) Butylenes.
xviii) Coke
xix) Activation carbon.
xx) Diamond.

PB3.5 a) Under what classification of compounds would you classify the chromosomes of animals? What are CENTRAL ATOMS in some selected ones?

b) Define Dissolution and Miscibility giving three examples each for them.

c) Distinguish between Solubilisation and Insolubilisation. Is it true that without Dissolution and Miscibilization, Solubilisation and Insolubilisation cannot take place? Use examples different from those given in the chapter.

d) Why is it for example that the same solvent cannot be used for dissolving or miscibilizing acrylamide and acrylonitrile?

PB3.6 What are reversible reactions? Use ammonia to provide the mechanisms for such reactions. Use another example of your choice to provide the mechanisms for such reactions.

PB3.7 What is molecular rearrangement of the first kind of the first type? How does it take place with the following compounds where possible-?

i) The butenes.
ii) Vinyl alcohol.
iii) Acrylamide.
iv) Acrylonitrile.
v) Acrylic acid.

PB3.8 a) What is molecular rearrangement of the first kind of the third type? Use only one example to explain the concept.

b) What are transfer species of the first kind of the first type and of the first kind of the third type? Of what relevance are they during polymerizations? Explain.

PB3.9 In the Petroleum industry, the fluid catalytic cracker is a very important process. Why the word "catalytic" when no catalyst is involved for the cracking of the hydrocarbon? The so-called catalysts are reservoirs for storing heat highly desired for the cracking. Give examples of the ones currently used in such industries. Provide the mechanisms for the cracking of propanes and butanes.

PB3.10 Greenhouse gases are of great concern to the environment when the Systems where they are emitted from are not closed. Distinguish between CLOSED and OPENED systems. Describe the mechanisms by which CO_2 and H_2S can be absorbed using a reagent which can be recycled. Should CO_2 belong to this family of compounds? Explain.

PB3.11 Provide the mechanisms for chlorination and fluorination of ammonia.

PB3.12 What are the significances of the use of the following elements as groups on CENTRAL ATOMS, such as H, Na for existence of some inorganic acids and salts- I, Br, Cl, F, N, S, P, and As? Identify those that can provide fuming acids with O and use one to show what is contained in the fuming acid. Why is O not in the least? What is the significance of H to all of them?

PB3.13 Explain why fuming sulfuric acid cannot be used as an oxidizing agent like fuming $HClO_4$ and HNO_3. Provide the mechanisms for the three of them for the oxidation of methane.

PB3.14 Distinguish between Combustion and Oxidation. Does a compound have to exist in Equilibrium state of existence before it can be combusted? If your answer is No, give an example of the combustion of such a compound.

PB3.15 Based on the considerations so far, answer the following questions.

a) What is Tautomerism?

b) What is Resonance Stabilization? Explain why it cannot take place chargedly.

PB3.16 If all that exist whether living or non-living are conglomerations of Atoms, then answer the following questions-

i) What is the essence of the FORCES of NATURE in all these things?

ii) What is essence of Boundaries and Limitations?

iii) Distinguish between Living and Non-living systems in terms of their abilities to use the FORCES of NATURE. Can they be used by atoms, elements and compounds?

iv) What is REAL and IMAGINARY?
v) What is the significance of CHEMISTRY as a discipline?

PB3.17 Can any discipline exist without the application of Mathematics which has been redefined as the Natural language of Communication? How do we apply Mathematics and Physics in the study of our Physical or Universal languages? What is the language of communication between atoms and compounds?

PB3.18 Distinguish between the NATURAL and PHYSICAL worlds. Which of them is more important or indeed Can the PHYSICAL world exist without the NATURAL world? What is the meaning of all these?

PB3.19 We see the role of MOTHER NATURE inside CHEMISTRY. How?

PB3.20 In NATURE, there are countless numbers of families of Compounds just as exists in our PHYSICAL world with Plants, Animals and Non-living things. Distinguish between each one of them.

PB3.21 In NATURE, the first member of all families of compounds is always different from all the other members in the family. Why and how does this apply in our world?

PB3.22 In NATURE, unknown to Present-day CHEMISTS, there are MALES and FEMALE compounds, just as exists in our PHYSICAL world. Distinguish between the kinds and types of the Males and Females in the two worlds.

PB3.23 In NATURE with Polymeric systems, it is the Female that diffuses with its positive end to a Male and form a couple when both are activated. Don't you think the same also applies in our world without our ability to see it? When the reverse is the case what do you expect to be the result of such coupling where possible? Use CHEMISTRY to seek for the solution.

PB3.24 Man by NATURE is a POLYGAMIST depending on THE OPERATING CONDITIONS, while Woman by NATURE is a PROSTITUTE depending on THE OPERATING CONDITIONS. Use CHEMISTRY to show the truth in these philosophical statements.

PB3.25 Distinguish between Male and Female radicals and charges and Male and Female compounds. How are they manifested in our PHYSICAL world?

PB3.25 Up to the present moment it has all along been thought that most chemical reactions take place IONICALLY while few take place radically. In fact, what CHARGES and RADICALS are have never be known by PRESENT-DAY SCIENCE. List the damages which the lack of these have done to humanity in all disciplines and our developments technologically or otherwise.

PB3.26 In NATURE with polymeric systems, one of the things we go against in our PHYSICAL world, takes place, except under different operating conditions. For example, females can add to themselves to give giant molecules only when no males exist in the system or that world and also under very harsh operating conditions. Why they exist in our world under different operating conditions unlike what exist in NATURE is questionable. Based on how NATURE operates, how can these problems be handled?

PB3.27 In NATURE, we have what are called Catalysts-active and passive. Identify what are identical to them in our PHYSICAL world.

PB3.28 In NATURE with polymeric systems, we have what are called INITIATORS. Identify what are identical to them in our PHYSICAL world.

PB3.29 In NATURE, regardless the operating conditions, many things which take place in our PHYSICAL world cannot take place in them. These you find everywhere including the applications of the ***Law discipline*** where their laws have exceptions. At the end of the day, we foolishly open our mouth and say "With GOD, all things are possible", when we humans are the one that make all things which should be impossible possible. Discuss using some Chemical reactions as examples.

PB3.30 Does Present-day Science have any classifications for Compounds, besides all the different types of obscure definitions for them. Compare their classifications with the new classification.

PB3.31 NATURE does not differentiate, but integrates, yet we differentiate in Mathematics. NATURE abhors a Vacuum, Non-linearity, Accidents, Discrimination and more, yet they exist in our world. Some should exist if inherently in NATURE, while some should never exist. Discuss the essence of all these.

PB3.32 What is TIME? Why does Combination mechanism take so short a time compared to Equilibrium and Decomposition mechanism? However, what is the best mechanism or hybrids of them a System like a Government should operate with for positive development. Discuss very broadly.

PB3.33 How can one use the different types of States of Existences and some of the new concepts to develop a viable model for the economic development of an Industry?

PB 3.34 Without H, nothing seems to exist in humanity. What is the significance of H?

References

1. S. N. E. Omorodion, "*New Classifications for Radicals and Their Impacts*", in Chapter 1 herein.
2. S. N. E. Omorodion, *"New Classifications for Bonds and Charges and Their Impacts"*, in Chapter 2 herein.
3. C. R. Noller, "*Textbook of Organic Chemistry*", W. B. Saunders Company, 1966, pg. 599.
4. C. R. Noller, "*Textbook of Organic Chemistry*", W. B. Saunders Company, 1966, pgs. 83-84.
5. C. R. Noller, "*Textbook of Organic Chemistry*", W. B. Saunders Company, 1966, pg. 581.
6. G. Odion, "Principles of Polymerization", McGraw-Hill Book Company, 1970, pgs. 355-357.
7. C. R. Noller, "*Textbook of Organic Chemistry*", W. B. Saunders Company, 1966, pgs. 67-69.
8. C. R. Noller, "*Textbook of Organic Chemistry*", W. B. Saunders Company, 1966, pg. 211.
9. C. R. Noller, "*Textbook of Organic Chemistry*", W. B. Saunders Company, 1966, pg. 369.
10. S.N.E.Omorodion, "Measurement of Molecular Weight Distribution Of Polyacrylamide by Turbidimetric Titration", M.Eng.Thesis, McMaster University, Hamilton, Ontario, Canada. (1976).
11. C. R. Noller, "*Textbook of Organic Chemistry*", W. B. Saunders Company, 1966, pg. 247.
12. E. de Barry Barnett, C. L. Wilson, "Inorganic Chemistry-A Text-Book for Advanced Students", ELBS and Longmans Green and Co Ltd. 1962, pgs. 369- 371.
13. Susumu Matsuzaki, Masaharu Taniguchi, Mizuka Sano, "Polymerization of Benzene occluded in Graphite-alkali metal intercalation compound", Synthetic Metals, Volume 16, Issue 3, (1986), pages 343-348.
14. S.N.E.Omorodion, "Size Exclusion Chromatography", PhD. Thesis, McMaster University, Hamilton, Ontario, Canada. (1980);
    S. N. E. Omorodion, A.E. Hamielec., "An Analytical Solution of a Newly Proposed Peak Broadening Equation and Extension of Polydispersities to Higher Molecular Weight Averages in Size Exclusion Chromatography (SEC)", Chromatographia Vol. 31, No. 5/6, March 1991.
15. Kyozi Kaeriyama, Mass-aki Sato, Kazuo Someno and Susumu Tanaka, J. Chem. Soc., Chem. Commun., (1984), 1199-1200.

16. Peter Kovacic, Alexander Kyriakis, "Polymerization of Benzene to p-Polyphenyl by Aluminium Chloride-Cupric Chloride", Journal of American Chemical Society, (1963) Vol. 85, Issue 4: Publisher: American Chemical Society, Pages 454-458.
17. Peter Kovacic, Fred. W. Koch, "Polymerization of Benzene to p-Polyphenyl by Ferric chloride", J. Org. Chem. (1963), 28(7), pp. 1864-1867.
18. Kovacic, P. and Wu, C. (1960), "Reaction of ferric chloride with Benzene", Journal of Polymer Science, Vol. 47, Issue 149, pages 45-54.
19. Kovacic, P., Koch, F. W. and Stephan, C. E., "Water catalysis in the polymerization of Benzene by Ferric chloride", Journal of Polymer Science Part A: General Papers, Vol.2, Issue 3, pages 1193-1203.
20. Peter Kovacic, Richard M. Lange, "Polymerization of Benzene to polyphenyl by Molybdenum Pentachloride", J. Org. Chem., (1963), 28(4), pp. 968-972.
21. Fumio Teraoka & Toshisada Takabashi, "Morphology of poly-p-phenylene formed during the polymerization of Benzene by molybdenum pentachloride or Ferric chloride", Journal of Macromolecular Science, Part B, Vol.18, Issue 1, (1980), pgs. 73-82.
22. Oprea, C. V. and Popa M. (1984)," Opening of aromatic rings in Mechanochemical polymerization of Benzene and Pyridine", Die Angewandte Makromoleculare Chemie, Vol.127, Issue 1, pages 49 -58.
23. R. H. Grubbs, C. B. Gorman, E. J. Ginsburg, Joseph W. Perry and Seth R. Marder, "New Polymeric Materials with Cubic Optical Non-linearities derived from Ring-Opening Metathesis Polymerization" Materials for Nonlinear Optica, Chapter 45, pp. 672-682.
24. C. R. Noller, "*Textbook of Organic Chemistry*", W. B. Saunders Company, 1966, pgs. 591-598.
25 Karin Roy, Christof Horn, Peter Frank, Vladislav G. Slutsky, Thomas Just, Symposium (International) on Combustion, Volume 27, Issue 1, 1998, Pages 329-336.
26. S. Nakra, R. J. Green, S. L. Anderson, "Thermal decomposition of JP-10 studied by micro flowtube pyrolysis-mass spectrometry", Combustion and Flame (2006), Volume: 144, Issue: 4, Pages 662-674.
27 Burcat, A. and Dvinyaninov, M. (1997), "Detailed Kinetics of cyclopentadiene decomposition studied in a shock tube",

International Journal of Chemical kinetics, Volume 29, Issue 7, pages 505-514.

28 Tokmakov, I. V., Moskaleva, L. V. and Lin, M. C. (2004) "Quantumchemical/vRRKM Study on the thermal decomposition of Cyclopentadiene" International Journal of Chemical kinetics, Volume 36, Issue 3, pages 139 – 151.

29 George, B. Bacskay and John, C, Mackie, "The pyrolysis of cyclopentadiene: quantum chemical and kinetic modelling studies of the acetylene plus propyne/allene decomposition channels" Phys. Chem. Chem. Phys., 2001, 3, 2467-2473.

30. Do Hyong Kim, Jeong-Kwon Kim, Seong-Ho Jang, James A. Mulholland, and Jae-Yong Ryu, "Thermal Formation of Polycyclic Aromatic Hydrocarbons from Cyclopentadiene (CPD)", Environ. Eng. Res. Vol. 12, No. 5, pp. 211-217, (2007).

31. McGivern, W. S., Mannion, J. A., Tsang, W., "Ring expansion reactions in the thermal decomposition of tert-butyl-1,3-cyclopentadiene", J. Phys. Chem A. 2006 Nov 30, 110 (47): 12822

32. Faina Dubnikokova and Assa Lifshitz, "Ring Expansion in Methycyclopentadiene Radicals. Quantum Chemical and Kinetics Calculations". J. Phys. Chem. A, 2002, 106 (35), pp. 8173-8183.

33. E. B. Rinker, S. S. Ashour, and O. C Sandall, *"Absorption of Carbon Dioxide into Aqueous Blends of Diethanolamine and Methyldiethanolamine",* Ind. Eng. Chem. Res. 2000, 39, 4346-4356.

34. B. P. Mandal, M. Guha, A. K. Biswas, S. S. Bandyopadhyay, *"Removal of carbon dioxide by absorption in mixed amines: modelling of absorption in aqueous MDEA/MEA and AMP/MEA solutions",* Chemical Engineering Science, 56, 2001, 6217-6224.

35. C. R. Noller, "*Textbook of Organic Chemistry*", W. B. Saunders Company, 1966, pgs. 282-283.

36. E. de Barry Barnett, C. L. Wilson, "Inorganic Chemistry-A Text-Book for Advanced Students", ELBS and Longmans Green and Co Ltd. 1962, pgs. 532- 540.

Appendix (I)
New Concepts on Inner Core of an Atom or Central atom of a Molecule (Nucleus)

The nucleus of an atom is a different world. Without the nucleus, there is no outer sphere of an atom. It is the nucleus that determines what can exist in the outer sphere. It is the nucleus that has made it impossible for humanity to indeed define an atom, since they know little or nothing about what takes place in the outer sphere of an atom, except via experimental observations. It is the nucleus that carries the source of ***life and living,*** through which HIS LAWS all set in MOTION link with via HIS rays and then to the Radicals. The nucleus has its positive and negative characters, ***the matter and antimatter,*** the real and imaginary respectively.

Modern investigation (as of today) of the atomic structure has shown and believe that protons, "Electrons" and neutrons appear to be fundamental constituents of atoms, while the existence of other particles such as positrons has also been clearly demonstrated. "Electrons" were thought to be in the outer sphere of the atom, while protons, neutrons and positrons (positive electrons!) are in the nucleus. "Electrons" are said to be negatively charged, while protons and positrons to be positively charged. ***While positrons are weightless, a proton has one atomic mass unit (i.e. has weight).*** Since "Electrons" have been identified to be radicals and to have both positive and negative characters in the New Frontiers, then what is an electron? What is a proton? What is a neutron? What is a positron? *Atomic Number of an atom is said to be the number of protons while Atomic weight is the sum of the atomic number and the number of neutrons in which each (neutron) has* ***one atomic mass unit*** *and no charge.* The numbers of "protons" are known to be equal to the numbers of "electrons".

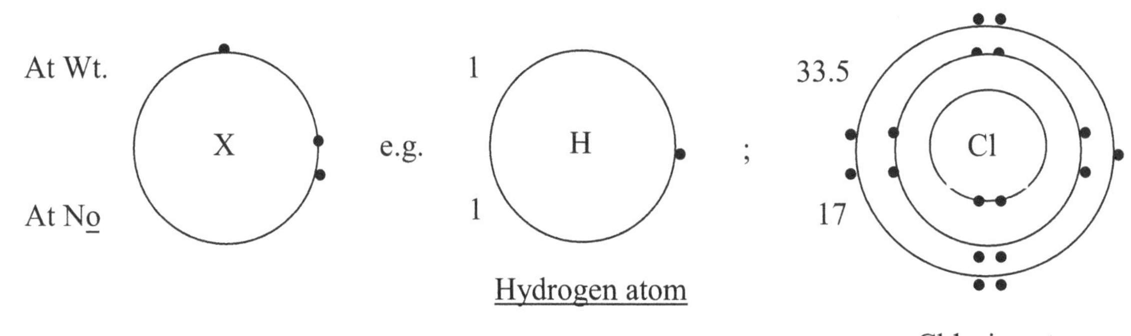

ApI. 1

In order words, the so-called "electrons", that is, the radicals are weightless. What is actually providing the weight of an atom, is embedded in the nucleus and not in the outer sphere up to the boundary. In order words, what is providing the weight in our solar system, a single MACRO-ATOM, is our SUN, wherein ***the planets (Radicals) surrounding it, such as the Earth, Jupiter, Mars, etc. are weightless relative to the SUN.***

While no change has, however, been suggested in the case of the established particles in the nucleus of importance to the Chemists and Physicists, no designation has been proposed for the "antiproton" or negative proton , one of the most recently discovered particles, as has been provided for "positrons" or positive electrons ($({}^{0}_{+1}e$ **or** ${}^{0}_{1}P$ *or* $e^{\oplus}$ -These are Symbols used in Present-day Science). This particle (antiproton) is known today to bear the same relation to proton as the positron does to the ordinary "electron". Its existence has been postulated for many years on theoretical grounds until recently. One is stated to have been observed in a Cosmic ray bombardment. In addition, there has been a report that *the bombardment of a copper target with 6200 Mev protons (very high energy) has produced particles which can only be identified as_antiprotons.* It is suggested that on collision with normal positive protons both types of particles must be annihilated, producing mesons (from Cosmic ray bombardment). It is open to speculation whether a type of matter may exist which is the inverse of normal matter, being composed of positrons, neutrons or "antineutrons," the nature of which is uncertain, and antiprotons. Indeed, without any doubt this is the basic foundation of the existence of so-called "quacks". What indeed are Cosmic rays?

Based on the nomenclatures or symbols for atoms in the world today (which are very much in order, ***except for something- absence of H sub-particles***), according to Equation ApI. 1, a "proton" (i.e. outside the nucleus) is a positive charge carried by the hydrogen center as shown below.

$$\overset{1}{\underset{1}{(H)}}{}^{\oplus} \equiv {}^{1}_{1}H^{\oplus} \equiv \text{Proton} \qquad \text{ApI. 2}$$

For the chlorine anion and chlorine molecule, the followings are obtained.

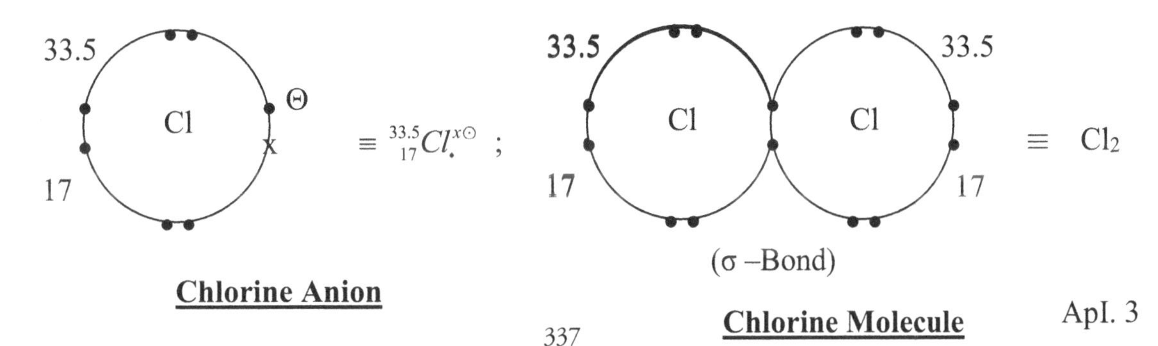

Notice that what happens in the outer sphere of the atom, does not affect the states of existence of the subparticles (protons + neutrons + positron + etc.), just as what goes on, on Earth or Mars or Jupiter etc., does not affect the state of existence of the SUN, the nucleus of our Solar system (a Macro-Atom). In order words, chemical reactions taking place on the outer sphere of an atom do not affect the state of existence of the nucleus, but dependent on the type of nucleus of an atom or central atom of a molecule.

However, whatever changes take place in the nucleus of an atom must affect the state of existence of the radicals in the outer sphere of the atom, just as what goes on in the SUN must affect the state of existence of Earth or Mars or Jupiter or other planets in the outer spheres of the SUN. The reason why this is so, is because first and foremost the number of radicals in the outer sphere whether female or male must always be equal to the number of "protons". For example, Uranium (U) cannot undergo a chemical reaction and be transformed to Lead (Pb); whereas it can undergo a nuclei disintegration reaction to form Lead. This is the basic foundation of so-called "Natural radioactivity" and "Artificial Atomic breakdown capture", domains of Chemists and Atomic Physicists.

With respect to ***Natural radioactivity***, two types of radiation, designated α-, β- rays were distinguished, their emission found to be unaffected by temperature, chemical combination or any other influence at the command of the Chemist. Thus, a radioactive substance may break down, emitting β-rays. These have been shown by their deflection in a magnetic field to be "negatively charged" and proved to be high-speed electrons, ejected from the nuclei of the atoms, and having speeds varying from 33 per cent to 99 per cent of that of light. ***Does this carry energy with it, since it is weight-less?***

(a)

$$\begin{matrix} \text{Radioactive substance} \\ \text{or atom} \end{matrix} \xrightarrow[\text{Down}]{\text{Breaks}} \begin{matrix} \beta-ray \\ \text{(From nucleus of atom)} \\ [-vely\ charged] \end{matrix} + \text{Another Atom} \qquad ApI.4$$

e.g.,

$$\begin{matrix} {}^{210}_{82}Pb \\ (Isotope) \\ \underline{Group\ IV\ A-Lead} \end{matrix} \longrightarrow \begin{matrix} {}^{210}_{83}Bi \\ (Isotope) \\ \underline{Group\ V\ A-Bismuth} \end{matrix} + \begin{matrix} {}^{0}_{-1}e \\ (\beta-ray) \\ \end{matrix} \qquad ApI.5$$

Based on the reaction above, apart from fully established presence of "neu-

trons" and "protons" in the nucleus, there is also the β-rays or electron (Recall "electron" in the outer sphere and electron in the inner part or core or nucleus). Why these confusions and fear of the unknown? So, in the nucleus we can so far identify the presence of "neutrons", "protons", "positrons" (positive electron- ${}^{0}_{1}P$ *or* ${}^{0}_{+1}e$) and β-rays (negative electron-${}^{0}_{-1}P$ *or* ${}^{0}_{-1}e$). An example of radioactive atom breaking down to produce positrons is that involving isotopic Nitrogen atom, noting that ***isotopes are atoms having same atomic number, but different atomic weights-Same atom.*** For example, Deuterium and Tritium are said to be isotopes of hydrogen atom.
(b)

$$\underset{\underline{Radioactive\ atom}}{\underset{(Isotope)}{{}^{13}_{7}N}} \longrightarrow \underset{(Isotope)}{\underset{Carbon}{{}^{13}_{6}C}} + \underset{"Positron"}{{}^{0}_{+1}e} \qquad ApI.6$$

The C obtained above is an isotope to the real C which is ${}^{12}_{6}C$.
N and C atoms here like Pb and Bi above are ***isobars (same atomic weight, but different At Nos-Two different atoms)***. Note that these reactions are *Decomposition mechanisms involving break-downs* and do not involve direct bombardment of Pb or N. Therefore, the β-ray and "positrons" are PRIMARY "particles" which exist in the nucleus of every atom. The "positron" itself is a type of radiation. In order words, two types have been identified so far- β-rays and positrons, ***wherein the atoms involved are isobars to each other***

The other important type of radiation produced by radioactive "matter" or indeed atom is the α-rays. ***{Note that the use of the word "matter" for atom is very much in place and acceptable, because anti-matter is imaginary and not real; it cannot be measured in our Solar system.}*** These α-rays, as their behaviour in a magnetic field shows, are positively charged, but they are now known not to be residues of the atoms breaking down, but to be made up of helium nuclei, that is, particles of mass 4 carrying two positive charges (${}^{4}_{2}He^{2\oplus}$). ***Why should this mass be neglected in Atomic weights of a matter? Is it because it has never been known to exist in the nucleus? Yet we are going to use it below.***
(c)

$$\underset{or\ atom}{Radioactive\ substance} \xrightarrow[Down]{Breaks} \underset{[+vely\ charged]}{\underset{(From\ nucleus\ of\ atom)}{\alpha - rays}} + Another\ Atom \qquad ApI.7$$

e.g.

$$\underset{\underline{Group\ II\ A - Radium}}{{}^{226}_{88}Ra} \longrightarrow \underset{\underset{(Inert\ gas)}{\underline{Group\,VIII\ A - Radon}}}{{}^{222}_{86}Rn} + \underset{(\alpha - rays)}{{}^{4}_{2}He} \qquad Apl.8$$

Can this case be said to be isotopic or isobaric? Yet, it is just a breakdown, just like in (a) and (b) which are isobaric. Note the absence of the +ve charges on the He center based on the so-called <u>***general rule***</u> of writing nuclear equations, since their primary concern is with the fundamental and deep-seated nuclear changes which is in order, but with something missing again. The reason for this is that the Scientist finds it "ambiguous to have the radon center (a stable inert gas) carrying negative charges in order to balance the equation. It has nothing to do with the difference between Chemical and Nuclear reactions, but something else.

$$\underset{\underline{Group\ II\ A - Radium}}{{}^{226}_{88}Ra} \not\longrightarrow \underset{\underset{(Inert\ gas)}{\underline{Group\,VIII\ A - Radon}}}{{}^{222}_{86}Rn} + \underset{(Helium\ cation)}{{}^{4}_{2}He^{2\oplus}} \qquad Apl.9$$

Indeed what these "particles" look like, based on the New Frontiers are not as so far represented and currently used universally. How can one have a helium **cation** inside the nucleus? How can one have negatively and positively charged electrons inside the nucleus, living with positively charged "protons" and probably negatively charged "protons"? Are these so-called charges indeed the same as charges carried outside the nucleus? Indeed, since no chemical reactions are involved, ***these are no charges, but positive and negative magnetic forces which can be said to be "nuclear charges".*** It is therefore no surprise that while positive and negative charges outside the Nucleus of an ATOM is within the domain of the Chemist, the positive and negative magnetic nuclear charges inside the Nucleus is within the domain of the Physicists. There is where the subject area of Magnetism in Physics grew from- The Nucleus of an ATOM. The α-ray which is a type of helium yet to be identified is also **a primary "particle"** unknown to the world subjectively. So far, three types of radiations have been identified, all existing in the nucleus of all atoms-β-rays (Matter), positrons (Anti-matter) and α-rays (Matter). Their removal from the nucleus is ***spontaneous (i.e. not by bombardment),*** and their removal depends on the nature of the atom in terms of Atomic numbers and Atomic weight and most importantly unknown to Science of today, ***the mode of transformation.*** That is, <u>*a radioactive atom which is metallic in character cannot disintegrate to give*</u>

<u>*another atom which is non-metallic in character.*</u> For example, consider the followings.

$$ {}^{23}_{11}Na \not\longrightarrow {}^{23}_{12}Mg + {}^{0}_{-1}e \; ; \; {}^{23}_{11}Na \not\longrightarrow {}^{19}_{9}F + {}^{4}_{2}He $$

(*Stable sodium*) (*Not an isotope of Mg*) – *Metal*; *Metal* → *Non – metal*

Metal

*ApI.*10

$$ {}^{24}_{11}Na \longrightarrow {}^{24}_{12}Mg + {}^{0}_{-1}e $$

(*Isotopic sodium*) (*Stable*) (β – *ray*) *ApI.*11*a*

$$ {}^{13}_{7}N \longrightarrow {}^{13}_{6}C + {}^{0}_{1}e $$

(*Isotopic nitrogen*) (*Isotopic carbon*) (*Anti* – β – *ray*) *Positron*

[*See Equation ApI.*6] *ApI.*11*b*

$$ {}^{24}_{11}Na \not\longrightarrow {}^{20}_{9}F + {}^{4}_{2}He $$

(*Isotopic sodium*) *Non – metal* *ApI.*12

It is important to note that a stable atom cannot disintegrate by itself and indeed to an unstable atom as shown by the first equation above. Note Equations ApI.11a and 11b where the emissions of β-ray and anti-β-ray (Positron) from unstable atoms were shown. When unstable atoms decompose, some natural laws must be obeyed, otherwise they will never decompose. ***Invariably what this implies, is that all atoms but one, are radioactive, their ability to release instantaneously nuclei "particles", depending on the type of atom, level of stability and application of natural laws.***

$$ {}^{127}_{53}I \; ; \; {}^{128}_{53}I \longrightarrow {}^{0}_{-1}e + {}^{128}_{54}Xe $$

Stable Iodine — *Unstable* [*Neither a metal nor a Non – metal*] — *Xenon* [*Neither a metal nor a Non – metal*] *ApI.*13

An atom is stable, when it cannot disintegrate and therefore largely in abundance.

Indeed, when atoms ***break down in a discharge tube***, they do so to give <u>cathode rays</u> and <u>positive rays.</u>

$$Atoms \xrightarrow[Discharge-Tube]{Break\ Down\ in} Cathode-rays + Positive-rays$$
$$(High-speed\ electron) \quad (Residue\ of\ atoms) \quad ApI14$$

Breakdown of Atom in a Discharge-tube

When a β-ray is emitted from an atom, the nucleus recoils, and the recoiled rays are said to correspond approximately to the positive rays of the discharge tube. This is a clear indication of the presence of some bonding forces in the nucleus, which in the discharge tube, the remaining opposite half is still exposed. <u>*The cathode-ray is not a primary particle*</u>, because it was forced out of the atom by the forces of the discharge tube. Hence, when cathode rays fall on matter, χ-rays are emitted, while when β-rays strike matter, γ-rays are produced.

Females:

$$Cathode-rays \xrightarrow[Matter]{Hits} \chi-rays\ are\ emitted$$
$$[Electromagnetic\ waves-10^{-10} metre\ wavelength]$$

$$\beta-rays \xrightarrow[Matter]{Hits} \gamma-rays\ are\ emitted$$
$$[Electromagnetic\ waves-10^{-12} metre\ wavelength] \quad ApI.15$$

χ-rays and γ-rays are electromagnetic waves similar to Light, but of very short wavelength. The only difference between them is that γ-rays are shorter than χ-rays, because the cathode rays which are essentially the same as β-rays are carrying added energy from the discharge tube. ***The χ-rays and γ-rays are SECONDARY particles which do not exist in the nucleus***, but obtained by nuclei reactions as secondary particles. γ-Rays are not only obtained from β-rays' **bombardments**, but also by "proton" or "neutron" **bombardments** as shown below.

$${}^{1}_{1}H + {}^{7}_{3}Li \longrightarrow \gamma-ray + {}^{8}_{4}Be$$
$$\underline{proton} \quad (Stable)-Lithium \quad (Unstable)-Beryllium \quad ApI.16$$

$${}^{1}_{0}H + {}^{127}_{53}I \longrightarrow \gamma-ray + {}^{128}_{53}I$$
$$\underline{neutron} \quad (Stable)-Iodine \quad (Unstable)-Iodine \quad ApI.17$$

<u>*BOMBARDMENT REACTIONS*</u>

Though these reactions look like capture, they are said to be bombardments. Notice that γ-rays do not carry any charge and are weightless (looks like ${}^{0}_{0}n \equiv {}^{0}_{0}H \equiv {}^{1}_{1}H^{\odot}$ an anion). The same also applies to the χ-rays. Most importantly, is the recognition of the fact that during nuclei bombardment, one moves either from ***Stable to Stable atom*** or ***Unstable to an Unstable atom*** or from ***Stable to an Unstable atom*** and never from Unstable to Stable atom. Hence, the phenomenon is unknowingly called "Artificial atomic breakdown". Notice the present new nomenclature or symbol for "proton" and a "neutron" in the equations above, using H (hydrogen) to show what the sub-atomic particles are with respect to 1 and 0 – the size of H. The "proton" cannot be said to be a CATION ($^{1}_{1}H^{\oplus}$) or H atom ($^{1}_{1}H{\bullet}e$) inside a nucleus!

So far, notice that "neutrons" and "protons" cannot be emitted spontaneously like β-rays and α-rays and are ***therefore no primary particles, but secondary particles.*** They can only be ejected by bombardment using other particles, in particular those that are "positively charged"-Males.

$$\underset{}{^{4}_{2}He} + \underset{(Stable)}{^{14}_{7}N} \longrightarrow \underset{(proton)}{^{1}_{1}H} + \underset{(Unstable)}{^{17}_{8}O} \qquad \underline{USING\ \alpha-rays} \qquad ApI.18$$

$$^{4}_{2}He + \underset{(Stable)}{^{9}_{4}Be} \longrightarrow \underset{(neutron)}{^{1}_{0}n\ (^{1}_{0}H)} + \underset{(Stable)}{^{12}_{6}C} \qquad \underline{USING\ \alpha-rays} \qquad ApI.19$$

$$\underset{\underline{Unstable}}{\underset{(Deuterium)}{^{2}_{1}D}} + \underset{a\ particle}{\underset{\underline{Deuteron}}{^{2}_{1}D}} \longrightarrow \underset{(proton)}{^{1}_{1}H} + \underset{\underline{Unstable}}{\underset{\underline{Tritium}}{^{3}_{1}T}} \qquad ApI.20$$

OR

$$\underset{\underline{Unstable}}{\underset{(Deuterium)}{^{2}_{1}D}} + \underset{a\ particle(Speed)}{\underset{\underline{Deuteron}}{^{2}_{1}D}} \longrightarrow \underset{\underline{Another\ ray}}{^{3}_{2}He} + \underset{(neutron)}{^{1}_{0}n\ (^{1}_{0}H)} \qquad \underline{USING\ Deuteron} \qquad ApI.21$$

$$^{4}_{2}He + \underset{(Stable)}{^{10}_{5}B} \longrightarrow \underset{(neutron)}{^{1}_{0}n\ (or\ ^{1}_{0}H)} + \underset{(Unstable)}{^{13}_{7}N} \qquad \underline{USING\ \alpha-rays} \qquad ApI.22$$

$$\underset{\underline{Deuteron}}{{}^{2}_{1}D} + \underset{(Stable)}{{}^{23}_{11}Na} \longrightarrow \underset{(proton)}{{}^{1}_{1}H} + \underset{(Unstable)}{{}^{24}_{11}Na}$$

$$\underline{USING\ Deuterons} \qquad ApI.23$$

"Proton" and "neutrons" can also be used for bombardments.

$$\underset{(proton)}{{}^{1}_{1}H} + \underset{\underline{Stable}}{{}^{7}_{3}Li} \longrightarrow \underset{\underline{\alpha - rays}}{2\,{}^{4}_{2}He} \quad OR \quad \underset{(Unstable)}{{}^{8}_{4}Be} + \gamma$$

$$[See\ Equation\ ApI.16]$$

$$\underline{USING\ \Pr otons} \qquad ApI.24$$

$$\underset{(proton)}{{}^{1}_{1}H} + \underset{\underline{Unstable}}{{}^{6}_{3}Li} \longrightarrow \underset{\underline{\alpha - rays}}{{}^{4}_{2}He} + \underset{(Another\ ray)}{{}^{3}_{2}He}$$

$$\underline{USING\ \Pr otons} \qquad ApI.25$$

$$\underset{(proton)}{{}^{1}_{1}H} + \underset{\underline{Unstable}}{{}^{11}_{5}B} \longrightarrow \underset{\underline{\alpha - rays}}{3\,{}^{4}_{2}He}$$

$$\underline{USING\ \Pr otons} \qquad ApI.26$$

Notice the major presence of α-rays as product with the use of proton.

$$\underset{(slow)}{{}^{1}_{0}n({}^{1}_{0}H)} + \underset{(Stable)}{{}^{238}_{92}U} \longrightarrow \underset{(Unstable)}{{}^{239}_{92}U}$$

$$(Neutron\ capture) \qquad ApI.27$$

$$\underset{(fast)}{{}^{1}_{0}n({}^{1}_{0}H)} + \underset{(Stable)}{{}^{238}_{92}U} \longrightarrow {}^{237}_{92}U + 2\,{}^{1}_{0}n({}^{1}_{0}H)$$

$$(Non-capture\ di\sin tegration) \qquad ApI.28$$

Notice that the type of products obtained, depends on certain operating conditions, of which *the energy carried by the bombarding particle, the speed of the particle and the type of atom being bombarded* are important (See Equations ApI. 20/21, Equations ApI.24 and Equations ApI. 27/28).
In Equation ApI.21 and Equation ApI.25, notice the presence of another type of ray (${}^{3}_{2}He$) a helium species of mass 3g carrying two positive charges

[${}_{0}^{1}He^{2\oplus}$ (*New Frontiers*) *versus* ${}_{2}^{3}He^{2\oplus}$ (*Present – day*)]. Equation ApI.1.21 is the origin of sourcing of ATOMIC ENERGY FROM HYDROGEN. The D we see is actually H and not D, since they are isotopes of H. The real symbol, for Deuteron particle and ${}_{2}^{3}He^{2\oplus}$ are not as shown. Deuterium and Deuteron cannot have the same symbol. Deuterium is not charged (${}_{1}^{2}D$) while deuteron is positively charged (${}_{0}^{1}D^{\oplus}$). So-called ${}_{2}^{3}He^{2\oplus}$ is indeed a subset of so-called ${}_{2}^{4}He^{2\oplus}$, and cannot exist in any atom but can be obtained by nuclear bombardment reactions. One used "so-called" above, because ${}_{2}^{4}He^{2\oplus}$ as shown does not exist as a member of the helium sub-particles.

From all considerations so far and based on the laws of Nature, the followings are worthy of note.

(i) The Nucleus does not carry radicals.

(ii) Bombardment reactions to produce other particles which exist in the Nucleus is limited to use of only positively nuclearly magnetically charged rays- ***"protons", "deuterons" and α-rays***. β-rays and "neutrons" cannot do these (See Equations Apl 14 to Apl 17). Instead, they are "captured" to produce another atom and γ types of rays. For "neutrons" they may in addition disintegrate the atom to produce more neutrons as if unstably reactive (See Equation ApI.28). In general, only males commence reactions (Nuclear or Chemical).

(iii) ***The H atom is the only atom that does not carry a "neutron", an α-ray, and therefore a β-ray. It carries only one "proton".***

(iv) Deuterium and Tritium isotopes of hydrogen, do not carry α-rays and therefore β-rays, but carry one "proton" and one and two neutrons respectively. They both look like combination between one proton and one and two neutrons respectively (Neutron Capture).

$${}_{1}^{1}H + {}_{0}^{1}H \longrightarrow {}_{1}^{2}D \;;\quad {}_{1}^{1}H + 2\,{}_{0}^{1}H \longrightarrow {}_{1}^{3}T \qquad ApI29$$

(ii) Nuclear bombardment reactions are movement of an atom from Order to Disorder or Order to Order or Disorder to Disorder, but never from Disorder to Order.

(iii) The balancings of Nuclear equations are independent of the magnetic charges carried by the particles, since the particles are no atoms or molecules. Balancing their equations has nothing to do with balancing of atoms, ***but with balancing of atomic weights and atomic numbers.*** Hence, mass and energy cannot be used interchangeably by application of Albert Einstein theory of

(i) relativity as used in present-day Science. It does not make any sense, since both of them have different units, unless the mass of sub-atomic sub-particles are dimensionless!

i.e.

$$Mass_1 \quad + \quad Energy_1 \quad = \quad Mass_2 \quad + \quad Energy_2$$
$$[Mass - energy\ equation] \qquad ApI.30a$$

where mass and energy are of the same units.

The helium particle of Equation ApI.21 can be a particular ray which carries a lot of energy with it, so large compared to that of deuteron, probably due to its high speed or something else. While the essence of Albert Einstein's theory of relativity is in place with respect to velocity of light, Einstein's equation ($E = mc^2$, c = velocity of Light, 10^{10}cm/sec) is not in place. The equation above is not in place, because what we are concerned with is not the atom or a radioactive atom in a molecule, but the sub-atomic particles in the nucleus of the atom. The fundamental laws of Conservation of Materials, Energy and Momentum must be obeyed. Mass on the LHS must be the same as for RHS. The same also applies with Energy and so on, whether inside or outside the Nucleus. The ways currently used universally for Nuclear reactions are all wrong. While Newton's Laws of motion with respect to energy ($E = \frac{1}{2}mv^2$) for example applies to the outer sphere of an atom (viz the surface of the Earth), Albert Einstein's theory of relativity ($E = mc^2$) applies to the inside of the nucleus of any atom with respect to the ***"mass"*** of sub-atomic particles. The mass highlighted is not mass, but something else, since these particles are almost massless (0, -1, 1, 4, -4). Note the absence of half (1/2) in the equation. The highest speed a particle made of matter and anti-matter from the nucleus can attain is the speed of Light. This is the origin of the concept of the "Black hole". Energy is only associated with a moving particle and not with the matter that is being bombarded or releasing.

(ii) However, the nucleus of the atom or matter being bombarded carries the largest form of energy one can ever envisage in human-ity, just like that of the SUN in our Solar system. Hence, the Law of Conservation of energy is never broken during Nuclear react-ions as shown in Figure ApI.1. Hence, indeed Energy can never be created or destroyed since it is already there inside the nucleus of every atom.

(a) RADIOACTIVITY

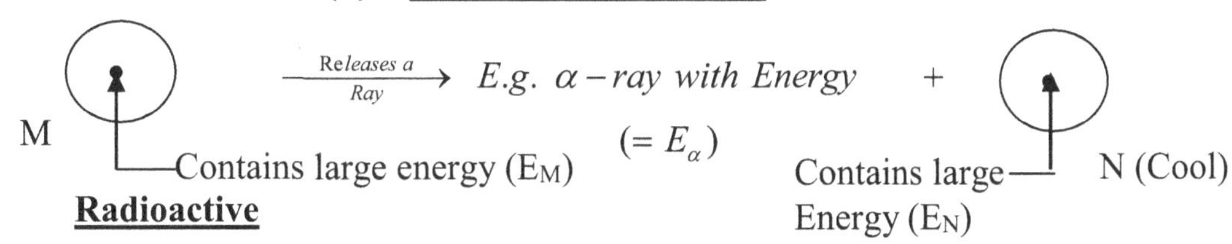

Material (Unstable)
Cool

Another atom

$E_M = E_\alpha + E_N$ (E_α cannot be > E_{AE} for same mass)

E_{AE} = Albert Einstein's energy (Maximum energy for particle)

(b) NUCLEAR BOMBARDMENT/RELEASE

(I)

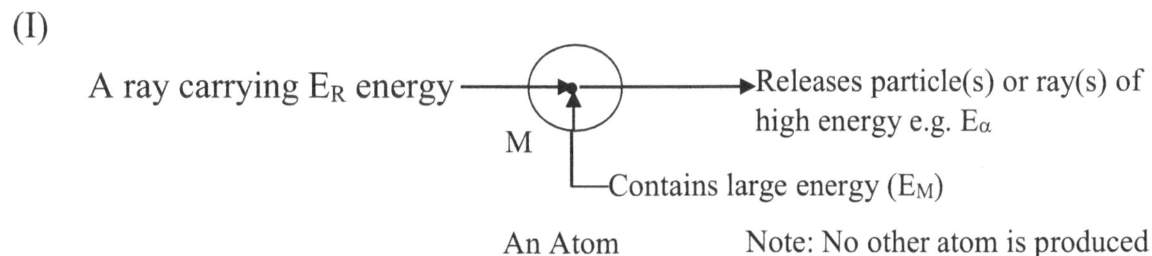

$E_R = E_\alpha - E_M$ (Source of Atomic Energy)

Since no atom is produced, $E_N = 0$, and $E_\alpha = E_R + E_M$.

(II)

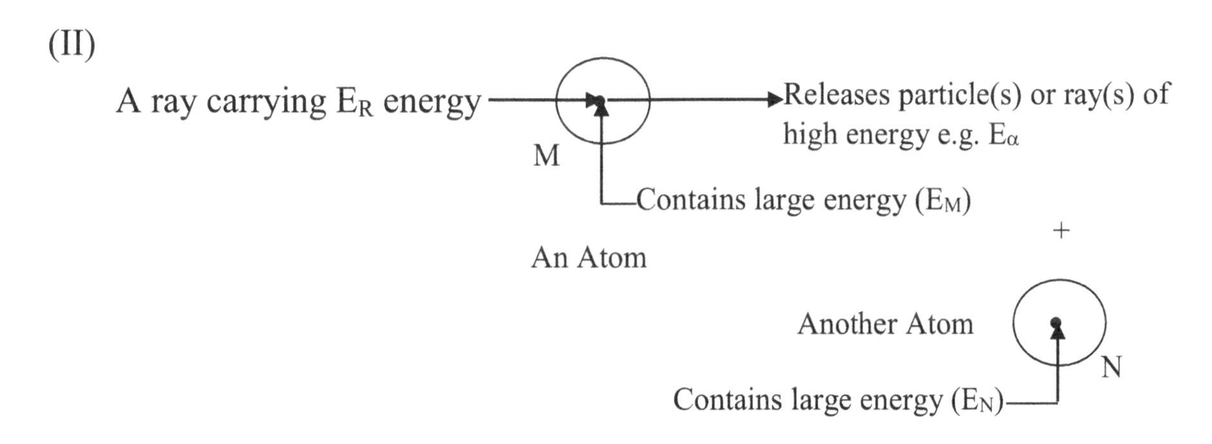

$E_R = (E_\alpha + E_N) - E_M$

(c) NUCLEAR BOMBARDMENT/CAPTURE

(I) Using the case of Equation Apl 17

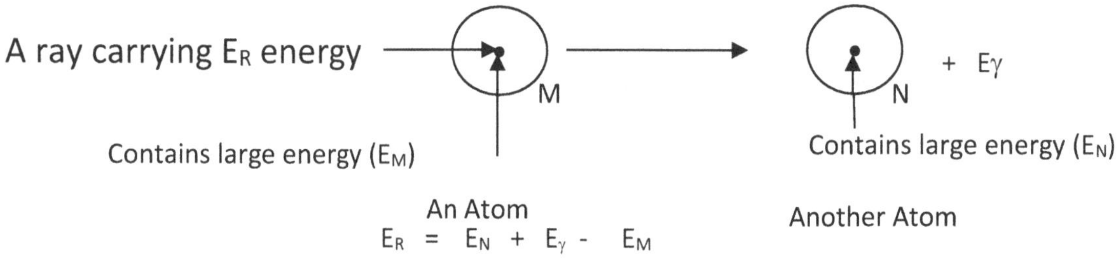

A ray carrying E_R energy → M → N + E_γ

Contains large energy (E_M) Contains large energy (E_N)

An Atom Another Atom

$E_R = E_N + E_\gamma - E_M$

Figure ApI.1 Conservation of Energy in Nuclear reactions

From the figures, one can observe the source of Hydrogen Atomic energy, a state in which no atom is produced during bombardment (b) (i) above. One can begin to very clearly see what an ATOM is. ***The energy inside the NUCLEUS of any ATOM sitting down there, is millions of times greater than the energy outside the Nucleus of an Atom whether for combustion or for any purpose.*** Basically, what this means is that when a sub-atomic particle is removed from the Nucleus of an atom, the energy content in the particle has little or nothing to do with the mass of the particle, but only with the speed at which that energy can be manifested. The higher the speed, the more the energy carried by the particle and the maximum energy content of the particle is realized when the speed is the speed of Light. Each particle has different maximum energy content. In order words, the Energy content of a particle is proportional to the square of the speed of the particle (p) as shown below.

$$E = Kp^2 \quad ; \quad E_{AL} = Kc^2 \equiv Energy\ content\ in\ particle\ at\ rest \qquad ApI.30b$$

$$(where\ K\ is\ a\ constant\ which\ has\ nothing\ to\ do\ with\ the\ mass\ of\ the\ particle)$$

This is the essence of Albert Einstein's theory of relativity, otherwise, why should a particle such as β-ray, anti-β-ray, and all particles with zero or negative mass carry energy with them? ***This energy is not Electrical, Mechanical or Chemical, but an Electromagnetic form of energy.*** Hence, it just like Electrical has nothing to do with the mass of the particle. It is the Equation that is wrong in Albert Einstein's theory, not the essence. The essence is extremely remarkable when the Energy content in the particle when it was sitting is attained or can be seen and felt to the fullest when the speed of the particle is the speed of Light. When the speed is zero, the Energy which is Electromagnetic is there, but not there. As the speed of the particle is increased, the Energy starts showing up increasing until the real Energy it was carrying manifests itself at the speed of Light. The constant K is a function of the type of sub-atomic particle (Quantum Physics). For example, the concept of **Fission** wherein a matter such as Uranium is split into two different atoms such as shown below when bombarded with neutrons or protons or deuteron or γ-rays or Cosmic rays, cannot yet be understood, because what these particles are and what they are carrying are not known.

$$_{92}U \longrightarrow {}_{56}Ba + {}_{36}Kr \qquad ApI.30c$$

If some of the particles do not have negative masses, the splitting of the Uranium into two or more different atoms would be impossible. Yet these particles do the splitting with this Energy. ***Though in the equation above, the splitting has been shown, the equation as written is meaningless.*** The entire positive rays both matter and antimatter do the splitting, while the primary sub-particles are released spontaneously. If neutrons alone are used, the splitting will take place limitedly. More than one sub-atomic particle must be involved, and this is only singly possible when Cosmic rays are used. *Protons, positrons, negatrons, and positive and negative mesons have all been detected in Cosmic radiation, as well as heavier particles believed to be the nuclei of atoms such as helium, carbon, nitrogen and oxygen.*

(vii) It looks as if the β-rays and α-rays borrow or give "neutrons" and "protons" from the atomic weights and atomic numbers of the original atom for their existence. Hence in an atom as it may seem, they remain as such without contributing to the mass and charge of the system. This particular point looks speculative, otherwise why should the atomic weight be limited only to the "neutrons" and "protons" alone when α-particles have weights twice as much as the others put together?

(viii) Since the α-rays are doubly "positively charged" and β-rays are singly "negatively charged", two β-rays are required to neutralize one α-ray. Indeed, in the nucleus, the α- and β- rays are paired as shown below. The charges as already said are not ionic or covalent or electrostatic or polar. ***These are called Nuclear Magnetic Electrostatic Charges (Real and Imaginary).*** The negatively charged center is real, while the positively charged center is imaginary analogous to the types used in conducting electric current in fluids in the outer sphere of the atom.

ELECTROSTATICALLY NUCLEARLY NEGATIVELY CHARGED-PAIRED BONDS

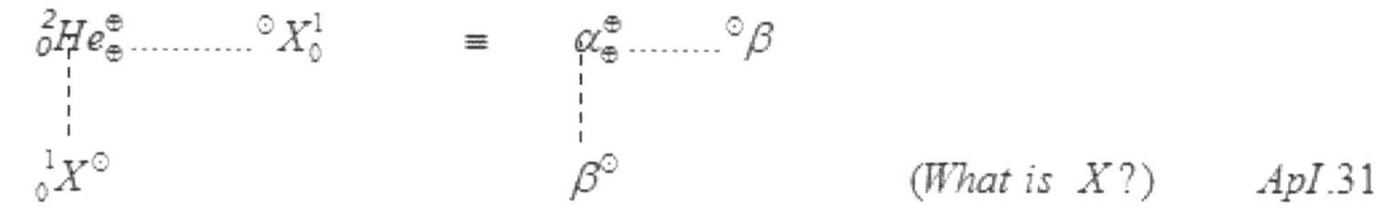

It should therefore be connected to the source of thunder and lightning. What indeed is X above which has no mass or neutron? In this manner, the α- and β- rays are placed in a neutral state. The number of β-rays equals the number of "protons". However, after neutralization of α- and β-rays, the presence of protons keeps the nucleus still positively charged on the Matter side. In view of the complex character displayed by the two rays above, hence only "protons" and "neutrons can be the only ones that can contribute to the atomic weight of an atom. The α-ray being imaginary cannot contribute to the mass of any atom. But it is there, just like the imaginary part of TIME. This is similar to the role of electrostatic and polar bonds (Imaginary bonds) outside the nucleus of an atom. They cannot be seen, but are there. But unlike outside the nucleus of an atom, these can be isolatedly placed when force is applied or when unstable.

(ix) Since the nucleus cannot be "charged", whether the outer sphere is charged or not, and since positrons and anti-protons have been identified, ***then the resultant "positive charge" is neutralized by an equal "negative charge" on the other side of the nucleus. That is, the nucleus is a combination of matter and anti-matter. While the matter has positive mass, the anti-matter has negative mass in all our gravitational systems in our Cosmos.*** In order words, anti-matter cannot be measured in any gravitational system whether here or in the moon or any planet inside our Solar system. It will still remain negative with respect to the Matter, since Anti-matter is the reverse of matter. So also, Matter will remain positive.

(x) Since the α-ray has been identified to be a helium particle and *since the mass of the other particles is either zero or positive or negative* ***unit mass,*** the other particles are nothing else than hydrogen particles of different types. Hence, one has been using H for β-ray, neutrons, protons and so on above in the equations. ***It is only with protons, H has been known to be universally used and not with the others.*** Just as the nucleus of our Solar system (The SUN) contains helium and hydrogen particles, so also exist for atoms. Incidentally, the two are the first members of Group 1A and Group VIIIA in the Periodic Table. That is, they are the only members in the FIRST PERIOD. Both of them do not have vacant orbitals. Helium is Polar while hydrogen is Non-polar outside the Nucleus. Inside the Nucleus, they are Non-polar, since ALL the radicals carried by them are not there. They are the only ones with one shell outside the Nucleus. But inside the Nucleus, they are shell-less, that is, they do not have any shell. One can see why amongst all known ATOMS, they are the only ones that can be inside the Nucleus, ***where as a rule the sub-atomic particles must be shell-less to reside inside the Nucleus.*** Hence, helium is nuclearly positively divalent, while hydrogen is nuclearly positively monovalent. If Na was to be used in place of H inside the Nucleus, then it will have a nuclearly positively charged valence state of 11!!!. ***This may be the case in another Solar system, but not ours, because of the SUN.*** Then, what are Cosmic rays where all of the nuclei of natural elements of the Periodic Table are said to be present in them in roughly the same proportion as they occur in our Solar system? They look like Cells. They are said to be present ***inside the nucleus of Cosmic rays.*** Obviously, these rays are no rays if other natural elements other than H and He exist inside them. They must be something else-***atoms without radicals,*** in which 90% are said to be H sub-particles, 9% He sub-particles and 1% the rest of other elements. Indeed, they exist and may be as such. They must be coming as already highlighted above from far away **Galaxy or Galaxies** outside our galaxy or from far away outside the galaxies in our Universe or from a **"Singularity".** Though the question has been partly answered, it may never be fully answered, because in it lies the only mystery in our planet Earth.

(xi) Since the "particles" in the nucleus can be observed to have their origin from H and He atoms, these are indeed sub-atomic particles, in which the He particle and neutron can be made to undergo further disintergration.

$$ {}^{4}_{2}He\,({}^{2}_{0}He^{2\oplus}) \longrightarrow {}^{3}_{2}He\,({}^{1}_{0}He^{2\oplus}) + {}^{1}_{0}H \quad \text{ApI.32a} $$
$$ (neutron) $$

$$ \underset{(neutron)}{{}^{1}_{0}H} \longrightarrow \underset{(\beta - rays)}{{}^{0}_{-1}H} + \underset{(proton)}{{}^{1}_{1}H} \quad \text{ApI.32b} $$

Hence indeed, the statement that ***"the atom is the smallest indivisible particle" is meaningless,*** whether it is the smallest of all things or not. In the first case, what atom amongst all known atoms is being referred to? Are all atoms the same? The so-called "quack" which is one of the subparticles of hydrogen, is not the smallest indivisible particle either, because ***all the sub-particles of H are indeed of the same size.*** It is what it is carrying that distinguishes them. Hence, the smallest indivisible particle is H and it is also the smallest atom. *Far more energies are required to remove anti-electrons (positrons) and anti-α-rays which are primary than electrons (β-rays) and α-rays which are also primary particles.* Just as some unstable atoms emit spontaneously β-rays, positron rays (See Equations ApI.11a and 11b) which are primary particles, so also the SUN emits some other rays such as γ (Gamma)-rays, χ-rays, Ultraviolet rays, Visible light ray, Infrared rays, Microwave, Radiowaves (Electromagnetic spectrum) and more.

Table ApI.1 below shows the types of sub-atomic particles in the Nucleus of an atom. Table ApI.2 shows the numbers and arrangement of these subparticles in the Nucleus of some atoms. From the first table, it can be observed that a "proton" which should be distinguished from the hydrogen atom, carries a "positive charge" (which has a unit mass) but has no atomic weight and atomic number. It should indeed be called a PROTONUCLEON. The electron inside the Nucleus carries a "negative charge" (which has a negative unit atomic mass) but has an atomic weight of one and no atomic number.

$$\underset{\substack{(proton)\\ Outside\ Nucleus}}{\underline{\underset{Cation}{{}_1^1H^{\oplus}}}} \quad versus \quad \underset{\substack{\text{"Proton"}\\ Inside\ Nucleus}}{\underline{\underset{Protonucleon}{{}_0^0H^{\oplus}}}} \quad ; \quad \underset{\substack{[Looks\ like\ \gamma-rays]\\ Not\ inside, Not\ Outside}}{\underline{\underset{Anion}{{}_1^1H_x^{\bullet\odot}}}} \quad versus \quad \underset{\substack{(Negatron)\\ Inside\ Nucleus}}{\underline{\underset{\beta-ray}{{}_0^1H^{\odot}}}} \qquad ApI.33$$

Table ApI.1 Sub-atomic particles in the Nucleus of an Atom.

	MATTER (VISIBLE)				ANTI-MATTER (INVISIBLE)			
Sub-atomic particle	α-rays	β-rays or Negatrons	Neutrons	Protons	Anti-protons	Anti-Neutrons	Positrons	Anti-α-rays
Type	Primary	Primary	Secondary	Secondary	Secondary	Secondary	Primary	Primary
Symbols Used Today universally	${}_2^4He^{2\oplus}$, ${}_2^4He$	${}_{-1}^{0}e$ e^-	${}_0^1n$	${}_1^1H$	None (But identified)	None	${}_{+1}^{0}e$, $e^{\oplus}$	None (Unknown)

NF Represen-tation	$^{2}_{0}He^{2\oplus}$ $(\equiv {}^{4}_{2}He)$	$^{1}_{0}H^{\odot}$ $(\equiv {}^{0}_{-1}H)$	$^{2}_{1}H^{\odot}$ $(\equiv {}^{1}_{0}H)$	$^{0}_{0}H^{\oplus}$ $(\equiv {}^{1}_{1}H)$	$^{0}_{0}H^{\odot}$ $(\equiv {}^{-1}_{-1}H)$	$^{-2}_{-1}H^{\oplus}$ $(\equiv {}^{-1}_{0}H)$	$^{-1}_{0}H^{\oplus}$ $(\equiv {}^{0}_{1}H)$	$^{-2}_{0}He^{2\odot}$ $(\equiv {}^{-4}_{-2}He)$
Mass of Sub-particle In Atomic Units	4.0....	0.0....	1.0.....	1.0....	-1.0.... (Negative Mass)	-1.0.... (Negative Mass)	0.0....	-4.0... (Negative Mass)
Nuclear Charge of Sub-particle	2⊕ charge Nuclearly	Θ charge Nuclearly	Neutral Nuclearly	⊕ charge Nuclearly	Θ charged Nuclearly	Neutral Nuclearly	⊕ charge Nuclearly	2Θ charge Nuclearly

Gamma rays- γ-rays $\equiv$ $^{1}_{1}H^{\odot}$ $(\equiv {}^{0}_{0}H)$, *neutral, mass* $= 0.0.....$
(These are Secondary particles not in the Nucleus)

χ-Rays are identical to γ-rays, Secondary particles not in the Nucleus.
Cathode rays are identical to β-rays (Negatrons) of higher speed.

Table ApI.2 Numbers and arrangements of Sub-atomic particles in the Nucleus of some Atoms [Electronic Configuration]

		MATTER				ANTI-MATTER			
# Atomic Number	Sub-atomic Particles Atom	α-rays	β-rays	Neutron	Proton	Anti-Proton	Anti-Neutron	Positron (Anti-β-rays)	Anti-α-rays
1	$^{1}_{1}H$ (99.98%)	0	0	0	1	1	0	0	0
	$^{2}_{1}H$ $({}^{2}_{1}D)$ (0.02%)	0	0	1	1	1	1	0	0
	$^{3}_{1}H$ $({}^{3}_{1}T)$ (0.00%)	0	0	2	1	1	2	0	0
2	$^{4}_{2}He$ (~100%)	1	2	2	2	2	2	2	1
	$^{3}_{2}He$ (.0001%)	1	1	2	1	1	2	1	1
3	$^{7}_{3}Li$ (92.5%)	1	3	4	3	3	4	3	1
	$^{6}_{3}Li$ (7.5%)	1	2	4	2	2	4	2	1
4	$^{9}_{4}Be$								

	(~100%)	2	4	5	4	4	5	4	2
	$^{8}_{4}Be$ (0.007%)	2	4	4	4	4	4	4	2
5	$^{10}_{5}B$ (18.2%)	2	5	5	5	5	5	5	2
	$^{11}_{5}B$ (81.2%)	2	5	6	5	5	6	5	2
6	$^{12}_{6}C$ (98.89%)	3	6	6	6	6	6	6	3
	$^{13}_{6}C$ (1.1%)	3	6	7	6	6	7	6	3
7	$^{14}_{7}N$ (99.635%	3	7	7	7	7	7	7	3
	$^{15}_{7}N$ (0.365%)	3	7	8	7	8	7	7	3
8	$^{16}_{8}O$ (99.759)	4	8	8	8	8	8	8	4
	$^{17}_{8}O$ (0.037%)	4	8	9	8	8	9	8	4
	$^{18}_{8}O$ (0.204%)	4	8	10	8	8	10	8	4
9	$^{19}_{9}F$ (~100%)	4	9	10	9	9	10	9	4
82	$^{207}_{82}Pb$ (22.6%)	41	82	125	82	82	125	82	41
	$^{208}_{82}Pb$ (52.3%)	41	82	126	82	82	126	82	41
	$^{206}_{82}Pb$ (23.6%)	41	82	124	82	82	124	82	41
	$^{204}_{82}Pb$ (1.5%)	41	82	122	82	82	122	82	41
83	$^{214}_{83}Bi$ (100%)	41	83	131	83	83	131	83	41
. . .									
89	$^{227}_{89}Ac$ (<100%)	44	89	138	89	89	138	89	44

. .									
92	$^{239}_{92}U$ (0.002%)	46	92	147	92	92	147	92	46
	$^{238}_{92}U$ (99.274%)	46	92	146	92	92	146	92	46
	$^{235}_{92}U$ (0.719%)	46	92	143	92	92	143	92	46
	$^{234}_{92}U$ (0.005%)	46	92	142	92	92	142	92	46

The neutron looks like deuterium carrying a "negative charge".

$$\underset{\underline{Deuterium}}{^{2}_{1}D} \quad versus \quad \underset{\underline{Deuteron}}{^{1}_{0}D^{\oplus}} \quad versus \quad \underset{\underline{Neutron}}{^{2}_{1}H^{\odot}\ (\equiv {^{1}_{0}H})} \qquad ApI.34$$

The α-ray is a helium center with atomic weight of two and atomic number of zero carrying two "positive charges" (with two unit masses). The γ-rays which are emitted when β-rays "hit" matter (electron capture) or protons or neutrons "hit" matter are as shown below. They look like hydrogen anion, but indeed neutral hydrogen with zero atomic weight and zero atomic number. It is not among the sub-atomic particles in the nucleus of any atom,

$$\underset{\beta - ray}{^{0}_{-1}H} \quad + \quad \underset{\underline{Curium}\ (Unstable)}{^{241}_{96}Cm} \quad \xrightarrow[Capture]{\beta-ray} \quad \gamma - rays \quad + \quad \underset{\underline{Americum}\ (unstable)}{^{241}_{95}Am} \qquad ApI.35$$

$$\gamma - rays \quad \equiv \quad {^{1}_{1}H^{\odot}} \ = \ {^{0}_{0}H} \qquad ApI.36$$

but it is one of sub-atomic particles of H. The "hit" above is not bombardment, but a COMBINATION mechanism system, that largely involving the Capture of a sub-particle. One can observe so far why H anion cannot exist in the outer sphere (H atom does not carry a neutron). It is important to note that anti-matter is a complete inverse of matter in almost all respects. Just as the β- and α- rays are paired electrostatically, so also are their antis paired.

ELECTROSTATICALLY NUCLEARLY POSITIVELY CHARGED PAIRED-BONDS

$$_{0}^{-1}H^{\oplus}..........{}_{\odot}^{\odot}He_{0}^{-2} \quad \text{versus} \quad {}_{0}^{2}He_{\oplus}^{\oplus}..........{}^{\odot}H_{0}^{1}$$

$$^{\oplus}H_{0}^{-1} \qquad\qquad {}_{0}^{1}H^{\odot}$$

$$\underline{Paired\ Anti-\alpha-\beta\ sub-particle} \qquad \underline{Paired\ \alpha-\beta\ sub-particle}$$

$$-Does\ not\ contribute\ to\ Mol.Wt. \qquad -Does\ not\ contribute\ to\ Mol.Wt.$$

$$-Invisible \qquad -Visible \qquad ApI.37$$

From Table ApI.2, one can explain all the observations which have been made so far with respect to radioactivity, nuclear bombardment and capture of sub-particles. The Table shows that H and its isotopes (which are no atoms in the Periodic Table) cannot carry β- and α-rays in their nucleus. The Table displays the Laws of duality of Nature-existence of matter and anti-matter in the nucleus of an atom. The Table shows that all elements or atoms apart from the first (H atom) are radioactive in character. H is the smallest and the only indivisible atom. Though it is permanent, the only thing permanent in our world is TIME an imaginary independent variable and not CHANGE. All first members must always show distinct characters different from all the other members in the same family. NATURE is too much to comprehend. The Table shows why the lighter atoms cannot display their radioactive character when stable; why the heavier metals do; when and how α- and β- particles are released and so much more. For example, consider Li.

$$\underset{(Stable)}{_{3}^{7}Li} \xrightarrow{Not\ Possible} \underset{(\beta-ray)}{_{-1}^{0}H} + \underset{(Unstable)}{_{4}^{7}Be} \quad ; \quad \underset{(Stable)}{_{3}^{7}Li} \xrightarrow{Not\ Possible} {}_{2}^{4}He + \underset{(Unstable)}{_{1}^{3}T}$$

$$\underset{(Stable)}{_{3}^{7}Li} \not\longrightarrow \underset{(Neutron)}{_{0}^{1}H} + \underset{(Unstable)}{_{3}^{6}Li} \quad ; \quad \underset{(Stable)}{_{3}^{7}Li} \not\longrightarrow {}_{1}^{1}H + \underset{(Unstable)}{_{2}^{6}He}$$

$$\underset{\substack{(Unstable)\\(Metal)}}{_{3}^{6}Li} \not\longrightarrow \underset{(Helium)}{_{2}^{4}He} + \underset{\substack{(Unstable)\\(Metallic\ Gas)}}{_{1}^{2}D} \qquad ApI.38$$

It is not possible for stable Li to even loose β-ray spontaneously, despite the fact that it has plenty of it, unlike U and Bi both of which have unstable isotopes. Only the unstable isotopes where they exist have the ability to loose it. The beryllium above does not exist. If an atom is an isotope, it must be such that is known to exist; otherwise indiscriminate rejection and addition will begin to take place. Though α-rays are largely emitted by heavy metals, the last equation above, has been used to show why it is not possible with small metals, though Li and Deuterium are ionic metals. The only problem and reason is that Li has given out not the smallest particle in her, but far more than what she is carrying, that which is against the laws of Nature. From the Tables, one can begin to explain some of the so-called "mysteries" of Life.

For the first time, a Table shows the numbers of α- and β-rays and their antis in every atom. For the first time, very new foundations have begun to be laid for Atomic Physicists and Chemists. ***These new foundations are not such that will be any source of threat to humanity, because Nuclear energy can only be a threat to humanity only and only when the mechanisms of Nuclear reactions are not understood as is the case in our world today.*** Any country in the world has the right to develop any form of Energy, provided the country understands the mechanisms of how that energy is being obtained. ***When the mechanisms are not understood, then the country becomes a threat more to humanity than to herself, because no country is an island to itself.*** Virtually every country using Nuclear energy as one of the sources of Energy in our world today, are threats not only to themselves, but to humanity, because none of them understand the mechanisms of Nuclear reactions. The situation is worsened by the fact that the same applies to all other forms of Energy (Electrical, Chemical, and even Mechanical) where the mechanisms of Combustion, Oxidation, flow of current, sound and much more are also unknown by the world. For lack of all these and much more, hence our environment has drastically changed, for which the need for Pollution Prevention is imminent. This again is just but the beginning based on the New Frontiers. The origin of the atomic structure of "matter" is lost in antiquity. For centuries, Philosophers (Most of which are basically Scientists) and Scientists have made great tremendous strides in trying to unfold the nature of the atom, because in it lies the key for advancement of knowledge for all disciplines. In it lies the "mysteries" of Life. It is when one knows what an atom is that one begins to "SEE" the ALMIGHTY INFINITE GOD IN HIS INFINITE CREATIONS AND WISDOM. All what were created in all the universes in the MIND OF THE ALMIGHTY INFINITE GOD from NOTHING were already there; for all we have been "seeing" unfortunately illusionarily are manifestations of a grain of sand of all what were already there, for which reason TIME has no beginning and no end. It is the single independent variable that is infinite.

An atom cannot be the smallest indivisible particle in our universe, when indeed there are many atoms and different types of elements for most or all atoms. ***An atom is the basic foundation of all things that exist in the universe, multi-spherical in character, composed of an inner core-the nucleus (which contains different types of helium and hydrogen sub-atomic particles with properties of matter and anti-matter) surrounded by a "burning" sphere (A very thin layer) like an outer core, then followed by a mantle and finally by an outer sphere containing radicals which are electrostatically and orderly arranged in orbitals and spherical shells distinguished by a domain and boundary- the last shell, that (i.e., the Boundary) on which all chemical and polymeric reactions take place, just like our life on the surface of the planet Earth.*** It is the Domain that in part determines the physical state of an atom or a central atom of a molecule. Everything that exists has a domain and boundary, because every atom has its own domain and boundary. Our Solar system has its own domain and boundary; and our Solar system is just one single macro-atom of so many macro-atoms in our Galaxy; and there are many galaxies. Figure ApI.2 shows the structure of the smallest atom-hydrogen. Figure ApI.3 shows the structure of the helium atom. These two atoms are the two most important atoms, followed by carbon, nitrogen and oxygen.

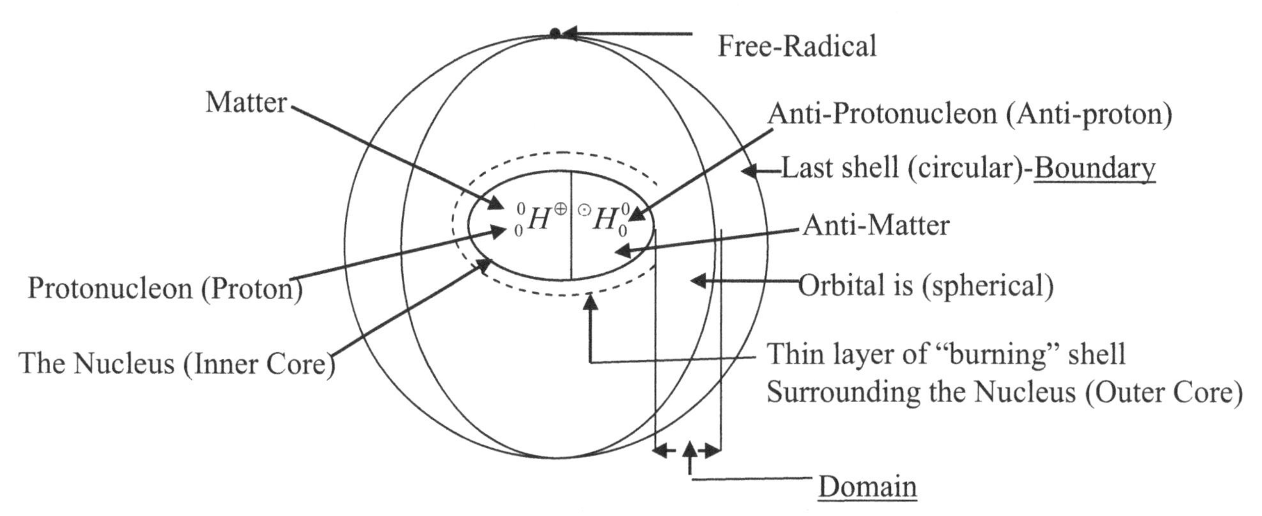

Figure ApI.2 Structure of the smallest atom of H-an element

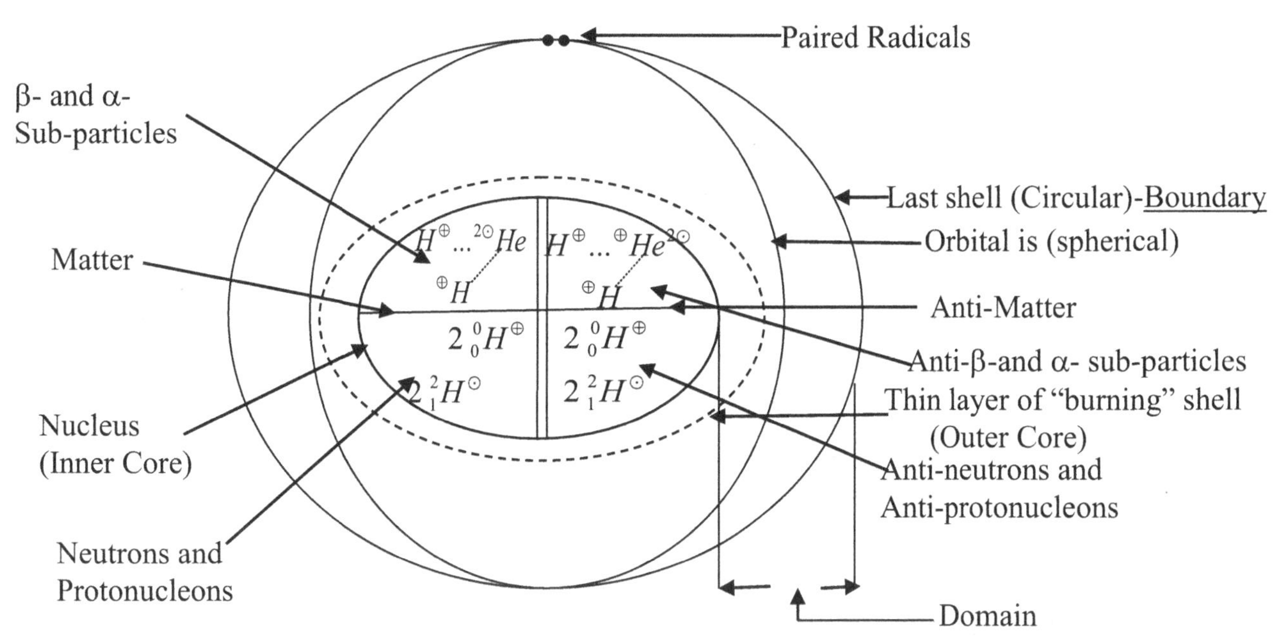

Figure ApI.3 Structure of the Helium atom, an inert element

The thin layer of burning shell is provided by the enormous amount of energy and mass located inside the nucleus. It is not combustion like that in the SUN, because of the size. There cannot be another nucleus in the center of a nucleus, because H sub-particle regardless the type is the smallest indivisible particle.

Concluding remarks

Based on one's foundations in many disciplines, in particular chemistry, physics, mathematics, chemical engineering and polymer engineering, there were just too many questions yet unanswered than answered for so many years. It has been the search for the solutions to these very large numbers of questions for which experimental data abundantly exist, but cannot yet be explained, that has given birth to the New Frontiers. Based on all considerations here, very little of which has been shown, one can imagine what great impact the New Frontiers will have on humanity for advancement of knowledge for all disciplines.

For years since one's writing (From Jan 1st 1992), colleagues, friends, and even immediate members of one's family have been asking many questions some of which include- When are you going to test your theories in the laboratory? When are you going to start publishing the Volumes? When are you going to stop writing your New Frontiers? Can I assist you as a co-author? While their concerns and criticisms are very much in order for which one cannot but thank them for the added inspiration to one's being, unknown to them is the important fact that, this was not just an ordinary assignment for human intervention, particularly with respect to those who wanted to assist one. The New Frontiers whose contents are original right from the beginning of Volume 1 to the end of Volume 12 (The first Series) to one author, are no "Accidental Inventions" or inventions, or textbooks or theories or hypotheses, but something else which is Divine in character. With so much abundance of data with little or no correct explanation, the need for one to go to the laboratory no longer arose when one started.

The foundation for all disciplines, center around what an ATOM is, in totality and knowing how ATOMS operate in Nature. Worthy of note herein intrinsically, Albert Einstein's theory has been redefined. The mass in the equation which is supposed to be the mass of the subatomic particle and not of an atom is not supposed to be there, because the Energy in question is not Mechanical or Chemical or even Electrical (all of which take place outside the Nucleus on the surface of an Atom), but Electromagnetic forms of energy, that which is inside the Nucleus. In Electrical and Electromagnetic forms of Energy, mass as the source of energy is irrelevant. On the other hand, based on what has been considered right from the beginning of this text, all three theories already proposed universally behind creations of our world and humanity-Big Bang theory, The Genesis, Darwin's theory are very much in place with only some missing links here and there, just as with the case of the outer sphere of an ATOM, as will be finally explained in the New Frontiers using the universal data so far collected so much in abundance. Yet we have only just begun.

Without mathematics, physics and chemistry, there is no other discipline. Hence academic orientations universally, will change very drastically following the New Frontiers. All books in Chemistry in particular, Physics and Mathematics will be thrown into the archive and rewritten again. Same will apply to other disciplines including History that deals with dates (Mathematics) of events. Based on the New Frontiers, all diseases new or old, genetic or non-genetic will have a cure; all industries (chemically or non-chemically based, refineries, etc.), will be forced to dismantle and rebuilt again in accordance to the Laws of Nature and what exist in Nature. Mental orientation and even Religion will completely take a new and final dimension, and so on. This does not mean that the world will eventually move into a perfect world because one cannot live in a world where good and bad, bitterness and sweetness, rich and poor, heaven and hell, matter and antimatter and so on can be separated, since both must coexist for any to exist.

**Major reference
E. de BARRY BARNETT, C. L. WILSON, "INORGANIC CHEMISTRY- A Text-Book for Advanced Students", The English Language Book Society and Longmans Green and Co Ltd., 1962.

We came from NOTHING with NOTHING that is REAL but IMAGINARY and we are going back to NOTHING with NOTHING that is REAL but IMAGINARY, and that which is IMAGINARY is the PAST, for without the PAST, there is no PRESENT and without the PRESENT, there is no FUTURE and without the FUTURE, there is no LIFE.

The Author.

INDE X

A

B

C

INDEX

D

E

INDEX

INDEX

Printed in the United States
By Bookmasters